Electromagnetics for Engineers:

With Applications to Digital Systems and
Electromagnetic Interference

Electromagnetics for Engineers:

With Applications to Digital Systems and Electromagnetic Interference

CLAYTON R. PAUL
Mercer University
and
University of Kentucky

WILEY

JOHN WILEY & SONS, INC.

Acquisitions Editor   *William Zobrist*
Marketing Manager   *Katherine Hepburn*
Senior Production Editor   *Norine M. Pigliucci*
Senior Designer   *Harry Nolan*
Production Management Services   *Suzanne Ingrao*
Senior Editorial Assistant   *Angie Vennerstrom*

This book was set in New Caledonia by The GTS Companies/York, PA Campus
and printed and bound by Donnelley Willard. The cover was printed by Lehigh Press.

This book is printed on acid free paper.   ∞

ISBN 0-471-27180-2
WIE ISBN 0-471-45245-9

Printed in the United States of America

10 9 8 7 6 5 4 3 2 1

*This textbook is dedicated to
the humane
and
compassionate treatment of animals*

Preface

Understanding and being able to apply electromagnetic (EM) principles and laws are among the most important skills that enable electrical and computer engineers (ECEs) to design modern electronic systems. The increasing frequencies of analog systems as well as the increasing speeds of digital systems require that designers have a fundamental understanding of the basic EM principles and laws that are contained in this text. An understanding and retention of these has become crucial in the design of modern electronic systems.

This text is intended for use as an introduction to electromagnetic principles and engineering applications for electrical and computer engineers. The principal purpose of the text is to provide electrical and computer engineering students with a lasting understanding of electromagnetic principles and their application to the design of modern high-speed digital and high-frequency analog devices and systems.

▶ THE PHILOSOPHY BEHIND THE PREPARATION OF THIS TEXT AND ITS KEY FEATURES

The proliferation of digital devices has had a major impact on undergraduate ECE curricula. The need to understand these new technologies has required that a substantial amount of new material and courses be fitted into an already crowded 4-year curriculum. In many ECE curricula, this has resulted in the required EM course sequence being reduced from two semesters to one semester. There is no indication that the ECE curriculum will be expanded beyond 4 years. Hence there is a continuing need to optimize the content and delivery of all courses in ECE programs. In addition, the frequency content of modern electronic systems is steadily increasing. Personal computers today operate with clock speeds on the order of 3 GHz (3×10^9 Hz). These clock signals contain, in addition to 3 GHz, frequencies at multiples of this fundamental frequency, for example, 6 GHz, 9 GHz, 12 GHz. Analog communication systems such as cell phones also operate at frequencies in the GHz range. These frequencies of operation will no doubt continue to increase. No longer can ECEs design these high-frequency, high-speed modern electronic systems without a fundamental understanding of the limitations imposed by EM principles and laws. This text grew out of those two needs.

An important guiding principle throughout the preparation and organization of the text was that the course it is intended to be used for will likely be the last course in electromagnetics that a majority of the ECE students will take. However, in future engineering design, that majority will have an increasing need to understand these EM principles in order to successfully design modern digital and analog systems and devices. This need is primarily being driven by the ever-increasing speeds of digital devices and the increasing requirement to use the GHz frequency spectrum in communication systems. Hence it is important that this first course cover these increasingly important aspects of EM that all ECEs will need in their future work. It is equally important that the material and its coverage be presented in a manner that will motivate the reader to give it serious study in order to promote long-term retention of these important principles and skills.

EM courses at the undergraduate level for which this text is intended have been traditionally divided into coverage of static (dc) field topics and those that apply to time-varying fields. Although there are numerous important applications of the static field principles, the majority of the important engineering applications deal with time-varying fields such as electromagnetic waves, transmission lines, and antennas. The time-varying topics are also becoming more and more important as the speeds of digital devices and the frequencies of analog devices continue to increase, seemingly without bound. As frequencies of excitation increase, electric circuits are becoming larger with respect to a wavelength, so that quasi-static field concepts such as lumped-circuit models and Kirchhoff's laws no longer apply and the time-varying principles must be used. The author has chosen to reduce the traditional coverage of static fields in order to provide an earlier and more thorough study of time-varying field principles and their important engineering applications. Early coverage of static fields has been retained since it gives the reader a simple introduction to and understanding of the meaning of the four primary EM variables of electric field intensity, electric flux density, magnetic field intensity, and magnetic flux density. Although the static field coverage has traditionally been contained in several chapters, the author has chosen to place the static field topics in one chapter (Chapter 3). The need to reduce the coverage of static fields has required that decisions be made as to what material to include in that chapter. Chapter 3 on static fields represents what the author feels is an appropriate balance. All ECE curricula contain several technical electives on EM in the undergraduate curriculum. Once the fundamentals have been covered in this first course, additional detail and topics that were omitted can be included and expanded on in these technical electives.

The required EM course has traditionally been viewed by students as being mathematically intensive. As such, many students fail to gain an appreciation and long-term retention of the important principles. In addition, many do not elect to take the EM elective courses, perhaps because of their view of EM as being primarily concentrated on mathematics. A study of EM requires dealing with vector quantities in space and time. Hence partial differential equations containing vector quantities underlie problem solutions. However, throughout the preparation of the text, the author has endeavored to reduce the mathematical detail where possible. Lengthy or complex derivations and vector manipulations are avoided where they do not substantially contribute to the student's learning of or ability to apply the basic principles. For example, in Chapter 2, which covers basic vector mathematics, only the vector calculus operations of line and surface integrals are discussed. The differential operations of divergence and curl are delayed until Chapter 4, where a study of time-varying fields begins. In Chapter 3, concerning static EM fields, problem solution methods stress visualization and the use of symmetry and trigonometry rather than vector manipulations in order to set up problem solutions. This stressing of visualization in setting up a problem solution has the added advantage of providing a lasting understanding of the concepts. The intent is to prevent overwhelming the student with mathematical detail. This should also have the advantage of motivating readers to deepen their understanding of EM by taking the follow-on EM elective courses. A certain amount of mathematical detail is unavoidable and, moreover, instructive. For students who choose to pursue EM in elective or graduate courses, the mathematical sophistication can be expanded once this solid foundation in understanding the fundamental principles has been established.

In order to motivate the reader toward a serious study of the text, the relevance of the material to his/her future engineering practice should be made clear. Hence the text should contain a significant number of demonstrations of how the theory applies to practical design of modern electrical and computer systems. The majority of those application examples in this text focus on three areas that all ECEs will inevitably be involved in: high-speed digital electronics, high-frequency analog electronics, and electromagnetic interference. These three topics are increasing in importance for both electrical and computer engineers due to the increasing speeds of digital devices and the increasing frequencies used in analog communication devices.

► ORGANIZATION OF TOPICS

Chapter 1 illustrates the need for understanding EM principles in order to design modern high-speed and high-frequency electronic devices. The important concepts of waves, wavelength, and electrical dimensions are briefly discussed, as is the concept of frequency (spectral) content of digital pulses. These concepts are later reinforced throughout the text. They are briefly discussed here to provide the reader with an awareness of the limitations of lumped-circuit models and Kirchhoff's laws in high-frequency, high-speed electronic design. Also provided are an overview of the history of the evolution of EM theory, a general understanding of the notion of a field and its sources, and some important motivational examples of practical applications. Chapter 2 discusses the basic orthogonal coordinate systems (rectangular, cylindrical, and spherical) whose understanding will be needed in all the later chapters. The dot and cross products of vectors are simple to comprehend and are also discussed. Finally, the operations of line and surface integrals are discussed to prepare the reader to understand the static field concepts of the next chapter. The author has chosen to discuss only the integral operations of line and surface integrals in Chapter 2, since they are sufficient for the static (dc) field analysis in the following chapter as well as the meaning of Maxwell's equations for time-varying fields in the later chapters. Discussion of the differential calculus topics of divergence and curl are deferred to Chapter 4, where time-varying fields are introduced.

The topics of static (dc) electric and magnetic fields are discussed in Chapter 3. Use of visualization and exploitation of symmetry in the solution of these static field problems are emphasized, and only the mathematical topics of line and surface integral are required. The static electric field topics of Coulomb's law, the meaning of the electric field intensity vector, and the electric flux density vector through Gauss' law are discussed. The additional topics of voltage, dielectric media, and capacitance are also covered. The static magnetic field is introduced through the use of the Biot-Savart law to compute the magnetic flux density vector. Ampere's law is used to compute the magnetic field intensity vector utilizing symmetry. Gauss' law for the magnetic field is discussed, as is the topic of magnetic fields in magnetic materials. Resistance and inductance are also discussed. The Lorentz force equation and its applications to the electric motor and generator (under Engineering Applications) are also discussed. Practical applications of the theory to low-frequency, quasi-static situations such as high-voltage transmission lines and electrostatic discharge (ESD), along with examples of 60-Hz interference and electrostatic shielding, are provided to motivate the student. Parasitic effects in components such as the effect of connection lead inductance and capacitance are also discussed. These aspects of the "hidden schematic" become important as the frequencies of excitation of the component increase. It was decided to place all static field concepts in this single chapter in order to emphasize which EM principles apply only to static fields.

Chapter 4 begins the discussion of Maxwell's equations for time-varying fields. The previously discussed integral forms for static fields are modified for time-varying fields and discussed. Next the point forms of these results, in terms of divergence and curl, are derived. Divergence and curl are delayed until this point to prevent overwhelming the student with mathematics. Power density and the Poynting vector are discussed, as are the boundary conditions and the method of images.

Chapter 5 introduces the student to the important concept of waves via the uniform plane wave. The discussion begins with uniform planes waves and their properties in lossless media and next is extended to lossy media. Power flow and the concept of skin depth are covered next. These ideas are extended to uniform plane waves with normal incidence on plane, material boundaries. Snell's laws for oblique incidence are briefly covered. Examination of the field vectors for oblique incidence on plane boundaries is contained in Appendix A for the instructor who wishes to include this topic. Several practical applications of the theory such as communication with submarines, shielding electronic equipment, design of radomes, fiber-optic cables, and microwave health hazards are presented.

Chapter 6 covers the topic of transmission lines. The concept of transverse electromagnetic (TEM) waves and the transmission-line equations are introduced. Time-domain (transient) solution of the transmission-line equations is discussed and is followed by the sinusoidal, steady-state (phasor) solution. Extensive discussion of the time-domain or transient solution for pulse waveforms is included to prepare the reader for understanding the problems imposed by high-speed digital devices. The Smith chart is briefly introduced as a computational aid, since it is used extensively in industry. Only its use in computing input impedance, reflection coefficient, and voltage standing wave radio (VSWR) on transmission lines is covered. Lumped-circuit approximate models of transmission lines and the restrictions on their applicability are discussed to provide a link to the lumped-circuit experience of the student. Throughout this chapter, SPICE (PSPICE) is used to provide solutions for both the transient and the phasor solutions to provide additional motivation for the student and to allow the investigation of practical digital systems wherein the line terminations may be digital logic gates and other digital devices. Several practical applications of transmission-line theory such as high-speed digital interconnect design, signal integrity in digital systems, microwave circuit components, the effect of antenna feed lines, and a brief discussion of crosstalk are discussed. The use of shielded wires and twisted wire pairs to reduce crosstalk is also briefly discussed. In this chapter, experimental results concerning digital signal integrity and crosstalk is provided to demonstrate the application of the EM principles. Students appreciate experimental confirmation of the theory.

Chapter 7 is the final chapter of the text and provides an introduction to antennas. The Hertzian (electric) dipole is introduced, and the properties of the far fields and radiation are examined. Next, the half-wave dipole and the quarter-wave monopole are examined by viewing their fields as the superposition of Hertzian dipoles along their length. A brief discussion of antenna arrays is given. The general properties of antennas such as patterns, gain, effective aperture (capture area), and the important Friis transmission equation are discussed. Finally, several important engineering applications of the theory are given. A simple model for predicting the radiation from transmission lines such as cables and printed circuit board lands is obtained and discussed in the context of complying with the governmental limits on the radiated emissions from digital devices. The use of shifting the phase of currents to antennas in an array in order to electronically steer the beam is discussed in the context of phased array radars. The design of a communication link for a specific signal-to-noise ratio is also discussed. A simple model to estimate the ability of a passing electromagnetic wave to couple to a transmission line is derived. This is used to illustrate the susceptibility of wires and PCB (printed circuit board) lands to incident fields.

▶ PEDAGOGICAL FEATURES

Example Problems Worked-out Example Problems are given after each new concept or law is introduced. These contain sufficient details of the solution process and are carefully chosen to illustrate the application of the concept or law.

Quick Review Exercises Quick Review Exercise problems with answers are also placed strategically throughout the text to allow the reader to self test his/her comprehension of the material.

End-of-Chapter Problems End-of-chapter problems are grouped according to the section of the text that they cover. The answers to selected end-of-chapter problems are given at the end of the question in brackets [ ]. For instructors who adopt the book, a summary of the answers to all of the end-of-chapter problems is provided at the beginning of the Solution Manual. These answers have been thoroughly checked by the author and are error free.

The IMPORTANT EQUATIONS are boxed in order to focus the attention of the student on their importance.

Use of SPICE (PSPICE) A unique feature of this text is the extensive use of the transmission-line model in the SPICE (PSPICE) circuit analysis program to analyze and confirm the results of both the time-domain and the frequency-domain solution of transmission-line problems. This also allows the analysis of practical digital systems wherein the line terminations are digital gates for which the analysis is more complex than for resistive terminations.

Engineering Applications Numerous examples of the application of the principles to engineering design are given at the end of each chapter under Engineering Applications. These Engineering Applications are placed at the end of each chapter in order to clearly separate the basic principles and theory from the applications to help the student distinguish between the two. These are chosen primarily to emphasize the application of the theory to high-speed and high-frequency systems and devices and to problems of electromagnetic interference. Careful selection of examples of the application of the concepts to design is also crucial to the student's acceptance of the material. These application examples are chosen to introduce the student to the important engineering applications of the principles that he/she will increasingly need in industry. Electromagnetic interference is becoming an increasingly important topic of ECEs as the speeds of digital devices and the frequencies of analog devices continue to increase. Hence electromagnetic compatibility (EMC) is rapidly becoming an important subdiscipline for ECEs. In addition, the text contains numerous photographs of practical engineering applications of the theory ranging from antennas and printed circuit boards of digital devices to shielded rooms used in EMC testing.

Chapter Learning Objectives Each chapter begins with a set of Chapter Learning Objectives to assist the student in seeing the "big picture."

Summary of Important Concepts and Formulae A Summary of Important Concepts and Formulae is provided at the end of each chapter. This provides a snapshot of what the student should retain from the chapter. Students are likely to be overwhelmed with formulae and are often unable to determine the minimum set of important skills and concepts that he/she must retain. The Chapter Learning Objectives and the Summary of Important Concepts and Formulae aid in helping the student focus on the important concepts and formulae that should be retained over the long term, which is one of the primary objectives of this text.

▶ OPTIONS IN COVERAGE OF THE MATERIAL

This text was designed for a one-semester (3 credit hours and 45 periods or 4 credit hours and 60 periods) or a two-quarter course. However, the text organization allows for sufficient flexibility in topic coverage.

There is a trend in teaching EM to begin with the topic of transmission lines and then go into the EM topics. It is entirely feasible to cover transmission lines in Chapter 6 before the traditional EM concepts. Chapter 1 provides a preliminary understanding of waves and the concept of electrical dimensions in wavelengths. Hence after a coverage of Chapter 1, it is possible to cover Chapter 6 on Transmission Lines and then begin the coverage of the EM concepts with Chapter 2. Chapter 6 on Transmission Lines is self-contained and does not substantially rely on prior coverage of Chapters 2–5.

Classic Outline	*Transmission Lines First*
Chapter 1, Introduction	Chapter 1, Introduction
Chapter 2, Calculation with Vectors	Chapter 6, Transmission Lines
Chapter 3, Static (DC) Electromagnetic Fields	Chapter 2, Calculation with Vectors
Chapter 4, Time-Varying Electromagnetic Fields	Chapter 3, Static (DC) Electromagnetic Fields
Chapter 5, Wave Propagation	Chapter 4, Time-Varying Electromagnetic Fields
Chapter 6, Transmission Lines	Chapter 5, Wave Propagation
Chapter 7, Antennas	Chapter 7, Antennas

A second option is to cover static field topics after the important time-varying engineering applications. Coverage of static fields in Chapter 3 was included in that early chapter primarily to give the student an understanding of the meaning and interpretation of the four important field vectors—electric field intensity, $\mathbf{E}$; electric flux density, $\mathbf{D}$; magnetic field intensity, $\mathbf{H}$; and magnetic flux density, $\mathbf{B}$—which the EM field laws contain whether the fields are static or time-varying. However, it would be possible to delay the discussion of static fields to the end of the course and go directly into the more important (from an engineering perspective) time-varying fields. Hence a second option would be

Classic Outline	*Static Fields Last*
Chapter 1, Introduction	Chapter 1, Introduction
Chapter 2, Calculation with Vectors	Chapter 2, Calculation with Vectors
Chapter 3, Static (DC) Electromagnetic Fields	Chapter 4, Time-Varying Electromagnetic Fields
Chapter 4, Time-Varying Electromagnetic Fields	Chapter 5, Wave Propagation
Chapter 5, Wave Propagation	Chapter 6, Transmission Lines
Chapter 6, Transmission Lines	Chapter 7, Antennas
Chapter 7, Antennas	Chapter 3, Static (DC) Electromagnetic Fields

▶ SUPPLEMENTS

A Solutions Manual containing the detailed solutions of the end-of-chapter problems is available to instructors who adopt the book for their course. The solution manual was prepared solely by the author and has been thoroughly checked for accuracy. Instructors should contact their Wiley representative for access to a website at Wiley to download an electronic copy. This and additional supplements are available for instructors at http//www.wiley.com/paul. These include Powerpoint slides of all figures in the text.

▶ ACKNOWLEDGMENTS

The author acknowledges the numerous discussions with his colleagues that have contributed markedly to shaping this book. Many of those colleagues are associated with the author's research in electromagnetic compatibility (EMC), which represents one of the significant applications of the EM principles of this text. A special acknowledgment is due Professor Robert G. Olsen of the Washington State University, who provided numerous extensive critiques of the manuscript as well as many suggestions for its improvement. The author also acknowledges the many helpful discussions with the editor William Zobrist.

Clayton R. Paul
Macon, Georgia

Contents

List of Engineering Applications

List of Tables

Electromagnetics for Engineers:

With Applications to Digital Systems and Electromagnetic Interference

CHAPTER 1

Introduction

Electromagnetic (EM) principles and laws govern all electrical and computer engineering (ECE) systems. It is therefore important for all electrical and computer engineers to understand these very basic EM principles in order to properly design modern electrical and computer systems and devices. This is becoming more important due to the increasing speeds of digital devices and the increased use of higher frequencies in modern communication systems. Some 20 years ago, digital computers operated with clock speeds in the low megahertz range, for example, 12 MHz. Digital computers and other imbedded processors are operating today with clock speeds in the low GHz (10^9 Hz) range. These digital pulses contain, in addition to the fundamental repetition frequency, multiples of this basic frequency. For example, a clock signal in a digital computer with a repetition rate of 1 GHz will contain, in addition to that frequency, sinusoidal components at 2 GHz, 3 GHz, 4 GHz, 5 GHz, etc. The presence of these increasingly higher frequencies means that electrical components such as capacitors, wires, printed-circuit-board (PCB) "lands," etc., will no longer behave like familiar lumped-circuit elements but will be governed by the EM laws that we will study in this text. Similarly, modern communication devices such as cell phones also operate at frequencies in the GHz range. Hence in order to design these circuits so that they will properly function will increasingly require an understanding of the EM principles that we will study in this text.

Chapter Learning Objectives

After completing the contents of this chapter you should be able to

▷ flawlessly convert units of measure from the English system to the metric system and vice versa,

▷ understand the concepts of a wave, and wavelength and phase shift,

▷ compute the electrical dimensions of an electronic device or component in wavelengths,

▷ determine when a lumped-circuit model and Kirchhoff's laws are invalid for an electric circuit,

▷ understand that a periodic, time-domain signal such as a digital computer clock contains, according to the Fourier series, sinusoidal components at frequencies that are multiples of the basic repetition frequency, and all of those frequency components go together to make up the time-domain wave shape,

▷ cite typical engineering applications of the EM principles.

▶ 1.1 UNITS AND UNIT CONVERSION

Units of measurement are an important aspect of scientific calculations. The basic set of units recognized throughout the world are the International System of Units or SI units. A subset of these is referred to as the MKSA system or the metric system. These names stem from the four basic units of measure. Length is measured in meters (m), mass is measured in kilograms (kg), time is measured in seconds (s), and current is measured in amperes (A). Other units are expressed in some combination of these basic units. For example, the units of the most basic quantity of EM is the unit of charge which is the coulomb (C). Current is the rate of flow of charge and hence charge in coulombs is expressed in amperes and seconds as $C = A \cdot s$. The common important powers of ten multipliers are listed in Table 1.1.

For example, 60 kilometers is written as 60 km or 0.06 Mm. Similarly, the value of a 5 microfarad capacitor is written as 5 μF or 5000 nF.

It is vitally important to realize that in the EM formulae that we will study, *the units of measure must be in SI units*. There are several universal constants in those formulae that are in SI units and hence other units in the formulae must necessarily also be in SI units. For example, one of the first laws we will study is Coulomb's law for the force between two point charges:

$$F = \frac{Q_1 Q_2}{4\pi \varepsilon_o R^2} \quad N$$

This gives the force F in newtons (N) between two charges, Q_1 and Q_2, which are separated a distance R in free space (approximately air). The charges must be given in coulombs and the distance must be in meters. The constant ε_o is the permittivity of free space and has units of farads per meter, F/m, or C^2/Nm^2.

Although the SI unit system is accepted around the world, some countries, including the United States, have not totally converted to this system. In the United States the predominant system of units used is the so-called English system where length is measured in inches (in.), or feet (ft), or yards (yd), or miles (mi). Radii of wires (circular cross section conductors) are commonly stated in mils where 1 mil = 1/1000 in. It appears that this will remain the case for the forseeable future, so there is a need to be able to flawlessly convert units in the English system to equivalent units in the SI system *before* inserting them into the EM formulae. The important conversions of length between the two systems are 1 in. = 2.54 cm, 12 in. = 1 ft, 1 yd = 3 ft, and 1 mi = 5280 ft. There are numerous other units in the English system that must be converted to SI but the length units will suffice. We must have a flawless way of converting between the two

TABLE 1.1   Unit Multipliers

Prefix	Multiplier	Symbol
Giga	10^9	G
Mega	10^6	M
Kilo	10^3	k
Centi	10^{-2}	c
Milli	10^{-3}	m
Micro	10^{-6}	μ
Nano	10^{-9}	n
Pico	10^{-12}	p

systems. A large number of mistakes that are made in scientific computation are made in unit conversion. Several years ago a spacecraft bound for Mars missed the planet due to an error in converting from English units to SI units by a U.S. contractor. A simple but effective way of converting between the two systems is to multiply by unit ratios. Cancellation of the unit names in this conversion avoids the improper multiplication (division) of a unity ratio when division (multiplication) should be used. For example, in order to convert 100 mi to km we perform the following multiplication and cancellation of unit names:

$$100 \text{ mi} \times \frac{5280 \text{ ft}}{1 \text{ mi}} \times \frac{12 \text{ in.}}{1 \text{ ft}} \times \frac{2.54 \text{ cm}}{1 \text{ in.}} \times \frac{1 \text{ m}}{100 \text{ cm}} \times \frac{1 \text{ km}}{1000 \text{ m}} = 160.93 \text{ km}$$

The reader should verify that $1 \text{ mil} = 1/1000 \text{ in.} = 2.54 \times 10^{-5} \text{ m}$.

▶ 1.2 THE NEED FOR AN UNDERSTANDING OF ELECTROMAGNETIC PRINCIPLES

ECE studies generally begin with a study of models of electric circuits and devices using lumped-element circuit models and Kirchhoff's laws. The circuit elements of resistance, R, inductance, L, and capacitance, C, are given values in those lumped-circuit models, for example, $1 \text{ k}\Omega$, $1 \text{ }\mu\text{H}$, 100 pF. How does one obtain the values of these elements? These elements are *models* of physical elements and hence the element values depend on the structure and dimensions of the physical element. The EM laws that we will study provide the ability to compute the values of the physical elements from their physical construction details. In order to develop mathematical models of physical devices, we must first understand the basic EM principles that govern the physical devices.

Although lumped-circuit models are simple to understand, they are *approximations* to the fundamental EM laws that we will study. Lumped-circuit models do not adequately explain the operation of numerous practical devices such as antennas, yet the EM laws we will study provide a clear explanation of their operation. As the frequency of excitation of electric circuits and devices increases, nonideal effects that are not predicted by the ideal lumped-circuit models become increasingly important. As the processing speeds of digital devices and operating frequencies of analog devices increase, it is becoming more important to understand these electromagnetic principles in order to design digital and analog devices that will operate properly at these increasingly higher frequencies. This text is intended to be an introduction to those fundamental EM principles and laws.

Lumped-circuit models do not predict several important phenomena. For example, when we construct lumped-circuit models we attach connection leads and assume that the current entering one lead attached to the element exits the other lead *immediately* as illustrated in Fig. 1.1. In other words, we *assume* that the length of the connection leads is unimportant. In fact there is a time delay associated with the physical connection leads of the device. Time delay is essentially the time it takes for a wave to propagate from one point to another. In free space (essentially air) this velocity of propagation is the speed of light $v_o = 2.99792458 \times 10^8 \text{ m/s}$ or approximately $v_o = 3 \times 10^8 \text{ m/s}$. The time delay in the propagation of a wave over a distance $\mathcal{L}$ is

$$T = \frac{\mathcal{L}}{v} \quad \text{s} \tag{1.1}$$

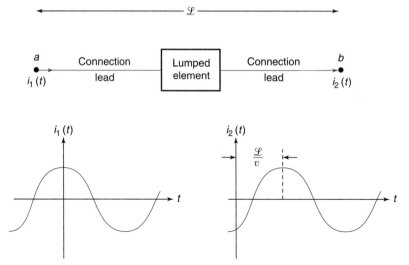

Figure 1.1 Illustration of the effect of element interconnection leads.

where v is the velocity of propagation of the wave. For example, the time delay for a wave propagating in free space (approximately air) over a distance of 1 m is approximately 3 ns or about 1 ns per ft.

The current along the leads is, in fact a *wave*. Suppose this current and the associated wave are sinusoidal. As we will see in Chapter 5, a sinusoidal propagating wave can be written as a function of time, t, and position, z, as (where we have chosen a cosine form)

$$i(t,z) = I \cos(\omega t - \beta z) \qquad (1.2)$$

where β is the phase shift constant in radians/m and $\omega = 2\pi f$ where f is the cyclic frequency. This is shown in Fig. 1.2a as a function of distance, z, for fixed times, t. As the wave propagates from one end of the connection lead, through the element and exits the other end of the connection lead it suffers a *phase shift* which is given in (1.2) by

$$\phi = \beta \mathscr{L} \qquad \text{radians} \qquad (1.3)$$

and $\mathscr{L}$ is the total length of the connection leads.

The phase shift is alternatively related to an important parameter, *wavelength*. Wavelength is denoted by λ and is the distance the wave must travel to change phase by 2π radians which is equivalent to $360°$ as shown in Fig. 1.2a. Hence the wavelength and the phase constant are related by

$$\beta \lambda = 2\pi \qquad \text{radians} \qquad (1.4)$$

Therefore, (1.2) can be written alternatively as

$$i(t,z) = I \cos\left(\omega t - \frac{2\pi}{\lambda} z\right) \qquad (1.5)$$

The wavelength of the wave is the distance between successive corresponding points such as the crest of the wave as shown in Fig. 1.2a. This is similar to observing waves in the ocean. Movement of the ocean wave is ascertained by observing the movement of the crest of the wave as shown in Fig. 1.2b. The water particles actually exhibit an up–down motion but the wave appears to move along the ocean surface. In order to track the movement of the wave we observe the movement of a common point on the wave. For

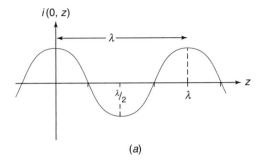

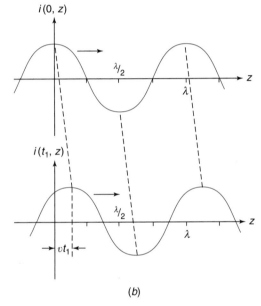

Figure 1.2 Wave propagation. (a) Wave propagation in space and wavelength. (b) Wave propagation as time progresses.

the sinusoidal wave in (1.2) this means that we track points where the argument of the cosine function, $\omega t - \beta z$, remains constant. For the wave in (1.2), as time t increases, position z must increase to keep this argument of the cosine function constant. Hence the wave in (1.2) is traveling in the $+z$ direction. We will also determine in Chapter 5 that the wavelength can be written in terms of the velocity of the wave propagation, v, and the cyclic frequency of the wave, f, as

$$\lambda = \frac{v}{f} \quad \text{m} \tag{1.6}$$

In the case of a connection wire where the surrounding medium is free space, that velocity of wave propagation is approximately 3×10^8 m/s. Table 1.2 gives the wavelength of sinusoidal waves propagating in free space for various frequencies of that wave. Combining (1.4) and (1.6) gives the velocity of propagation in terms of the phase constant and frequency of the wave as

$$
\begin{aligned}
v &= \lambda f \\
&= \frac{\omega}{\beta}
\end{aligned}
\tag{1.7}
$$

TABLE 1.2 Frequencies of Sinusoidal Waves and Their Corresponding Wavelengths

Frequency (f)	Wavelength (λ)
60 Hz	3107 miles (5000 km)
3 kHz	100 km
30 kHz	10 km
300 kHz	1 km
3 MHz	100 m
30 MHz	10 m
300 MHz	**1 m**
3 GHz	10 cm
30 GHz	1 cm
300 GHz	1 mm

Substituting (1.7) into (1.2) gives

$$
\begin{aligned}
i(t,z) &= I \cos\left(\omega\left(t - \frac{\beta}{\omega}z\right)\right) \\
&= I \cos\left(\omega\left(t - \frac{z}{v}\right)\right)
\end{aligned}
\tag{1.8}
$$

The result in (1.8) illustrates that *phase shift of the wave is equivalent to time delay* which is given by z/v seconds.

From (1.3) and (1.4) as the current propagates along the connection leads a distance of one wavelength, $\mathcal{L} = \lambda$, it suffers a phase shift of $\phi = \beta\lambda = 2\pi$ radians or 360°. In other words, if the total length of the connection leads is one wavelength, the current entering the connection leads and the current exiting those leads are in phase but have changed phase 360° in the process of transiting the element. On the other hand, if the total length of the connection leads is one-half wavelength ($\mathcal{L} = \lambda/2$), then the current suffers a phase shift of 180° so that the current entering the connection leads and the current exiting those leads are completely *out of phase*. If the length of the connection leads is 1/10 of a wavelength the current suffers a phase shift of 36°. Over a distance of 1/20 of a wavelength it suffers a phase shift of 18°, and over a distance of 1/100 of a wavelength it suffers a phase shift of 3.6°. If the effects of the connection leads are to be unimportant as is assumed by the lumped-circuit model, then the total length of the connection leads must be such that this phase shift is negligible. There is no fixed criterion for this but we will asume that the phase shift is negligible if the lengths are smaller than, say, 1/10 of a wavelength at the excitation frequency of the source. For some situations the phase shift must be smaller than this to be negligible. Physical dimensions are not as important as *electrical dimensions* in determining the behavior of an electric circuit or device. Electrical dimensions are the physical dimensions in wavelengths. A physical dimension that is smaller than 1/10 of a wavelength is said to be *electrically small* in that the phase shift as a wave propagates across that dimension may be ignored. These concepts give rise to the rule of thumb that lumped-circuit models of circuits are an adequate representation of the physical circuit so long as the largest *electrical dimension* of the physical circuit is less than, say, 1/10 of a wavelength. Table 1.3 gives the frequencies and corresponding wavelengths for various applications.

TABLE 1.3 Frequencies and Corresponding Wavelengths of Electronic Systems

Frequency Band[a]	Wavelength	Uses
EHF (30–300 GHz)	1 cm–1 mm	Radar, remote sensing, radio astronomy
SHF (3–30 GHz)	10 cm–1 cm	Radar, satellite communication, remote sensing, microwave electronic circuits, aircraft navigation, **digital systems**
UHF (300–3000 MHz)	1 m–10 cm	Radar, TV, microwave ovens, air navigation, cell phones, military air traffic control communication and navigation
VHF (30–300 MHz)	10 m–1 m	TV, FM broadcasting, police radio, mobile radio, commercial air traffic control (ATC) communication and navigation
HF (3–30 MHz)	100 m–10 m	Short wave radio (ham), citizens band
MF (300–3000 kHz)	1 km–100 m	AM broadcasting, maritime radio, ADF direction finding
LF (30–300 kHz)	10 km–1 km	Loran long-range navigation, ADF radio beacons, weather broadcasting
VLF (3–30 kHz)	100 km–10 km	Long-range navigation, sonar
ULF (300–3 kHz)	1 Mm–100 km	Telephone audio range
SLF (30–300 Hz)	6214 mi–621 mi	Communication with submarines, commercial power (60 Hz)
ELF (3–30 Hz)	62,137 mi–6214 mi	Detection of buried metal objects

[a]E = extra, S = super, U = ultra, V = very, H = high, M = medium, L = low, F = frequency.

Today's technology utilizes ever-increasing frequencies. Cell phones, radar systems, and satellite broadcast systems utilize frequencies in the GHz (10^9 Hz) range where a wavelength is on the order of centimeters. Lumped-circuit models of these circuits are only valid when the circuit dimensions are in the millimeter range. Digital computers and other imbedded processors operate with clock speeds (repetition rates of the pulses) at GHz speeds. These waveforms are in the form of trapezoidal pulses at the clock rate with very fast rise and fall times of the leading and falling edges (picoseconds) as shown in the *time domain* in Fig. 1.3a. This time-domain pulse train can be viewed, alternatively, in the *frequency domain* using the Fourier series as being composed of sinusoidal components whose sum gives the time-domain pulse shape. Hence the time-domain waveform having period P can be represented as

$$v(t) = V_0 + V_1 \cos(\omega_o t + \theta_1) + V_2 \cos(2\omega_o t + \theta_2) + V_3 \cos(3\omega_o t + \theta_3) + \cdots \quad (1.9)$$

where $\omega_o = 2\pi f_o$ and $f_o = 1/P$ is the fundamental repetition frequency of the pulse train. The frequency content of the pulses consists of a dc term (the average value), V_0, as well as sinusoids at the fundamental frequency, f_o, plus all its higher-frequency harmonics which are integer multiples of this fundamental frequency. In Fig. 1.3b we have illustrated this by showing the amplitudes of the sinusoidal components of the pulse in the *frequency domain*. Figure. 1.3c shows a *spectrum analyzer* that is used to display the frequency components of a periodic waveform. A spectrum analyzer is essentially a radio receiver with a bandpass filter that is swept in time. At a particular point in the sweep, those frequency components that fall within the bandwidth of the bandpass filter are displayed. Figure 1.3c shows the spectrum of a 1-V, 1-MHz trapezoidal waveform having a 50% duty cycle, and rise/fall times of 12.5 ns. For a 50% duty cycle, the even

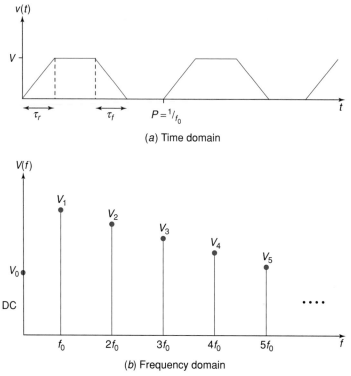

Figure 1.3 Digital clock signal. (a) In the time domain. (b) In the frequency domain. (c) A spectrum analyzer for determining the frequency (spectral) content of a periodic signal (reprinted from C.R. Paul, *Introduction to Electromagnetic Compatibility*, John Wiley Interscience, 1992, p. 378, courtesy of Agilent Technologies ©2002).

harmonics have, ideally, zero amplitude. The spectrum analyzer plot shows that these even harmonics are not zero although they are significantly below the amplitudes of the odd harmonics. Hence the frequency (spectral) content of digital signals is spread out from the basic repetition frequency to much higher frequencies that are multiples of the basic repetition frequency. *In order for lumped-circuit models to be valid, the electrical dimensions of the circuit must be small at all of these frequencies that are significant in forming the pulse spectrum.* For example, a digital computer clock waveform shown in Fig. 1.3a at 600 MHz having a level of 5 V, rise/fall times of 500 ps and a duty cycle of 50% will contain frequency components (sinusoidal components) shown in Table 1.4.[1]

Hence, there are significant frequency components of this pulse well into the GHz range and the electrical dimensions must be small *at all these significant spectral components* in order to correctly analyze the processing of this signal by a circuit using a lumped-circuit model of the circuit. For example, in order to successfully use lumped-circuit models to predict the processing of the seventh harmonic at 4.2 GHz, the maximum dimension of the circuit must be less than some 0.71 cm or 7.1 mm!

When the dimensions of a circuit are no longer electrically small, we must use the EM laws which are called Maxwell's equations. Although use of these EM laws in modeling a structure is mathematically more difficult than the use of lumped-circuit models, *we have no alternative!* Using a lumped-circuit model to analyze a circuit or

---

[1]C.R. Paul, *Introduction to Electromagnetic Compatibility*, John Wiley Interscience, 1992.

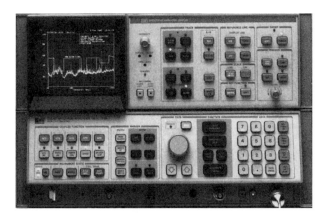

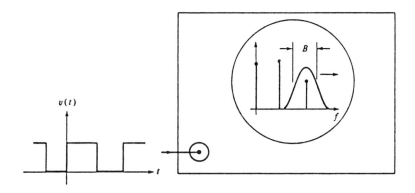

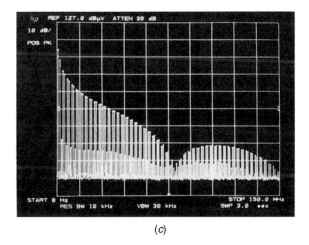

(c)

Figure 1.3 *(Continued)*

structure that is electrically large is a waste of time since the answers derived from it will be incorrect.

This text is devoted to a study of the basic electromagnetic principles and laws and their application to electrical and computer engineering problems. EM can be inherently intensive, mathematically. However, there has been a conscious attempt throughout this text to minimize that mathematical detail where possible. A certain amount of mathe-

TABLE 1.4   Spectral (Frequency) Components of a 5-V, 600-MHz, 50% Duty Cycle, 500-ps Rise/Fall Time Digital Clock Signal

Harmonic	Frequency	Wavelength	Level (V)
1	600 MHz	50 cm	2.73 V
2	1.2 GHz	25 cm	0
3	1.8 GHz	16.7 cm	116 mV
4	2.4 GHz	12.5 cm	0
5	3 GHz	10 cm	135 mV
6	3.6 GHz	8.3 cm	0
7	4.2 GHz	7.1 cm	21 mV
⋮			

matical detail is unavoidable. However, any mathematics that is not directly necessary will be avoided or delayed. It is important for the reader to adopt the following philosophy. Virtually all EM problems that can be solved by hand have been solved. Those that are solvable by hand must possess some symmetry and fit one of the three coordinate systems that we will describe (rectangular, cylindrical, or spherical). It is a serious error to expect that we can solve EM problems in the same fashion as lumped-circuit problems, namely, plug numbers into equations and grind out an answer. In order to be successful at solving the EM equations, we must generally do two things: visualize and *sketch* the problem in two or three dimensions, and anticipate the *form of the solution* wherein we can determine the unknown constants in that form so that they satisfy the EM field laws. If this attitude is adopted, EM problems can become virtually *trivial* to solve. If the attitude of "plug and chug" is chosen, the EM problems can become very difficult to solve.

▶ 1.3 A BRIEF HISTORY OF ELECTROMAGNETICS

Although an awareness of electromagnetism extends as far back as the early Greek civilizations, the major laws were discovered, experimentally, in the nineteenth century. For a thorough and entertaining discussion of this history, the reader is referred to R.S. Elliot, Electromagnetics, McGraw-Hill, 1966. It is important to recognize that although these basic laws were discovered in the 1800s via seemingly crude (in today's context) experiments, they have remained unchanged for some 200 years. This is a credit to the ingenuity and careful construction of experimental apparatus by those early EM experimentalists.

One of the most important and fundamental aspects of electromagnetism is the inverse square law. The gravitational attraction between two bodies varies directly as the product of the masses of the bodies and inversely as the square of the distance between them. Similarly, the force exerted by two electric charges on each other varies as the product of the charges and inversely as the square of the distance between them. This is called Coulomb's law and will be the first EM law that we will study. The fact that the force of attraction or repulsion varies inversely with distance squared (not distance to the 1.999 power, not distance to the 2.0001 power but *precisely* distance to the power 2.000000...) is extremely important. Without that precise variation, the EM laws would not make sense. The inverse square law was discovered by John Mitchell (1724–1793) in 1750 and later confirmed for electric charges by Charles deCoulomb (1736–1806) in his now famous Coulomb's law which we will first study in Chapter 3. The experiments of Mitchell and Coulomb involved stationary distributions of charge. The next important discovery was that a steady flow of charge, a dc current, could also produce another type of electromagnetic field, the magnetic field. In order to produce a current one needed a battery. Batteries to produce such a current were not in existence until Alessandro Volta (1745–1827) produced the first one in 1800 (which he called the voltaic pile). Hans Christian Oersted (1777–1851) discovered in 1820 that a wire carrying a current could deflect the needle of a compass. André Ampère (1775–1836) soon after discovered that two current-carrying wires would create forces on each other of either repulsion or attraction depending on the relative current directions. At about the same time Jean Baptiste Biot (1774–1862) and Felix Savart (1791–1841) formulated the famous Biot-Savart law giving the magnetic flux density from a current element. The Biot-Savart law is, in certain respects, the counterpart to Coulomb's law in that it is also an inverse square law for currents, whereas Coulomb's law is an inverse square law for static charges. Karl Frederich Gauss (1777–1855) discovered Gauss' law of electrostatics which provides that knowing the number of electric flux lines leaving a closed surface allows one to determine the net positive charge enclosed by that surface.

At this point the electric field (due to stationary charges) and the magnetic field (due to the steady flow of electric charge, i.e., dc current) seemed independent of each other. A dc current had been shown to produce a magnetic field. Michael Faraday (1791–1867) discovered the converse to this in 1831 that a *time-changing* magnetic field would produce an electric field which produces a current in a closed loop and obtained one of the more important of Maxwell's equations, Faraday's law. This showed that for time-changing currents, the electric and magnetic fields are not independent but are related. Faraday wound two coils on an iron ring and attached a battery to one coil and a galvanometer (essentially a voltmeter of that time) to the terminals of the other coil. He discovered that when he connected the battery a deflection of the needle of the galvanometer occurred, and when he disconnected the battery the galvanometer deflected in the other direction. While the battery was connected (and the current in the first coil was steady) no deflection of the galvanometer occurred. From this he reasoned that a *time-changing* magnetic field (due to the time-changing current) caused a time-changing electric field which is his celebrated law. Until Faraday's experiment it was thought that the electric field was caused solely by the static distribution of charges. Joseph Henry (1797–1878) is also said to have made this discovery at about the same time as Faraday, but he did not publish his work so Faraday is credited with the discovery.

It remained for a Scottish mathematician, James Clerk Maxwell (1831–1879) to unify the previous discoveries and to make an important addition to Ampere's law in his famous treatise, *Electricity and Magnetism*, in 1873. These famous equations predicted the existence of electromagnetic *waves* which Heinrich Hertz (1857–1894) demonstrated, experimentally, in 1887. This led to modern radio, radar, television transmission, and numerous other wireless communication devices. It is rather remarkable that the majority of these important laws were discovered in the 1800s with seemingly crude experimental apparatus yet they have remained *unchanged* for some 200 years. This is an extraordinary tribute to the ability of these early experimenters to construct precise experimental equipment and to their scientific ability to translate their experimental observations into the mathematical laws that govern all electromagnetic phenomena.

▶ 1.4 AN OVERVIEW OF ELECTROMAGNETIC FIELDS

From the previous historical discussion we see that charges produce forces on other charges. Hence we can envision an invisible *force field* around stationary charges. This force field will be one of our primary EM quantities that we will be interested in determining and is called the *electric field*. Currents, or the flow of charge, also produce forces on other currents. Again we can envision an invisible force field in the space around currents. This force field will be the second primary EM quantity that we will be interested in determining and is called the *magnetic field*.

Each of these force fields has a magnitude and direction of effect and hence is characterized, mathematically, as a *vector*. The EM laws are therefore stated, mathematically, in terms of vector quantities. The field vectors will be dependent on not only time, t, but also their location in space, x,y,z. On the other hand, the currents and voltages in lumped-circuit models are independent of position and depend only on time. Hence the EM field quantities will be functions of time and spatial location, and the governing EM equations will be partial differential vector equations. These partial differential equations are somewhat more difficult to solve than the ordinary differential equations associated with lumped circuits. An example of another type of vector field having similar governing equations is fluid flow in which we depict with arrows (vectors) the direction and rate-of-flow of the fluid at various locations in the fluid. Heat transfer is another example whose governing equations are also vector partial differential equations. Our task in solving EM problems and in understanding the physical principles embodied in the laws will be to mathematically determine these vectors throughout space for each problem. The vector aspect of the field equations makes them difficult to solve for the most general cases. Fortunately, we will not be so ambitious as to try to solve them for every conceivable case. We will only attempt to solve them for special cases that exhibit some symmetry, fit a coordinate system, and are useful in modeling practical problems. Although we can interpret in words these EM laws (which will be important to do), we cannot quantify the solution unless we can determine the mathematical solution of them. Lord Kelvin, expressed this important thought as

> When you can measure what you are speaking about and express it in numbers you know something about it; but when you cannot measure it, when you cannot express it in numbers your knowledge is of meagre and unsatisfactory kind; it may be the beginning of knowledge but you have scarcely progressed in your thoughts to the stage of science whatever the matter may be.

▶ 1.5 ENGINEERING APPLICATIONS OF ELECTROMAGNETICS

1.5.1 Transmission Lines

Consider a pair of parallel wires connecting a source to a load shown in Fig. 1.4. This structure is called a *transmission line* which will be studied in Chapter 6. Suppose a pulse source is switched onto the line. Based on our previous electric circuit courses, we might expect that the pair of wires has no effect and the pulse would immediately arrive at the load. In actuality, a wave will be sent down the line and will require a time delay of $T = \mathscr{L}/v$ to reach the load where v is the velocity of propagation on the line

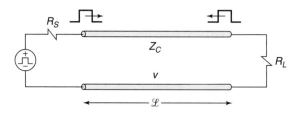

Figure 1.4 A transmission line illustrating propagating waves and reflections at the terminations.

and $\mathscr{L}$ is the total line length. The voltage and current are waves of the sinusoidal components of this pulse and will be of the form

$$V(z,t), I(z,t) = A\cos(\omega t - \beta z)$$

$$= A\cos\left(\omega\left(t - \frac{z}{v}\right)\right)$$

If the medium surrounding the wires is air, that velocity of propagation is approximately 3×10^8 m/s. The consequence of this is that in today's high-speed digital computers, the time delay of propagation can affect timing of the logic signals. For example, a typical printed circuit board (PCB) constructed of glass-epoxy is used to support the electronic components and their interconnections. The components are connected by "lands" which are thin copper strips of thickness 35 μm (1.4 mils where 1 mil is 0.001 in.) and width on the order of 127 μm (5 mils). These lands form transmission lines. The velocity of propagation on a PCB is influenced by the board material and is less than in free space being on the order of 1.8×10^8 m/s. Hence a typical land length of some 18 cm (some 7 in.) can have a delay of 1 ns. Modern high-speed digital computers rely on precise time intervals for passing and interpreting data and instructions. This time delay of 1 ns is becoming on the order of those time intervals and hence the time delay is becoming intolerable. A PCB in a laser printer is shown in Fig. 1.5. The thin copper lands interconnect the modules on the board and constitute transmission lines.

Equivalently, the wave suffers a phase shift as it propagates along the wires. If the physical length of the wires is electrically small, for example, $\mathscr{L} \ll 1/10\ \lambda$, this phase shift may be approximately ignored and the behavior of the line can be approximately described by Kirchhoff's laws and lumped-circuit models.

In addition, we will find in Chapter 6 that another parameter of a transmission line, the characteristic impedance, will determine how much of the incoming wave at the load is reflected back to the source. This could also affect the performance of digital computer components that are interconnected by these lands. The logic levels must be within

Figure 1.5 A printed circuit board (PCB) used in a laser printer illustrating the interconnection lands.

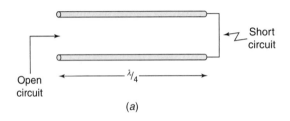

(a)

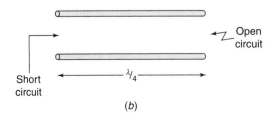

(b)

Figure 1.6 Illustration of the nonideal effects caused by an electrically long transmission line. (a) A short-circuited line that is one-quarter of a wavelength appears to be an open circuit at its input. (b) An open-circuited line that is one-quarter of a wavelength appears to be a short circuit at its input.

certain voltage levels in order to be correctly interpreted as a logic 0 or a logic 1. The tolerances for transistor-transistor logic (TTL) are 2.4 V–5 V for a logic 1 and 0 V–0.4 V for a logic 0. If the voltage at the load (the input to a digital gate) is between these, an error may occur because the level may be incorrectly interpreted. If the input to the load gate (represented by R_L) does not equal the characteristic impedance of the line, Z_C, then a portion of the incoming wave will be reflected and the level of the total voltage will not equal that sent out by the source, for example, 5 V. Hence a logic error may be created. In Chapter 6 we will learn how to determine these total voltages at the source and the load. Lumped-circuit models of the line may not be sufficiently accurate to determine this, and another model, the transmission-line equations, must be used in order to correctly calculate the load voltage. Increasing speeds of digital computers are requiring the use of the transmission-line model for ascertaining proper timing.

The transmission line also has some features in the frequency domain that are not predicted with lumped models of the line. For example, suppose a transmission line is excited by a single-frequency sinusoidal source. If the line is λ/4 in electrical length at that frequency and is terminated in a short circuit, the input to the line will appear to be an open circuit and vice versa as shown in Fig. 1.6. Once again, lumped circuit models of the line will not predict this important result.

1.5.2 Antennas

Consider the dipole antenna shown in Fig. 1.7. Antennas will be studied in Chapter 7. The antenna consists of two wire segments, and a sinusoidal source is attached to the input terminals. We will find that the electric and magnetic fields "radiated" by antennas will also be of the form of waves:

$$E(r,t), H(r,t) = \frac{A}{r} \cos(\omega t - \beta r)$$

$$= \frac{A}{r} \cos\left(\omega\left(t - \frac{r}{v}\right)\right)$$

where r is the radial distance from the antenna to the point of interest. Again, these waves will suffer a time delay r/v representing the fact that changes occurring in the antenna current will not be felt at some distant point until after a time delay of $T = r/v$. Since the top and bottom ends of the wires are not connected, lumped-circuit theory and Kirchhoff's

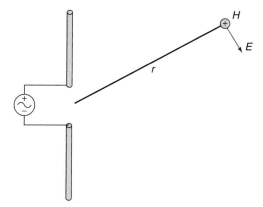

Figure 1.7 A dipole antenna illustrating its radiated electromagnetic fields.

laws would predict that the current along the wires is zero. How then would this antenna "radiate" a signal to another antenna? The answer is that Kirchhoff's laws only consider one type of current, conduction current, on the wires. The EM field equations include a second form of current, displacement current, which has an effect similar to conduction current. Displacement current in the air between the wires "completes the circuit" and allows current to flow along the wires. It is displacement current between the plates of a capacitor that explains how a capacitor can conduct "current" across an open circuit between the capacitor plates. But Kirchhoff's laws do not directly include this; we "fix them up" to include this in the same manner that we include mutual inductance in our list of lumped-circuit elements in order to model elements that are not directly connected.

Another aspect of an antenna not predicted by lumped-circuit models is that power will be delivered to the antenna terminals. Where does it go? Part of that power (a small part) is dissipated in the resistance of the antenna wires but another part of that power (the major portion) leaves the antenna and radiates into space never to return. The EM field laws successfully predict all these features of an antenna.

Photographs of various types of antennas are shown in Fig. 1.8.

1.5.3 Electromagnetic Compatibility (Interference)

Interference in electronic systems is becoming an increasing concern in today's high-density, high-speed electronic systems. A branch of electrical engineering known as electromagnetic compatibility (EMC) deals with the design of electronic systems so that unintended coupling of signals from one system to another will not cause an incorrect or unintended response of that system. Figure 1.9 shows some examples of how such interference can occur. Figure 1.9a illustrates the phenomenon of *crosstalk*. Two pairs of parallel wires (transmission lines) are routed in close proximity (perhaps in a common bundle). A pulse sent down one pair of wires will cause electric and magnetic fields about those wires. These fields will interact with the second pair of wires and induce a signal (similar to the one sent down the first pair) in that circuit so that a "noise" voltage will appear across the terminations of that second circuit. This is unintended, and if the induced signal is of sufficient magnitude or frequency content, the terminations at the ends of the second pair may interpret this as a signal being transmitted from the other end and an error may occur. This is similar to talking on the telephone and hearing another conversation faintly in the background. *Crosstalk* has occurred. Metallic connections are intended to guide a signal from a source to a load and nowhere else. This unintended coupling essentially provides an unintended path between two circuits almost as though they were intentionally connected. In high-speed digital computers

(a)

Figure 1.8 Typical antennas. (a) A parabolic antenna used for satellite communication (courtesy of Scientific-Atlanta, Inc.). (b) A horn antenna used for testing electronic devices for susceptibility to electromagnetic waves (courtesy of ETS-Lindgren, Inc.). (c) A log-periodic antenna used for electromagnetic compatibility (EMC) testing (courtesy of ETS-Lindgren, Inc.) (d) A biconical antenna used for electromagnetic compatibility (EMC) testing (courtesy of ETS-Lindgren, Inc.).

the digital logic pulses contain high-frequency components extending well into the GHz range. Typically, the higher the frequency, the more easily signals will "couple" from one pair of lands on a printed circuit board (PCB) to another pair creating crosstalk. Logic errors on PCBs in digital computers that are due to crosstalk between the lands of the PCBs are becoming an increasing concern and will continue to increase in importance as the speeds of those computers increase.

Figure 1.9b illustrates another potential interference problem. The air is filled with a wide range of signals, for example, AM radio transmissions, FM radio transmissions, cell phone transmissions, and radar transmissions. Even though some of these are not digital, if they inadvertently couple into a digital device they can cause signals to be induced in that device and cause it to be upset. A typical example is an airport surveillance radar. A digital device (computer, typewriter, laser printer, etc.) in a home near an airport

(b)

Figure 1.8 *(Continued)*

is illuminated by the airport's radar as the beam sweeps past the house. The radar signal may be strong enough to induce signals into the digital device to cause it to malfunction. Manufacturers of digital devices routinely test their products for susceptibility to radar signals by illuminating them with typical radar waveforms to insure that the device will operate properly in the presence of these signals. Conversely, a digital computer can cause interference in a radio or a TV. For example, suppose someone is using their digital computer in an apartment, and in an adjacent apartment someone is watching their TV. The computer is constantly emanating electromagnetic signals (unintended)

(c)

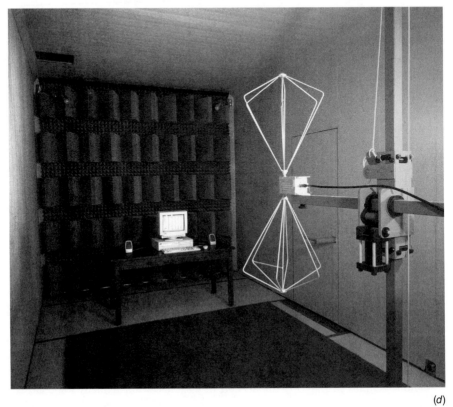

(d)

Figure 1.8 *(Continued)*

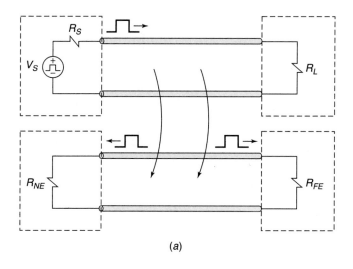

(a)

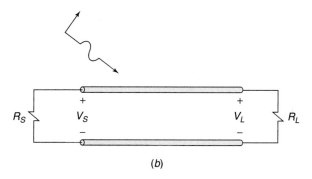

(b)

Figure 1.9 Illustration of various situations of electromagnetic interference. (a) Crosstalk between two transmission lines. (b) Susceptibility of a transmission line to an incident electromagnetic field.

and some portion of those signals may be picked up by the adjacent TV causing degradation of the picture. This is also becoming an increasing problem with the proliferation of high-speed computing devices. The Federal Communications Commission (FCC) in the U.S. instituted regulations in 1979 that require that all digital devices sold in the United States must be tested to determine their radiated emissions from 30 MHz to well over a GHz in order to prevent interference with wire and radio communications. The FCC requires that those emissions be below certain limits or else *it is illegal to sell that device in the United States*. Most other countries throughout the world have similar regulations for a similar purpose of preventing digital devices from causing electromagnetic interference. Hence it does little good to design a digital device that performs some unique function if the resulting product fails to meet the FCC limits; it cannot be sold! Manufacturers of digital devices understand this so that EMC has become as important a design criterion as cost. Figure 1.10 shows a semianechoic chamber that is used to test a digital device for compliance to the FCC radiated emission limits.

Figure 1.11a illustrates another coupling situation that is predicted by the EM laws. A potentially susceptible electronic device is housed in a metallic box ("shielded"). Yet a hole or slot is cut in that box to, for example, allow air flow for cooling. An incident wave from, for example, an airport radar may couple through that slot and be picked up on the sensitive electronics circuit inside causing interference. And finally, Fig. 1.11b illustrates another important EMC problem that is becoming of increasing concern in today's digital electronics. Many of us have experienced the small shock caused by walking along a nylon carpet and touching a metallic doorknob. What has happened is that as we walk along the carpet, charge is deposited on our body. When we touch the

Figure 1.10 Illustration of a typical test of a digital device for its radiated emissions as required by various governmental regulations. (Courtesy of ETS-Lindgren, Inc.)

doorknob, that charge is "discharged" resulting in a miniature lightning bolt between our hand and the doorknob. This is not a rare occurrence and the level of discharge can be on the order of 10 kV. (Discharges of less that 3 kV are usually not felt.) This is called electrostatic discharge (ESD) and can either upset or permanently damage digital electronics. Manufacturers of digital devices routinely test their products for ESD susceptibility by discharging ESD "guns" to the device to simulate what will happen in an office environment and to insure that the device is immune to this phenomenon.

In today's dense configuration of sensitive electronics, it would do irreparable harm to a manufacturer's name to produce a device (digital or otherwise) that is susceptible to the above interference sources. The company would suffer a devastating loss of consumer confidence in the company's ability to produce a quality product. Hence electromagnetic compatibility (EMC) is becoming an important design criterion. Today's

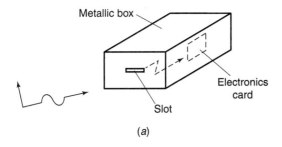

(a)

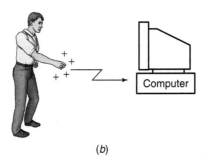

(b)

Figure 1.11 Illustration of additional situations of electromagnetic interference. (a) Coupling of an incident electromagnetic field through openings in a shielded enclosure. (b) Effect of an electrostatic discharge (ESD) on digital devices.

design of digital devices require that designers have some understanding of EMC and good EMC design practices. The problem in not having that knowledge is that the designer does not "see" the "hidden schematic" causing this unintended coupling and hence does not design the system to prevent it. Study of the EM laws and principles of this text will not make designers experts in EMC but at least they will be aware of when they need to consider it. An analogy is that it is not a tragedy to not know how to spell every word one encounters, but one should know when he/she does not know the correct spelling of a word and look it up in a dictionary.

1.5.4 Design of Communication Systems

Wireless communications such as cell phones rely on having an efficient transmission medium between the transmitter and the receiver. This transmission medium generally takes the form of a transmitting antenna, a propagation path (usually air), and a receiving antenna. We want to focus the transmitted energy in the direction of the receiving antenna because any energy that is directed in other directions is wasted. This is one property of an antenna which is called its *gain*. Another property of an antenna is to provide a smooth transition between the source (which generates the signal to be transmitted) and the propagation path (which is generally air). The antenna provides a "matching" between the source (whose impedance is typically 50 Ω) and the propagation path through air (whose intrinsic impedance is 377 Ω) so that little of the source power is reflected back to the source and the majority of the source signal is sent to the antenna and radiated. Antennas will be covered in Chapter 7.

Figure 1.12 shows a parabolic antenna used in a line-of-sight communication link.

1.5.5 Design of High-Speed Digital Electronics

From the previous discussion, the reader should have gained an appreciation of the EM problems that exist in the design of high-speed computers and other digital devices. Simply writing software to perform a task is of no use unless the machine (the computer)

Figure 1.12   Antennas in a line-of-sight communication link. (Courtesy of Scientific-Atlanta, Inc.)

can reliably implement that in electronic form. Time delay of propagation paths, re-flections at terminations, and interference are all critical aspects of the design of high-speed computers and other digital devices. Without attention being given to these as-pects, the digital electronics will encounter errors and will not faithfully carry out the software instructions that are intended.

▶ SUMMARY OF IMPORTANT CONCEPTS AND FORMULAE

1. **Unit conversion:** multiply by unity ratios between the two systems and cancel unit names to avoid the improper multiplication (division) where division (multiplication) is intended.

2. **Powers of ten multipliers:** 10^9 giga (G), 10^6 mega (M), 10^3 kilo (k), 10^{-2} centi (c), 10^{-3} milli (m), 10^{-6} micro (μ), 10^{-9} nano (n), 10^{-12} pico (p).

3. **Time delay of wave propagation:** $T = \mathcal{L}/v$ s, $v = 3 \times 10^8$ m/s (in air).

4. **Sinusoidal traveling wave:** $i(z,t) = I \cos(\omega t - \beta z) = I \cos(\omega(t - z/v))$.

5. **Wavelength:** $\lambda = v/f$ m, $v = 3 \times 10^8$ m/s (in air).

6. **Phase shift of sinusoidal wave:** $\phi = \beta \mathcal{L}$ radians.

7. **Periodic, time-domain pulses (digital clock signals):** are composed of sinusoidal com-ponents at multiples of the basic repetition frequency.

8. **Electrical dimensions and lumped-circuit models:** If the largest electrical dimension (in wavelengths) of an electric circuit or structure is greater than about $1/10$ λ, then lumped-circuit models and Kirchhoff's laws no longer apply and will give incorrect answers.

9. **Electromagnetic field:** An electromagnetic field is an invisible force field caused by charges and their movement (current). One of our primary objectives in this text is to determine (visu-alize and sketch) that field.

▶ PROBLEMS

SECTION 1.1 UNITS AND UNIT CONVERSION

1.1.1. Express the following values of resistance, capacitance, and inductance in terms of the multipliers in Table 1.1:

a. $25 \times 10^4 \, \Omega$ [250 kΩ]

b. $0.035 \times 10^4 \, \Omega$

c. 0.00045F [450 μF]

d. 0.003×10^{-7}F

e. 0.005×10^{-2}H [50 μH]

1.1.2. Convert the following dimensions to those indicated:

a. 30 miles to km [48.3 km]

b. 1 ft to mils

c. 100 yds (length of a U.S. football field) to meters [91.44 m]

d. 5 mm to mils

e. 20 μm (micron) to mils [0.7874 mils]

f. 880 yds (race distance) to m

SECTION 1.2 THE NEED FOR AN UNDERSTANDING OF ELECTROMAGNETIC PRINCIPLES

1.2.1. Determine the wavelength at the following frequencies in SI and in English units:

a. LORAN C long-range navigation 90 Hz [3333.3 km, 2071.2 mi]

b. Submarine communication 1 kHz

c. Automatic direction finder in aircraft 350 kHz [857.14 m, 0.533 mi]

d. AM radio transmission 1.2 MHz

e. Amateur radio 35 MHz [8.57 m, 28.12 ft]

f. FM radio transmission 110 MHz

g. Instrument landing system 335 MHz [89.55 cm, 2.94 ft]

h. Satellite 6 GHz

i. Remote sensing 45 GHz [6.67 mm, 262.5 mils]

1.2.2. Determine the following physical dimensions in wavelengths, that is, their electrical dimension:

a. A 50-mile length of a 60-Hz power transmission line $[1/62 \, \lambda]$

b. A 500-ft AM broadcast antenna broadcasting at 500 kHz

c. A 4.5-ft FM broadcast antenna broadcasting at 110 MHz $[0.5 \, \lambda]$

d. A 2-in. land on a printed circuit board (assume a velocity of propagation of 1.5×10^8 m/s) at 2 GHz

1.2.3. A sinusoidal current wave is described below. Determine the velocity of propagation and the wavelength. If the wave travels a distance d determine the time delay and phase shift.

a. $i(t,z) = I_o \cos(2\pi \times 10^6 t - 2.2 \times 10^{-2} z)$, $d = 3$ km. $[v = 2.856 \times 10^8$ m/s, $\lambda = 285.6$ m, $T = 10.5$ μs, $\phi = 3781.5°]$

b. $i(t,z) = I_o \cos(6\pi \times 10^9 t - 75.4 z)$, $d = 4$ in.

c. $i(t,z) = I_o \cos(30\pi \times 10^7 t - 3.15 z)$, $d = 20$ ft. $[v = 2.99 \times 10^8$ m/s, $\lambda = 1.995$ m, $T = 20.4$ ns, $\phi = 1100.2°]$

d. $i(t,z) = I_o \cos(6\pi \times 10^3 t - 0.126 \times 10^{-3} z)$, $d = 50$ mi.

CHAPTER 2

Calculation with Vectors

In this chapter we will study the basic methods for manipulating vector mathematical quantities. The reason we need to study vectors is that the electromagnetics (EM) laws will be stated in terms of these vector operations. In order to understand what the laws are telling us, we must understand the vector operations. The algebra of vectors (addition, subtraction, dot product, and cross product) will be studied first. Next we introduce the three important coordinate systems (rectangular, cylindrical, and spherical) that will allow us to quantify our solutions. Finally we will study the two important vector calculus ideas of line integrals and surface integrals. These mathematical operations constitute the "language" of electromagnetics in that the EM field laws will be stated in mathematical terms using these ideas. Certain other vector calculus operations will be introduced later when they are needed.

Chapter Learning Objectives

After completing the contents of this chapter you should be able to

▷ understand the meaning of vectors,

▷ compute the vector algebra operations of addition, subtraction, dot product, and cross product in rectangular, cylindrical, and spherical coordinate systems,

▷ compute the line integral of a vector field in each of the three coordinate systems,

▷ compute the surface integral of a vector field in each of the three coordinate systems, and

▷ sketch a vector field in three dimensions.

▷ 2.1 VECTORS

The EM quantities of interest, electric field intensity, electric flux density, magnetic field intensity, and magnetic flux density, are vector quantities. A vector is denoted as a line with an arrow at one end and contains two items of information, a direction of effect and a magnitude of that effect. The length of the vector denotes its magnitude and the arrow denotes the direction of effect. As a simple example of the use of vectors, consider Fig. 2.1. In attempting to move an object along a surface we apply a force denoted as a vector $\mathbf{F}$. A vector is denoted as boldface in this text and its magnitude is the symbol not in boldface, that is, $|\mathbf{F}| = F$. If the force is not directed parallel to the surface, only a portion of that force (the projection of $\mathbf{F}$ along the surface), $F\cos(\theta)$, will be useful in moving the block.

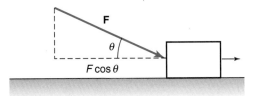

Figure 2.1   Illustration of components of a force vector in moving an object.

▶ 2.2 ADDITION AND SUBTRACTION OF VECTORS

Two vectors are added by displacing one vector to the tip of the other. The resulting vector formed by this chain is their sum, **A** + **B** as shown in Fig. 2.2. Two vectors are subtracted as **A** − **B** = **A** + (−**B**) by reversing the direction of **B** to give the vector −**B** and adding this to **A**. Note that the order of addition does not matter: **A** + **B** = **B** + **A**.

▶ 2.3 THE DOT PRODUCT OF VECTORS

Vectors cannot be simply "multiplied" together. There are two useful definitions of the product of two vectors, the dot product and the cross product. The necessity for these two product definitions is that the EM field laws are mathematically stated in terms of them. Hence if we are to understand the EM laws we must understand the meanings of these two products.

The first vector product is the *dot product*. The dot product of two vectors is the product of their magnitudes and the cosine of the angle between them:

$$\mathbf{A} \cdot \mathbf{B} = AB \cos(\theta) \tag{2.1}$$

as shown in Fig. 2.3. This can also be stated as the product of the magnitude of **A** and the projection of **B** *onto* **A** ($B \cos\theta$) or the product of the magnitude of **B** and the projection of **A** *onto* **B** ($A \cos\theta$). Observe that *the dot product gives a scalar (no direction information) as the result*. Note that the order of the two vectors does not matter: **A** · **B** = **B** · **A**. The dot product has many uses. For example, the dot product of a vector with itself yields the magnitude squared of the vector:

$$\mathbf{A} \cdot \mathbf{A} = |\mathbf{A}|^2 = A^2 \tag{2.2}$$

Note that the magnitude of a vector can be denoted one of two ways: $|\mathbf{A}|$ or A. Similarly, we can determine whether two vectors are perpendicular when it may not be obvious.

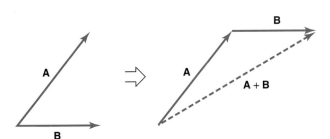

Figure 2.2   Addition of two vectors.

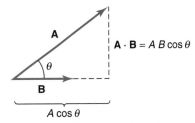

Figure 2.3   The dot product of two vectors.

If the two vectors are perpendicular then the included angle is $\theta = 90°$ and $\mathbf{A} \cdot \mathbf{B} = 0$. The dot product is distributive:

$$\mathbf{A} \cdot (\mathbf{B} + \mathbf{C}) = \mathbf{A} \cdot \mathbf{B} + \mathbf{A} \cdot \mathbf{C} \qquad (2.3)$$

▶ 2.4 THE CROSS PRODUCT OF VECTORS

The next vector product is the *cross product*. Unlike the dot product, *the cross product gives a vector as the result:*

$$\boxed{\mathbf{A} \times \mathbf{B} = AB \sin(\theta)\mathbf{a}_n} \qquad (2.4)$$

Hence the cross product has a magnitude that is the product of the magnitudes and the sine of the angle between them. The direction of this resulting vector is determined by the *right-hand rule* as shown in Fig. 2.4. If we curl the fingers of our right hand *from* $\mathbf{A}$ *to* $\mathbf{B}$, the thumb gives the direction of the resulting *vector* $\mathbf{A} \times \mathbf{B}$. This result is *perpendicular to the plane containing* $\mathbf{A}$ *and* $\mathbf{B}$. The notation $\mathbf{a}_n$ is a *unit vector* whose length is unity. Note that $\mathbf{A} \times \mathbf{B} = -\mathbf{B} \times \mathbf{A}$ and the order matters in the cross product unlike the dot product. Hence, the angle θ in (2.4) is measured *from* vector $\mathbf{A}$ *to* vector $\mathbf{B}$ according to the right-hand rule. The cross product, like the dot product, is distributive:

$$\mathbf{A} \times (\mathbf{B} + \mathbf{C}) = \mathbf{A} \times \mathbf{B} + \mathbf{A} \times \mathbf{C} \qquad (2.5)$$

The cross product of two vectors that are parallel is zero, $\mathbf{A} \times \mathbf{B} = 0$, since $\theta = 0°$.

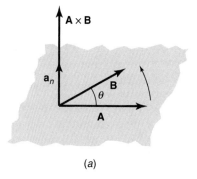

(a)

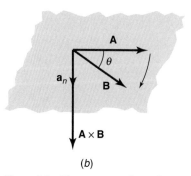

(b)

Figure 2.4 The cross product of two vectors and the right-hand rule for determining the direction of the resultant.

▶ 2.5 THE RECTANGULAR COORDINATE SYSTEM

In order to perform computations with vector quantities and obtain numerical results we require coordinate systems. The primary such coordinate system is the *rectangular coordinate system* shown in Fig. 2.5a. The coordinate system consists of three mutually orthogonal axes, x,y,z, and a point in space is denoted as $P(x_1,y_1,z_1)$. The location of the point is defined by the intersection of three planes as shown in Fig. 2.5b; one for x constant as $x = x_1$, one for y constant as $y = y_1$, and one for z constant as $z = z_1$. Unit vectors at this point are denoted as $\mathbf{a}_x$, $\mathbf{a}_y$, and $\mathbf{a}_z$ and point in the direction of increasing values of the respective coordinate axes. The three axes are labeled x, y, and z. These are ordered with the right-hand-rule such that $\mathbf{a}_x \times \mathbf{a}_y = \mathbf{a}_z$. Note that, for example, $\mathbf{a}_z \times \mathbf{a}_y = -\mathbf{a}_x$. A vector in this coordinate system can be written, using these unit vectors, as

$$\mathbf{A} = A_x\mathbf{a}_x + A_y\mathbf{a}_y + A_z\mathbf{a}_z \qquad (2.6)$$

where A_x, A_y, and A_z are the magnitudes of the projections of the vector on the respective coordinate axes. Addition of these vectors is obtained by adding the components since

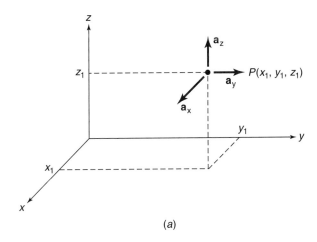

(a)

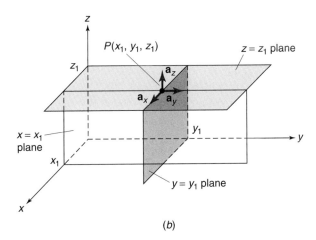

(b)

Figure 2.5 The rectangular coordinate system. (a) The axes of the coordinate system and the unit vectors. (b) Location of a point as the intersection of three constant-coordinate planes.

the unit vectors for each of the corresponding vector components are parallel:

$$\boxed{\mathbf{A} + \mathbf{B} = (A_x + B_x)\mathbf{a}_x + (A_y + B_y)\mathbf{a}_y + (A_z + B_z)\mathbf{a}_z}$$ (2.7)

The dot product is simply the sum of the products of the components:

$$\boxed{\mathbf{A} \cdot \mathbf{B} = A_xB_x + A_yB_y + A_zB_z}$$ (2.8)

This can be demonstrated by directly forming the result, using (2.3), as

$$\mathbf{A} \cdot \mathbf{B} = (A_x\mathbf{a}_x + A_y\mathbf{a}_y + A_z\mathbf{a}_z) \cdot (B_x\mathbf{a}_x + B_y\mathbf{a}_y + B_z\mathbf{a}_z)$$
$$= A_xB_x(\mathbf{a}_x \cdot \mathbf{a}_x) + A_xB_y(\mathbf{a}_x \cdot \mathbf{a}_y) + A_xB_z(\mathbf{a}_x \cdot \mathbf{a}_z) + \cdots$$

The terms involving the dot product of two identical unit vectors is unity because of (2.2), and the dot product of two different unit vectors is zero because these unit vectors are perpendicular. Note once again that *the dot product produces a scalar and not a vector.*

The cross product can be likewise obtained in terms of components but the result is a bit more involved. Using the result in (2.5), the cross product of the two vectors is

$$\mathbf{A} \times \mathbf{B} = (A_x\mathbf{a}_x + A_y\mathbf{a}_y + A_z\mathbf{a}_z) \times (B_x\mathbf{a}_x + B_y\mathbf{a}_y + B_z\mathbf{a}_z)$$
$$= A_xB_x(\mathbf{a}_x \times \mathbf{a}_x) + A_xB_y(\mathbf{a}_x \times \mathbf{a}_y) + A_xB_z(\mathbf{a}_x \times \mathbf{a}_z) + \cdots$$

The terms involving the cross product of two identical unit vectors is zero because the unit vectors are parallel, and the cross product of two different unit vectors yields a unit vector that is along the other coordinate axis, for example, $\mathbf{a}_x \times \mathbf{a}_y = \mathbf{a}_z$, $\mathbf{a}_z \times \mathbf{a}_y = -\mathbf{a}_x$. The result is

$$\boxed{\mathbf{A} \times \mathbf{B} = (A_yB_z - A_zB_y)\mathbf{a}_x + (A_zB_x - A_xB_z)\mathbf{a}_y + (A_xB_y - A_yB_x)\mathbf{a}_z}$$ (2.9)

This seemingly complicated result can nevertheless be formed very quickly using the following observation. Observe that each component is of the form

$$(A_\beta B_\gamma - A_\gamma B_\beta)\mathbf{a}_\alpha$$ (2.10)

Remembering the *cyclic* ordering of the axis labels as $x \rightarrow y \rightarrow z \rightarrow x \rightarrow y \rightarrow \cdots$ allows this rapid construction. If we are forming the x component we set $\alpha = x$. The next axis label in the sequence is y followed by z so we assign $\beta = y$ and $\gamma = z$ and subtract the product of the components with the labels swapped to give the x component. If we are forming the y component, the next axis label in the sequence is z followed by x so we assign $\alpha = y$, $\beta = z$, and $\gamma = x$. An alternative method of forming the cross product is the following mnemonic. Evaluate the following determinant, pretending that the unit vectors are scalars:

$$\mathbf{A} \times \mathbf{B} = \begin{vmatrix} \mathbf{a}_x & \mathbf{a}_y & \mathbf{a}_z \\ A_x & A_y & A_z \\ B_x & B_y & B_z \end{vmatrix}$$ (2.11)

which gives the result in (2.9). It is important to observe in (2.11) that the unit vectors (in the order $x \rightarrow y \rightarrow z$) form the first row and the components of $\mathbf{A}$ form the second row while the components of $\mathbf{B}$ form the third row. In computing $\mathbf{B} \times \mathbf{A}$, the components of $\mathbf{B}$ form the second row while the components of $\mathbf{A}$ form the third row.

▶ **EXAMPLE 2.1**

Determine the dot product and cross product of the following vectors.

$$\mathbf{A} = 2\mathbf{a}_x + 3\mathbf{a}_y - 4\mathbf{a}_z$$
$$\mathbf{B} = -1\mathbf{a}_x - 5\mathbf{a}_y + 6\mathbf{a}_z$$

SOLUTION The dot product is

$$\mathbf{A} \cdot \mathbf{B} = (2)(-1) + (3)(-5) + (-4)(6)$$
$$= -41$$

The cross product $\mathbf{A} \times \mathbf{B}$ is

$$\mathbf{A} \times \mathbf{B} = [(3)(6) - (-4)(-5)]\mathbf{a}_x + [(-4)(-1) - (2)(6)]\mathbf{a}_y + [(2)(-5) - (3)(-1)]\mathbf{a}_z$$
$$= -2\mathbf{a}_x - 8\mathbf{a}_y - 7\mathbf{a}_z$$

or, alternatively

$$\mathbf{A} \times \mathbf{B} = \begin{vmatrix} \mathbf{a}_x & \mathbf{a}_y & \mathbf{a}_z \\ 2 & 3 & -4 \\ -1 & -5 & 6 \end{vmatrix}$$
$$= [(3)(6) - (-4)(-5)]\mathbf{a}_x - [(2)(6) - (-4)(-1)]\mathbf{a}_y + [(2)(-5) - (3)(-1)]\mathbf{a}_z$$
$$= -2\mathbf{a}_x - 8\mathbf{a}_y - 7\mathbf{a}_z$$

The cross product $\mathbf{B} \times \mathbf{A}$ is

$$\mathbf{B} \times \mathbf{A} = [(-5)(-4) - (6)(3)]\mathbf{a}_x + [(6)(2) - (-4)(-1)]\mathbf{a}_y + [(-1)(3) - (-5)(2)]\mathbf{a}_z$$
$$= 2\mathbf{a}_x + 8\mathbf{a}_y + 7\mathbf{a}_z$$
$$= -\mathbf{A} \times \mathbf{B}$$

or, alternatively

$$\mathbf{B} \times \mathbf{A} = \begin{vmatrix} \mathbf{a}_x & \mathbf{a}_y & \mathbf{a}_z \\ -1 & -5 & 6 \\ 2 & 3 & -4 \end{vmatrix}$$
$$= [(-5)(-4) - (6)(3)]\mathbf{a}_x - [(-1)(-4) - (6)(2)]\mathbf{a}_y + [(-1)(3) - (-5)(2)]\mathbf{a}_z$$
$$= 2\mathbf{a}_x + 8\mathbf{a}_y + 7\mathbf{a}_z$$
$$= -\mathbf{A} \times \mathbf{B}$$

▷ **QUICK REVIEW EXERCISE 2.1**

Determine the dot product and cross product of the following vectors.

$$\mathbf{A} = -2\mathbf{a}_x + 5\mathbf{a}_y + 4\mathbf{a}_z$$
$$\mathbf{B} = 6\mathbf{a}_x - 3\mathbf{a}_y + \mathbf{a}_z$$

ANSWERS $\mathbf{A} \cdot \mathbf{B} = -23, \mathbf{A} \times \mathbf{B} = 17\mathbf{a}_x + 26\mathbf{a}_y - 24\mathbf{a}_z.$

Vector calculus operations involving vector field quantities that are expressed in this coordinate system require differential path lengths, surface areas, and volumes. These are shown in Fig. 2.6. Differential changes along the coordinate axes yield changes in those directions of dx, dy, and dz. A differential change in path length is represented as a vector in the direction of the change and is formed as the sum of the differential changes along the coordinate axes as

$$d\mathbf{l} = dx\mathbf{a}_x + dy\mathbf{a}_y + dz\mathbf{a}_z \qquad (2.12)$$

A differential volume is the product of the differential changes as

$$dv = dxdydz \qquad (2.13)$$

Differential surfaces are obtained as the product of the two differential sides. It turns out that we will need to ascribe a "direction" to these differential surfaces in the EM

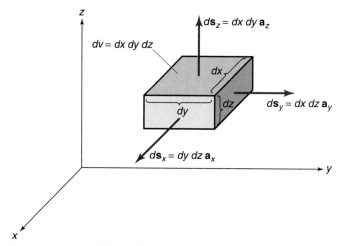

Figure 2.6 The differential surfaces in a rectangular coordinate system.

field laws so we define them as

$$\begin{array}{c} d\mathbf{s}_x = dydz\mathbf{a}_x \\ d\mathbf{s}_y = dxdz\mathbf{a}_y \\ d\mathbf{s}_z = dxdy\mathbf{a}_z \end{array} \qquad (2.14)$$

Observe that the vector direction of a differential surface is perpendicular to it in the direction of increasing coordinate value of that axis.

▶ **EXAMPLE 2.2**

Suppose we wish to compute the distance between the two points $P_1(2,-3,4)$ and $P_2(5,1,-6)$. Determine the distance between these two points directly and by integration.

SOLUTION   The direct computation is obtained from the square root of the sum of the squares of the distances along the coordinate axis as

$$D = \sqrt{(5 - 2)^2 + (1 - (-3))^2 + ((-6) - 4)^2} = \sqrt{125}$$

Alternatively we could integrate a differential length between these two points as

$$D = \int_{P_1}^{P_2} dl$$

We cannot simply integrate $\int_{x_1}^{x_2} dx + \int_{y_1}^{y_2} dy + \int_{z_1}^{z_2} dz$ since this would not constrain the path to a straight line between the points but instead would determine the sum of the distances along the x, y, and z axes which would be a "zigzag" path. In order to constrain the movement along a straight line between the two points we must incorporate the equation of a straight line between the two points into the integration. To do this we write dl in terms of one axis variable, say, x, as

$$dl = \sqrt{dx^2 + dy^2 + dz^2}$$

$$= dx\sqrt{1 + \left(\frac{dy}{dx}\right)^2 + \left(\frac{dz}{dx}\right)^2}$$

We first must write the relation between y and x and between z and x *along the path*. These are written using the coordinates of the point in the form of $y = mx + b$ and $z = nx + c$ as

$$y = \underbrace{\left[\frac{1 - (-3)}{5 - 2}\right]}_{m} x + [1 - m5]$$

$$= \frac{4}{3}x - \frac{17}{3}$$

Similarly we obtain

$$z = \underbrace{\left[\frac{(-6) - 4)}{5 - 2}\right]}_{n} x + [(-6) - n5]$$

$$= -\frac{10}{3}x + \frac{32}{3}$$

From these results we obtain

$$dy = \frac{4}{3}dx$$

$$dz = -\frac{10}{3}dx$$

and

$$dl = dx\sqrt{1 + \left(\frac{4}{3}\right)^2 + \left(-\frac{10}{3}\right)^2}$$

$$= \frac{\sqrt{125}}{3}dx$$

Hence

$$D = \int_{x=2}^{5} \frac{\sqrt{125}}{3}dx$$

$$= \sqrt{125}$$

as before.

▶ **QUICK REVIEW EXERCISE 2.2**

Determine the distance between the two points $P_1(-1,2,3)$ and $P_2(4,-3,-2)$ directly and by integration.

ANSWER $D = \sqrt{75}$.

▶ **2.6 THE CYLINDRICAL COORDINATE SYSTEM**

The (circular) cylindrical coordinate system is formed by three surfaces. One is a plane for constant z, $z = z_1$. The next surface is a cylinder centered on the z axis of radius $r = r_1$, and the third surface is a plane perpendicular to the xy plane and rotated about the z axis by angle $\phi = \phi_1$ *in the direction from x to y* as shown in Fig. 2.7. Observe that $0 \leq \phi \leq 2\pi$. A point is defined as the intersection of these three surfaces, $P(r_1,\phi_1,z_1)$. Unit vectors $\mathbf{a}_r$, $\mathbf{a}_\phi$, and $\mathbf{a}_z$ point in the direction of increasing coordinate value. Note that the unit vectors $\mathbf{a}_r$ and $\mathbf{a}_\phi$ change direction from point to point unlike

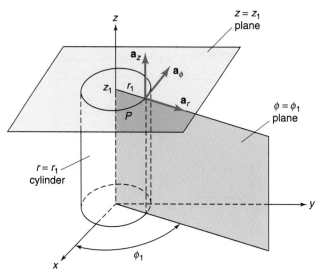

Figure 2.7 The cylindrical coordinate system illustrating the unit vectors and the location of a point as the intersection of three constant-coordinate surfaces.

the three unit vectors of a rectangular coordinate system which are fixed in direction. The three axes are labeled r, ϕ, and z and these are ordered with the right-hand-rule such that $\mathbf{a}_r \times \mathbf{a}_\phi = \mathbf{a}_z$. Note that, for example, $\mathbf{a}_r \times \mathbf{a}_z = -\mathbf{a}_\phi$.

Two vectors expressed in this coordinate system at point P can be added at that point since the unit vectors of the corresponding components are parallel *at that point*. Hence for

$$\mathbf{A} = A_r\mathbf{a}_r + A_\phi\mathbf{a}_\phi + A_z\mathbf{a}_z$$
$$\mathbf{B} = B_r\mathbf{a}_r + B_\phi\mathbf{a}_\phi + B_z\mathbf{a}_z$$

the sum is

$$\mathbf{A} + \mathbf{B} = (A_r + B_r)\mathbf{a}_r + (A_\phi + B_\phi)\mathbf{a}_\phi + (A_z + B_z)\mathbf{a}_z \qquad (2.15)$$

The dot and cross products are very similar to those for the rectangular coordinate system:

$$\mathbf{A} \cdot \mathbf{B} = A_rB_r + A_\phi B_\phi + A_zB_z \qquad (2.16)$$

$$\mathbf{A} \times \mathbf{B} = (A_\phi B_z - A_zB_\phi)\mathbf{a}_r + (A_zB_r - A_rB_z)\mathbf{a}_\phi + (A_rB_\phi - A_\phi B_r)\mathbf{a}_z \qquad (2.17)$$

Observe that the cyclic ordering of the axis labels, $r \rightarrow \phi \rightarrow z \rightarrow r \rightarrow \phi \rightarrow \cdots$ can, like the rectangular coordinate system, be used to quickly form the cross product using the rule in (2.10). Alternatively the cross product can be formed using the determinant mnemonic device in (2.11).

Differential path lengths, volumes, and surface areas are formed in a fashion similar to the rectangular coordinate system. There is one important difference. In the cylindrical coordinate system, r and z have the dimensions of length but ϕ is an angle whose units are not length. Hence, a differential length in the ϕ direction is $r d\phi$. This can be shown as in Fig. 2.8 in the xy plane. For a fixed r and a differential change in ϕ of $d\phi$, the arc length is $r \sin(d\phi) \cong r d\phi$ using the small angle approximation for the

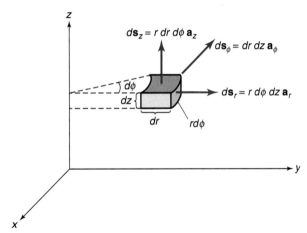

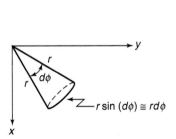

Figure 2.8 Illustration of differential arc length in a cylindrical coordinate system.

Figure 2.9 The differential surfaces in a cylindrical coordinate system.

sine function. Hence a differential path length is

$$d\mathbf{l} = dr\mathbf{a}_r + rd\phi\,\mathbf{a}_\phi + dz\mathbf{a}_z \qquad (2.18)$$

A differential volume is formed by taking differential changes in each of the coordinate axes, as shown in Fig. 2.9, giving

$$dv = (dr)(rd\phi)(dz) \\ = rdr\,d\phi\,dz \qquad (2.19)$$

Again the differential surface areas are needed as vector quantities so that

$$d\mathbf{s}_r = (rd\phi)(dz)\mathbf{a}_r \\ = rd\phi\,dz\,\mathbf{a}_r \\ d\mathbf{s}_\phi = (dr)(dz)\mathbf{a}_\phi \\ = drdz\,\mathbf{a}_\phi \\ d\mathbf{s}_z = (dr)(rd\phi)\mathbf{a}_z \\ = rdr\,d\phi\,\mathbf{a}_z \qquad (2.20)$$

▶ **EXAMPLE 2.3**

Determine the volume enclosed by a cylinder of radius R and length L as well as the surface area of that volume.

SOLUTION To determine the volume enclosed we integrate

$$V = \int_v dv$$

$$= \int_{r=0}^{R} \int_{\phi=0}^{2\pi} \int_{z=0}^{L} \underbrace{rdr\,d\phi\,dz}_{dv}$$

$$= \pi R^2 L$$

To determine the surface area we integrate over the sides, the bottom, and the top:

$$S = \underbrace{\int_{\phi=0}^{2\pi} \int_{z=0}^{L} (r = R)d\phi dz}_{\text{sides}} + \underbrace{\int_{\phi=0}^{2\pi} \int_{r=0}^{R} r d\phi\, dr}_{\text{bottom}} + \underbrace{\int_{\phi=0}^{2\pi} \int_{r=0}^{R} r d\phi\, dr}_{\text{top}}$$

$$= 2\pi RL + 2\pi R^2$$

◁

▷ QUICK REVIEW EXERCISE 2.3

Determine the volume enclosed by the surface defined by $0 \le r \le 2, 0 \le \phi \le 60°, 2 \le z \le 4$ as well as the surface area of that volume. (Hint: sketch the volume and surface.)

ANSWERS $4\pi/3, 8\pi/3 + 8$.

◁

▷ 2.7 THE SPHERICAL COORDINATE SYSTEM

The spherical coordinate system is also formed by three surfaces. One surface is a cone about the z axis of half angle θ. The next surface is a sphere centered on the origin of radius r, and the third surface is a plane perpendicular to the xy plane and rotated about the z axis by angle ϕ *in the direction from x to y* as shown in Fig. 2.10. Observe that $0 \le \phi \le 2\pi$ and $0 \le \theta \le \pi$. A point is defined as the intersection of these three surfaces, $P(r_1, \theta_1, \phi_1)$. Unit vectors $\mathbf{a}_r$, $\mathbf{a}_\theta$, and $\mathbf{a}_\phi$ point in the direction of increasing coordinate value. Note that all three unit vectors change direction from point to point unlike the three unit vectors of a rectangular coordinate system which are fixed in direction. The three axes are labeled r, θ, and ϕ and these are ordered with the right-hand-rule such that $\mathbf{a}_r \times \mathbf{a}_\theta = \mathbf{a}_\phi$. Note that, for example, $\mathbf{a}_r \times \mathbf{a}_\phi = -\mathbf{a}_\theta$. The same symbol r is used in both the cylindrical and the spherical coordinate systems but they have different meaning.

Two vectors expressed in this coordinate system at point P can be added at that point since the unit vectors of the corresponding components are parallel at that point.

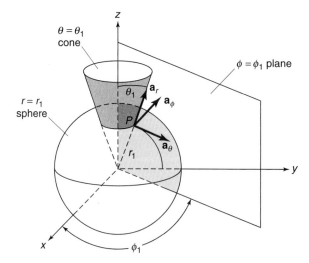

Figure 2.10 The spherical coordinate system illustrating the unit vectors and the location of a point as the intersection of three constant-coordinate surfaces.

Hence for

$$\mathbf{A} = A_r\mathbf{a}_r + A_\theta\mathbf{a}_\theta + A_\phi\mathbf{a}_\phi$$
$$\mathbf{B} = B_r\mathbf{a}_r + B_\theta\mathbf{a}_\theta + B_\phi\mathbf{a}_\phi$$

the sum is

$$\mathbf{A} + \mathbf{B} = (A_r + B_r)\mathbf{a}_r + (A_\theta + B_\theta)\mathbf{a}_\theta + (A_\phi + B_\phi)\mathbf{a}_\phi \qquad (2.21)$$

The dot and cross products are very similar to those for the rectangular coordinate system:

$$\mathbf{A} \cdot \mathbf{B} = A_rB_r + A_\theta B_\theta + A_\phi B_\phi \qquad (2.22)$$

$$\mathbf{A} \times \mathbf{B} = (A_\theta B_\phi - A_\phi B_\theta)\mathbf{a}_r + (A_\phi B_r - A_r B_\phi)\mathbf{a}_\theta + (A_r B_\theta - A_\theta B_r)\mathbf{a}_\phi \qquad (2.23)$$

Observe that the cyclic ordering of the axis labels, $r \to \theta \to \phi \to r \to \theta \to \phi \to \cdots$ can, like the rectangular coordinate system, be used to quickly form the cross product using the rule in (2.10). Alternatively the cross product can be formed using the determinant mnemonic device in (2.11).

Differential path lengths, volumes, and surface areas are formed in a fashion similar to the rectangular coordinate system. As in the cylindrical coordinate system there are some differences. In the spherical coordinate system, r has the dimensions of length but θ and ϕ are angles whose units are not length. Hence, a differential length in the θ direction is $rd\theta$. This is illustrated in Fig. 2.11a in the yz plane. A differential change in ϕ will depend on the value of θ as shown in Fig. 2.11b. The projection of this change in the xy plane $\theta = 90°$ is $rd\phi$ but for $\theta = 0°$ the change is zero. Thus the change in arc length

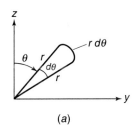

(a)

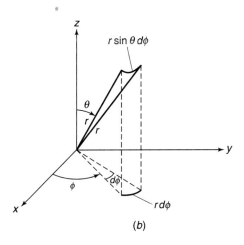

(b)

Figure 2.11 Illustration of differential arc lengths in a spherical coordinate system. (a) Differential arc length for a constant ϕ. (b) Differential arc length for a constant θ.

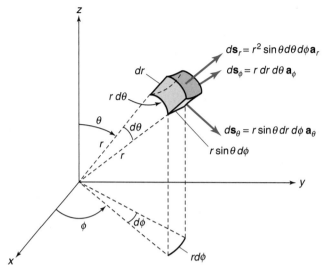

Figure 2.12   The differential surfaces in a spherical coordinate system.

due to a differential change in ϕ is $r\sin(\theta)d\phi$. Hence a differential path length is

$$d\mathbf{l} = dr\mathbf{a}_r + rd\theta\,\mathbf{a}_\theta + r\sin(\theta)d\phi\,\mathbf{a}_\phi \qquad (2.24)$$

A differential volume is formed by taking differential changes in each of the coordinate axes, as shown in Fig. 2.12, giving

$$\begin{aligned} dv &= (dr)(rd\theta)(r\sin\theta\,d\phi) \\ &= r^2\sin\theta\,drd\theta\,d\phi \end{aligned} \qquad (2.25)$$

Again the differential surface areas are needed as vector quantities so that

$$\begin{aligned} d\mathbf{s}_r &= (rd\theta)(r\sin\theta\,d\phi)\mathbf{a}_r \\ &= r^2\sin\theta\,d\theta\,d\phi\,\mathbf{a}_r \\ d\mathbf{s}_\theta &= (dr)(r\sin\theta\,d\phi)\mathbf{a}_\theta \\ &= r\sin\theta\,drd\phi\,\mathbf{a}_\theta \\ d\mathbf{s}_\phi &= (dr)(rd\theta)\,\mathbf{a}_\phi \\ &= rdr\,d\theta\,\mathbf{a}_\phi \end{aligned} \qquad (2.26)$$

▶ **EXAMPLE 2.4**

Determine the volume enclosed by a sphere of radius R as well as the surface area of that sphere.

SOLUTION   To determine the volume enclosed we integrate

$$V = \int_v dv$$

$$= \int_{r=0}^{R}\int_{\theta=0}^{\pi}\int_{\phi=0}^{2\pi} r^2\sin\theta\,dr\,d\theta d\phi$$

$$= \frac{4}{3}\pi R^3$$

To determine the surface area we integrate

$$S = \oint_s ds_r$$

$$= \int_{\theta=0}^{\pi} \int_{\phi=0}^{2\pi} (r = R)^2 \sin\theta \, d\theta \, d\phi$$

$$= 4\pi R^2$$

▷ **QUICK REVIEW EXERCISE 2.4**

Determine the volume enclosed by the surface defined by $0 \leq r \leq 3, 0 \leq \theta \leq 90°$, and $0 \leq \phi \leq 60°$ as well as the surface area of that volume. (Hint: sketch the volume and surface.)

ANSWERS $3\pi, 9\pi$.

▷ **2.8 THE LINE INTEGRAL**

This and the next section of this chapter cover the two important vector calculus operations: the line and surface integrals. The necessity for these is due to the fact that the EM laws are expressed mathematically in terms of these integrals. Hence if we are to understand what the laws are telling us, we must understand these mathematical operations.

A common case where the line integral occurs is in determining the work required to move an object along a path as illustrated in Fig. 2.13. In moving an object from point a to point b along a designated path, only the portion of the force that is tangent to the path is used in moving the object. This tangential component of the force is $F \cos\theta$ and hence the incremental work done in moving the object through a distance dl is $F \cos\theta \, dl = \mathbf{F} \cdot d\mathbf{l}$. Thus the work required to move the object from a to b is

$$W = \int_a^b \mathbf{F} \cdot d\mathbf{l}$$

$$= \int_a^b F \cos\theta \, dl \qquad (2.27)$$

This is an example of the *line integral*. The line integral sums the components that are tangent to the path. Performing the line integral in the various coordinate systems is quite simple. First form $\mathbf{F} \cdot d\mathbf{l}$ in those coordinate systems as

$$\mathbf{F} \cdot d\mathbf{l} = F_x dx + F_y dy + F_z dz \quad \text{(rectangular coordinates)} \qquad (2.28a)$$
$$\mathbf{F} \cdot d\mathbf{l} = F_r dr + F_\phi r d\phi + F_z dz \quad \text{(cylindrical coordinates)} \qquad (2.28b)$$
$$\mathbf{F} \cdot d\mathbf{l} = F_r dr + F_\theta r d\theta + F_\phi r \sin\theta \, d\phi \quad \text{(spherical coordinates)} \qquad (2.28c)$$

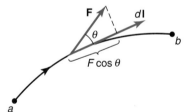

Figure 2.13 Illustration of the line integral; determination of the component of a vector along the path.

Next we simply integrate these components from a to b as

$$\int_a^b \mathbf{F} \cdot d\mathbf{l} = \int_{x_a}^{x_b} F_x dx + \int_{y_a}^{y_b} F_y dy + \int_{z_a}^{z_b} F_z dz \quad \text{(rectangular coordinates)} \qquad (2.29a)$$

$$\int_a^b \mathbf{F} \cdot d\mathbf{l} = \int_{r_a}^{r_b} F_r dr + \int_{\phi_a}^{\phi_b} F_\phi r d\phi + \int_{z_a}^{z_b} F_z dz \quad \text{(cylindrical coordinates)} \qquad (2.29b)$$

$$\int_a^b \mathbf{F} \cdot d\mathbf{l} = \int_{r_a}^{r_b} F_r dr + \int_{\theta_a}^{\theta_b} F_\theta r d\theta + \int_{\phi_a}^{\phi_b} F_\phi r \sin\theta \, d\phi \quad \text{(spherical coordinates)} \qquad (2.29c)$$

In other words, the line integral sums the products of the components of $\mathbf{F}$ that are *tangent* to the path and the differential path lengths.

This is a simple and straightforward process. The only complication is when the vector component contains one of the coordinate variables which is not the one we are integrating with respect to. For example, suppose we are integrating the component of (2.29a) $\int_{y_a}^{y_b} F_y dy$. Suppose F_y contains x or z. We simply write x or z in terms of y, the integration variable, *using the equation of the path*. The following example illustrates this simple remedy.

▶ **EXAMPLE 2.5**

Perform the line integral of $\mathbf{F} = x\mathbf{a}_x + z\mathbf{a}_y + \mathbf{a}_z$ along a straight-line path from point a at $x = 1, y = 2, z = 3$ to point b at $x = -2, y = 4, z = 5$ shown in Fig. 2.14.

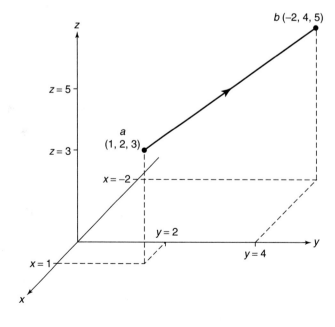

Figure 2.14   Example 2.5.

SOLUTION We form (2.29a):

$$\int_a^b \mathbf{F} \cdot d\mathbf{l} = \int_{x=1}^{-2} x\,dx + \int_{y=2}^{4} z\,dy + \int_{z=3}^{5} dz$$

Observe that the second integral contains z which is not the variable of integration for this integral. Hence we must write z in terms of the variable of integration, y, using the equation of the path. We write that equation in the form of a straight line since the path is a straight line:

$$z = my + k$$

From the coordinates of the points we determine

$$m = \frac{5-3}{4-2} = 1$$

and

$$k = 5 - m(4) = 1$$

Hence, the desired equation is $z = y + 1$. Substituting this into the second integral gives

$$\int_a^b \mathbf{F} \cdot d\mathbf{l} = \int_{x=1}^{-2} x\,dx + \int_{y=2}^{4} (y+1)\,dy + \int_{z=3}^{5} dz$$

$$= \frac{x^2}{2}\Big|_1^{-2} + \left(\frac{y^2}{2} + y\right)\Big|_2^4 + z\Big|_3^5$$

$$= \frac{23}{2}$$

▶ **EXAMPLE 2.6**

Evaluate the line integral of $\mathbf{F} = 3\mathbf{a}_r + 2r\mathbf{a}_\phi + \mathbf{a}_z$ along a circular path of radius 2 from point a at $x = 2, y = 0, z = 0$ to point b at $x = 0, y = 2, z = 0$ shown in Fig. 2.15.

SOLUTION The path fits a circular coordinate system where $r = 2$. The vector is also in cylindrical coordinates. Hence we choose a cylindrical coordinate system and form

$$\mathbf{F} \cdot d\mathbf{l} = 3(dr) + 2r(rd\phi) + 1(dz)$$
$$= 3\,dr + 2r^2\,d\phi + dz$$

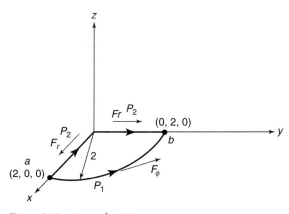

Figure 2.15 Example 2.6.

Observe that along this path $r = 2$. Hence

$$\int_{P_1} \mathbf{F} \cdot d\mathbf{l} = \underbrace{\int_{r=2}^{2} 3dr}_{0} + \int_{\phi=0}^{\pi/2} 2(r=2)^2\, d\phi + \underbrace{\int_{z=0}^{0} dz}_{0}$$

$$= 4\pi$$

To demonstrate that the path is important we alternatively perform this computation along path P_2 as shown in Fig. 2.15. Along these two segments we obtain

$$\int_{P_2} \mathbf{F} \cdot d\mathbf{l} = \int_{r=2}^{0} 3dr + \underbrace{\int_{\phi=0}^{0} 2r^2 d\phi}_{0} + \underbrace{\int_{z=0}^{0} dz}_{0}$$

$$+ \int_{r=0}^{2} 3dr + \underbrace{\int_{\phi=\pi/2}^{\pi/2} 2r^2 d\phi}_{0} + \underbrace{\int_{z=0}^{0} dz}_{0}$$

$$= 0$$

Observe that along each portion of path P_2 only one of the components of $\mathbf{F}$ is tangent to it as indicated in Fig. 2.15. Hence the contributions from the other two components are zero and we need not set up the integrals in order to determine this. It is important to visualize the problem and, as in this example, determine the components of $\mathbf{F}$ that are tangent to it in order to reduce the computation. ◀

▶ EXAMPLE 2.7

Evaluate the line integral of $\mathbf{F} = 2\mathbf{a}_r + 3r\mathbf{a}_\theta + r\mathbf{a}_\phi$ along a path from point a at $x = 0, y = 0, z = 3$ to point b at $x = 3, y = 0, z = 0$ along the path shown in Fig. 2.16.

SOLUTION The path fits a spherical coordinate system where $r = 3$ The vector is also in spherical coordinates. Hence we choose a spherical coordinate system and form

$$\mathbf{F} \cdot d\mathbf{l} = 2(dr) + 3r(rd\theta) + r(r\sin(\theta)d\phi)$$
$$= 2dr + 3r^2 d\theta + r^2 \sin(\theta)d\phi$$

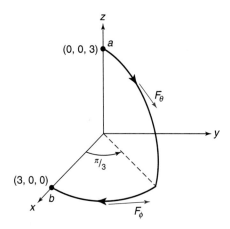

Figure 2.16   Example 2.7.

Observe that along this path $r = 3$. Hence

$$\int_P \mathbf{F} \cdot d\mathbf{l} = \underbrace{\int_{r=3}^{3} 2dr}_{0} + \underbrace{\int_{\theta=0}^{\pi/2} 3(r = 3)^2 d\theta}_{} + \underbrace{\int_{\phi=\pi/3}^{\pi/3} (r = 3)^2 \sin(\theta)d\phi}_{0}$$

$$+ \underbrace{\int_{r=3}^{3} 2dr}_{0} + \underbrace{\int_{\theta=\pi/2}^{\pi/2} 3(r = 3)^2 d\theta}_{0} + \underbrace{\int_{\phi=\pi/3}^{0} (r = 3)^2 \sin\left(\theta = \frac{\pi}{2}\right)d\phi}_{1}$$

$$= \frac{21\pi}{2}$$

As in the previous example only one of the components of $\mathbf{F}$ is tangent to each portion of the path as indicated in Fig. 2.16. Hence the contributions from the other two components are zero and we need not set up the integrals in order to determine this. It is important to visualize the problem and, as in this example, determine the components of $\mathbf{F}$ that are tangent to it in order to reduce the computation. ◁

▷ QUICK REVIEW EXERCISE 2.5

Evaluate the line integral of $\mathbf{F} = 2y\mathbf{a}_x + 3x\mathbf{a}_y + \mathbf{a}_z$ along a straight-line path from point a at $x = 0, y = 0, z = 0$ to point b at $x = 1, y = 2, z = 3$.

ANSWER 8. ◁

▷ QUICK REVIEW EXERCISE 2.6

Evaluate the line integral of $\mathbf{F} = 2\mathbf{a}_r + 3zr\mathbf{a}_\phi + \mathbf{a}_z$ along a curved, circular path from $r = 2, \phi = 0, z = 3$ to $r = 2, \phi = 90°, z = 3$. (Hint: Sketch the path and observe which components of $\mathbf{F}$ are tangent to it.)

ANSWER 18π. ◁

▷ QUICK REVIEW EXERCISE 2.7

Evaluate the line integral of $\mathbf{F} = 2\mathbf{a}_r + 3r\mathbf{a}_\theta + 2\sin\theta\mathbf{a}_\phi$ along constant radius, curved circular paths from point a at $x = 0, y = 0, z = 2$ to $x = 2, y = 0, z = 0$ to point b at $x = 0, y = 2, z = 0$. (Hint: Sketch the path and observe which components of $\mathbf{F}$ are tangent to it.)

ANSWER 8π. ◁

The above examples and Quick Review Exercise problems show an important idea in computing line integrals. When the path is along only one of the coordinate axes of the coordinate system, only the component of $\mathbf{F}$ that is *tangent to that path* (in the direction of that coordinate axis) contributes to the integral so the other two components of (2.29) do not contribute and need not be evaluated. If we sketch a picture of the problem, this will become readily apparent. This is not always the case as in Example 2.5, but when it is, we should take advantage of it.

▶ 2.9 THE SURFACE INTEGRAL

The second vector calculus operation in the statement of the EM laws is the *surface integral*:

$$\psi = \int_s \mathbf{F} \cdot d\mathbf{s}$$

$$= \int_s \mathbf{F} \cdot \mathbf{a}_n \, ds \qquad (2.30)$$

$$= \int_s F \cos\theta \, ds$$

where $\mathbf{a}_n$ is a unit vector normal to the differential surface. This is said to give the net *flux* of the vector $\mathbf{F}$ through the open surface s as shown in Fig. 2.17. The use of the term *flux* to describe this result stems from the analogy to light flux through an opening. Observe that the component of $\mathbf{F}$ that is tangent to the surface does not contribute to flux exiting the surface; only the component of $\mathbf{F}$ that is perpendicular to the surface contributes. Observe the important distinction between the line integral in (2.27) and the surface integral in (2.30). The line integral in (2.27) requires that we obtain the components of the vector that are *parallel to the path,* whereas the surface integral in (2.30) requires that we obtain the components that are *perpendicular to the surface.* There are two sides to this *open surface* so we must be clear as to the desired direction. If the surface is *closed*, we will take the flux to be *out of the surface* and indicate it with a circle on the integral:

$$\psi = \oint_s \mathbf{F} \cdot d\mathbf{s} \qquad (2.31)$$

The computation of surface integrals is very simple *if the surface involved fits one of the coordinate surfaces of the coordinate system being used.* If not, the computation can be quite involved. Fortunately, our uses of the surface integral will be for surfaces that fit the coordinate system surfaces, for example, a sphere or a portion of a sphere in a spherical coordinate system. The key to evaluating such integrals is to *sketch a picture of the problem and determine the components of* $\mathbf{F}$ *that are perpendicular to the sides*

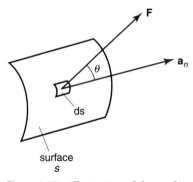

Figure 2.17 Illustration of the surface integral; determination of the component of a vector perpendicular to the surface.

of that surface. Then the above integrals are evaluated using only that component of **F**. The integrands of the surface integral for the various coordinate systems are

$$\mathbf{F} \cdot d\mathbf{s} = F_x \, dy \, dz + F_y \, dx \, dz + F_z \, dx \, dy \quad \text{(rectangular coordinate system)} \tag{2.32a}$$

$$\mathbf{F} \cdot d\mathbf{s} = F_r \, r \, d\phi \, dz + F_\phi \, dr \, dz + F_z \, r \, d\phi \, dr \quad \text{(cylindrical coordinate system)} \tag{2.32b}$$

$$\mathbf{F} \cdot d\mathbf{s} = F_r \, r^2 \sin\theta \, d\phi \, d\theta + F_\theta \, r \sin\theta \, dr \, d\phi + F_\phi \, r \, dr \, d\theta \, \text{(spherical coordinate system)} \tag{2.32c}$$

These are simple to form using the differential surfaces (and visualizing) given in (2.14), (2.20), and (2.26). The simplest way to evaluate surface integrals is to *evaluate the integral over each individual side of the surface* as shown in the following examples.

▷ **EXAMPLE 2.8**

Determine the flux of the vector $\mathbf{F} = 4x\mathbf{a}_x + 5y\mathbf{a}_y + 6\mathbf{a}_z$ out of the rectangular surface bounded by $x = 1$, $y = 2$, and $z = 3$ shown in Fig. 2.18.

SOLUTION The surface has six sides and we draw the component of **F** that is perpendicular to each side on the diagram. The flux *out of* the surface over the front side is

$$\psi_{\text{front}} = \int\limits_{y=0}^{2} \int\limits_{z=0}^{3} \underbrace{4x}_{F_x} \underbrace{dy \, dz}_{ds_x}$$

Over this surface, $x = 1$ so we substitute that constraint into the integrand giving

$$\psi_{\text{front}} = \int\limits_{y=0}^{2} \int\limits_{z=0}^{3} 4(x = 1) dy \, dz = 24$$

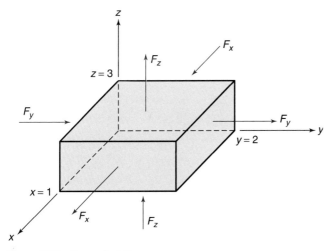

Figure 2.18 Example 2.8.

Over the back side $x = 0$ and F_x is pointing into the volume so that

$$\psi_{\text{back}} = -\int_{y=0}^{2}\int_{z=0}^{3} \underbrace{4(x=0)}_{F_x}\underbrace{dydz}_{ds_x} = 0$$

Note the minus sign which symbolizes that this flux is *into* the volume because F_x is pointing into the volume. Over the right side, we similarly obtain

$$\psi_{\text{right}} = \int_{x=0}^{1}\int_{z=0}^{3} \underbrace{5(y=2)}_{F_y}\underbrace{dxdz}_{ds_y} = 30$$

and over the left side

$$\psi_{\text{left}} = -\int_{x=0}^{1}\int_{z=0}^{3} \underbrace{5(y=0)}_{F_y}\underbrace{dxdz}_{ds_y} = 0$$

Over the top we have

$$\psi_{\text{top}} = \int_{x=0}^{1}\int_{y=0}^{2} \underbrace{6\,dxdy}_{F_z\ ds_z} = 12$$

and over the bottom we have

$$\psi_{\text{bottom}} = -\int_{x=0}^{1}\int_{y=0}^{2} \underbrace{6\,dxdy}_{F_z\ ds_z} = -12$$

giving

$$\psi = 24 + 0 + 30 + 0 + 12 - 12 = 54 \qquad \triangleleft$$

▶ **EXAMPLE 2.9**

Determine the flux of the vector $\mathbf{F} = 6\mathbf{a}_r + 5\mathbf{a}_\phi + 4z\mathbf{a}_z$ out of the closed surface bounded by $r = 3, 0 \le z \le 2$, and $0 \le \phi \le 90°$.

SOLUTION The surface is sketched in Fig. 2.19 and has five surfaces. The surface fits a cylindrical coordinate system and the vector field is also given in a cylindrical coordinate system. The components of $\mathbf{F}$ that are perpendicular to each surface are sketched on the diagram. Over the front, curved surface, $r = 3$ and F_r is the only component perpendicular to

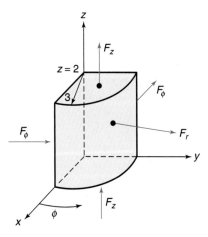

Figure 2.19 Example 2.9.

that surface. Hence

$$\psi_{\text{front}} = \int_{z=0}^{2} \int_{\phi=0}^{\pi/2} \underbrace{6(r=3)d\phi dz}_{F_r \quad ds_r} = 18\pi$$

Over the top, $z = 2$ and F_z is the only component perpendicular to that surface. Hence

$$\psi_{\text{top}} = \int_{r=0}^{3} \int_{\phi=0}^{\pi/2} \underbrace{4(z=2)rd\phi \, dr}_{F_z \quad ds_z} = 18\pi$$

Over the bottom, $z = 0$ and F_z is the only component perpendicular to that surface but it is pointing into the surface. Hence

$$\psi_{\text{bottom}} = -\int_{r=0}^{3} \int_{\phi=0}^{\pi/2} \underbrace{4(z=0)rd\phi \, dr}_{F_z \quad ds_z} = 0$$

Over the right rear side, $\phi = 90°$ and F_ϕ is the only component perpendicular to that surface. Hence

$$\psi_{\text{rear}} = \int_{r=0}^{3} \int_{z=0}^{2} \underbrace{5 \, drdz}_{F_\phi \, ds_\phi} = 30$$

Over the left side, $\phi = 0°$ and F_ϕ is the only component perpendicular to that surface but that is pointing into the surface. Hence

$$\psi_{\text{left}} = -\int_{r=0}^{3} \int_{z=0}^{2} \underbrace{5 \, drdz}_{F_\phi \, ds_\phi} = -30$$

Hence the total flux out of the surface is

$$\psi = 18\pi + 18\pi + 0 + 30 - 30 = 36\pi$$

◀

EXAMPLE 2.10

Determine the flux of $\mathbf{F} = 2\mathbf{a}_r + 4\mathbf{a}_\theta + 3r\mathbf{a}_\phi$ out of the closed surface bounded by $r = 2, 0 \le \theta \le 90°$, and $0 \le \phi \le 90°$.

SOLUTION The surface is sketched in Fig. 2.20 and there are four sides. The surface fits a spherical coordinate system and the vector field is also given in a spherical coordinate system. The components of the vector that are perpendicular to each surface are sketched on the diagram. Over the curved front side $r = 2$ and only F_r is perpendicular to that surface. Hence

$$\psi_{\text{front}} = \int_{\theta=0}^{\pi/2} \int_{\phi=0}^{\pi/2} \underbrace{2(r=2)^2 \sin(\theta)d\phi d\theta}_{F_r \quad ds_r} = 4\pi$$

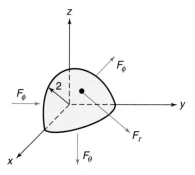

Figure 2.20 Example 2.10.

Over the right rear side $\phi = \pi/2$ and only F_ϕ is perpendicular to that surface. Hence

$$\psi_{\text{rear}} = \int_{\theta=0}^{\pi/2} \int_{r=0}^{2} \underbrace{3r}_{F_\phi} \underbrace{rdr\, d\theta}_{ds_\phi} = 4\pi$$

Over the left side $\phi = 0$ and only F_ϕ is perpendicular to that surface. But F_ϕ points into the surface. Hence

$$\psi_{\text{left}} = -\int_{\theta=0}^{\pi/2} \int_{r=0}^{2} \underbrace{3r}_{F_\phi} \underbrace{rdr\, d\theta}_{ds_\phi} = -4\pi$$

Over the bottom side $\theta = \pi/2$ and only F_θ is perpendicular to that surface. Hence

$$\psi_{\text{bottom}} = \int_{\phi=0}^{\pi/2} \int_{r=0}^{2} \underbrace{4\,r\,\sin(\theta = 90°)dr\, d\phi}_{F_\theta \qquad\qquad ds_\theta} = 4\pi$$

Hence the total flux out of the surface is

$$\psi = 4\pi + 4\pi - 4\pi + 4\pi = 8\pi$$

▷ **QUICK REVIEW EXERCISE 2.8**

Determine the flux of $\mathbf{F} = -2\mathbf{a}_x + y\mathbf{a}_y - \mathbf{a}_z$ out of the closed surface bounded by $-1 \le x \le 1$, $-2 \le y \le 2$, and $0 \le z \le 3$.

ANSWER 24. ◁

▷ **QUICK REVIEW EXERCISE 2.9**

Determine the flux of $\mathbf{F} = 2r\mathbf{a}_r + \mathbf{a}_\phi - 3\mathbf{a}_z$ out of the closed surface bounded by $r = 3$, $-90° \le \phi \le 90°$, and $0 \le z \le 2$.

ANSWER 36π. ◁

▷ **QUICK REVIEW EXERCISE 2.10**

Determine the flux of $\mathbf{F} = 3r\mathbf{a}_r + 2\mathbf{a}_\theta - 3\mathbf{a}_\phi$ out of the closed surface bounded by $r = 2$, $0 \le \theta \le 90°$, and $0 \le \phi \le 90°$.

ANSWER 14π. ◁

▶ 2.10 ELECTROMAGNETIC FIELDS

The laws governing the electromagnetic quantities are stated in terms of the above vector ideas. Our first goal is to understand what the laws are telling us in qualitative terms. The second goal is to determine numerical answers for specific problems again using the above vector ideas. In order to accomplish the first goal we must be able to *visualize* the electromagnetic field vectors. This also greatly aids in evaluating the mathematical relations for a specific problem. As an example of these EM laws, the first such law that we will study in the next chapter governs the electric field intensity vector for static (not varying with time) cases and is

$$\oint_c \mathbf{E} \cdot d\mathbf{l} = 0$$

In plain words this law provides that the line integral of the electric field intensity vector, **E**, around a closed path *c* will yield a result of zero for every static electric field. In other words, if we add up the products of the components of **E** that are *tangent to the path* and the differential path lengths, we will always obtain a result of zero for any static **E** and any closed path. Another example of an EM law that we will study in the next chapter governs the electric flux density vector for static (not varying with time) cases and is

$$\oint_s \mathbf{D} \cdot d\mathbf{s} = Q_{\text{enclosed}}$$

In plain words this law provides that the surface integral of the electric flux density vector, **D**, over a closed surface *s* will yield the net positive charge enclosed by that surface. In other words, if we add up the products of the components of **D** that are *perpendicular to the closed surface* and the differential surface areas, we will always obtain a result that is the net positive charge enclosed by that surface for any **D** and any closed surface.

    This is the type of qualitative understanding of the EM laws that we want to obtain. Our understanding of the meaning of line and surface integrals gives us that valuable insight. We can further broaden that insight as well as help solve specific problems by visualizing the electromagnetic field vectors in three-dimensional space. We have been using the term *field* to describe the EM quantities. Essentially there are two types of fields. The first is a *scalar field* and the other is a *vector field*. A scalar field describes with a mathematical equation the values of a scalar quantity throughout some region of space. An example of a scalar field is a drawing of contours of constant elevation on a topographical map as illustrated in Fig. 2.21a. A mathematical function EL(*x,y*) would

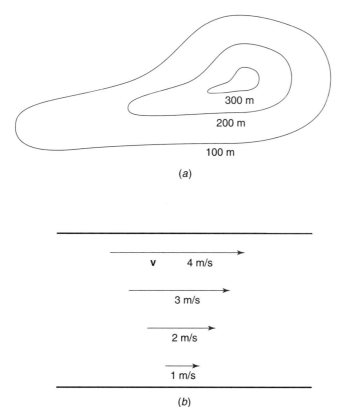

300 m

200 m

100 m

(a)

v   4 m/s

3 m/s

2 m/s

1 m/s

(b)

Figure 2.21 Illustration of the two types of fields. (a) A scalar field (magnitude only) as in a topographical map. (b) A vector field (magnitude and direction) as in the flow of a fluid in a pipe.

give the elevation at a particular location x,y over the earth's surface. A scalar field has magnitude only and does not possess a direction. On the other hand, a vector field has both magnitude and direction information. An example might be the illustration of the flow rate at various points in a pipe as illustrated in Fig. 2.21b. The vector field $\mathbf{v}(x,y)$ would give the magnitude and direction of the velocity of the fluid. The EM fields that we will be interested in are either scalar fields or vector fields.

Visualizing a particular scalar or vector field is a very important aspect of understanding what the resulting mathematical equations mean in physical terms. Hence, we should develop the ability (and willingness) to visualize the EM fields that we are dealing with in order to develop a physical understanding of them and not just a mathematical manipulative ability. In order to depict a scalar field we usually draw contours where the field has a constant value as in the case of a topographical map. In the case of vector fields there are two ways to depict the field. The first way is to show, at a sufficient number of points in the space, the magnitude and direction of the field vector at that point as illustrated in Fig. 2.22a. The relative lengths of the vectors give their relative magnitudes. This method was used to sketch the field in Fig. 2.21b. The other method is with streamlines where the streamlines indicate the vector directions and their transverse density (closeness) indicates the relative magnitudes of the fields as illustrated in Fig. 2.22b. We will use both these methods for different occasions.

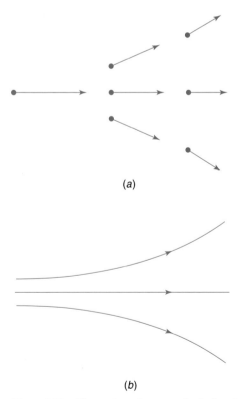

(a)

(b)

Figure 2.22 Illustration of two methods for sketching a vector field. (a) A "quiver plot" where the length of the vectors are proportional to the magnitude. (b) A "steamline plot" where the density of the lines is proportional to the magnitude of the vector field.

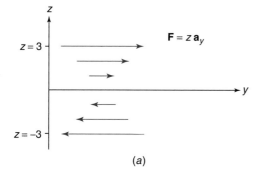

(a)

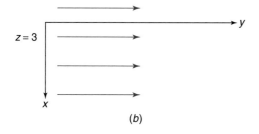

(b)

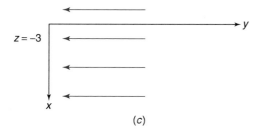

(c)

Figure 2.23 Example 2.11.

▶ EXAMPLE 2.11

Sketch the vector field $\mathbf{F} = z\mathbf{a}_y$.

SOLUTION First examine the field vector equation and determine its direction. The field vectors are pointing in the y direction. Next examine the equation for each component to determine how the magnitude of the component varies. In this case, the magnitude varies directly with z. At $z = 3$, its magnitude is 3 and at $z = -3$ its magnitude is 3 but the negative value reverses the direction. A field plot in the yz plane is shown in Fig. 2.23a. The vector magnitude increases for increasing z (positive and negative z). The component is independent of x and y. Hence a view in the xy plane will show no variation in the vector magnitude but the direction will depend on the value of z for which this view is taken as illustrated in Fig. 2.23b and c. ◀

▶ EXAMPLE 2.12

Sketch the vector field $\mathbf{F} = r\mathbf{a}_\phi$ which is given in a cylindrical coordinate system.

SOLUTION First examine the field to determine the direction. In this case the vectors are pointing in the ϕ direction everywhere. Next examine the field to determine the variation with the coordinate values. In this case the variation is with r so that the field magnitude is constant for

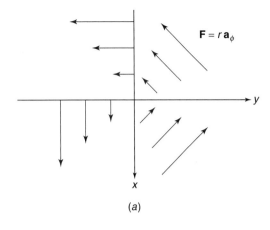

(a)

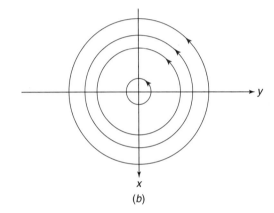

(b)

Figure 2.24   Example 2.12.

contours located at a fixed radius from the z axis (the location from which the radius vector is directed in a cylindrical coordinate system). Hence we show the field plot in the xy plane since the field vector is independent of z. This is shown two ways in Fig. 2.24.   ◁

▷ **EXAMPLE 2.13**

Sketch the vector field $\mathbf{F} = 1/r\,\mathbf{a}_\theta$ which is given in a spherical coordinate system.

SOLUTION   First examine the field to determine the direction. In this case the field vectors are directed in the θ direction everywhere. Next observe that the magnitudes of the field decay inversely with distance from the origin of the coordinate system. Also note that the magnitude of the field is independent of ϕ and θ. This means that we can plot the field in the yz or xz plane and it is rotationally symmetric about the z axis. The two methods of plotting the field are shown in Fig. 2.25.   ◁

▷ **QUICK REVIEW EXERCISE 2.11**

Sketch the field $\mathbf{F} = \sin(\theta)\mathbf{a}_\theta$ which is given in a spherical coordinate system.

ANSWER   The field vectors are directed in the θ direction and are rotationally asymmetric about the z axis. The magnitudes are zero along the positive $\theta = 0°$ z axis and along the negative $\theta = 180°$ z axis. The magnitudes are unity in the xy plane, $\theta = 90°$.   ◁

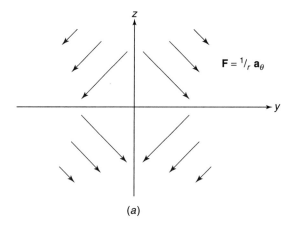

(a)

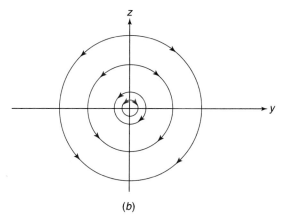

(b)

Figure 2.25 Example 2.13.

▶ SUMMARY OF IMPORTANT CONCEPTS AND FORMULAE

1. **Vector:** denotes a magnitude (length) and direction of effect (arrow) of some physical quantity.

2. **Addition and subtraction of vectors:** two vectors defined at the same point in space can be added by displacing the tail of one to the tip of the other and forming the resultant vector.

3. **Dot product:** dot product of two vectors gives a scalar (no direction of effect) that is the product of the length of one vector and the length of the projection of the other vector on the first: $\mathbf{A} \cdot \mathbf{B} = AB \cos(\theta)$. Vectors that are perpendicular have zero dot product since $\theta = 90°$.

4. **Cross product:** cross product of two vectors gives a vector as the result which is $\mathbf{A} \times \mathbf{B} = AB \sin(\theta)\mathbf{a}_n$. The vector result is perpendicular to the plane containing $\mathbf{A}$ and $\mathbf{B}$ according to the right-hand-rule as signified by the unit vector $\mathbf{a}_n$. Vectors that are parallel have zero cross product since $\theta = 0°$.

5. **Coordinate systems:** it is important to choose the coordinate system to fit the problem dimensions. Vector operations and differential elements in the three coordinate systems are (in order to quickly form the differential elements it is important to visualize changes in the coordinate axes in that coordinate system)

Rectangular (x,y,z)

Addition	$\mathbf{A} + \mathbf{B} = (A_x + B_x)\mathbf{a}_x + (A_y + B_y)\mathbf{a}_y + (A_z + B_z)\mathbf{a}_z$
Dot product	$\mathbf{A} \cdot \mathbf{B} = A_x B_x + A_y B_y + A_z B_z$
Cross product	$\mathbf{A} \times \mathbf{B} = (A_y B_z - A_z B_y)\mathbf{a}_x + (A_z B_x - A_x B_z)\mathbf{a}_y + (A_x B_y - A_y B_x)\mathbf{a}_z$

Path length	$d\mathbf{l} = dx\mathbf{a}_x + dy\mathbf{a}_y + dz\mathbf{a}_z$
Volume	$dv = dxdydz$
Surface areas	$d\mathbf{s}_x = dydz\mathbf{a}_x,\ d\mathbf{s}_y = dxdz\mathbf{a}_y,\ d\mathbf{s}_z = dxdy\mathbf{a}_z$

Cylindrical (r,ϕ,z)

Addition	$\mathbf{A} + \mathbf{B} = (A_r + B_r)\mathbf{a}_r + (A_\phi + B_\phi)\mathbf{a}_\phi + (A_z + B_z)\mathbf{a}_z$
Dot product	$\mathbf{A} \cdot \mathbf{B} = A_r B_r + A_\phi B_\phi + A_z B_z$
Cross product	$\mathbf{A} \times \mathbf{B} = (A_\phi B_z - A_z B_\phi)\mathbf{a}_r + (A_z B_r - A_r B_z)\mathbf{a}_\phi + (A_r B_\phi - A_\phi B_r)\mathbf{a}_z$
Path length	$d\mathbf{l} = dr\mathbf{a}_r + rd\phi\,\mathbf{a}_\phi + dz\mathbf{a}_z$
Volume	$dv = rdr\,d\phi\,dz$
Surface areas	$d\mathbf{s}_r = rd\phi\,dz\,\mathbf{a}_r,\ d\mathbf{s}_\phi = drdz\mathbf{a}_\phi,\ d\mathbf{s}_z = rdr\,d\phi\,\mathbf{a}_z$

Spherical (r,θ,ϕ)

Addition	$\mathbf{A} + \mathbf{B} = (A_r + B_r)\mathbf{a}_r + (A_\theta + B_\theta)\mathbf{a}_\theta + (A_\phi + B_\phi)\mathbf{a}_\phi$
Dot product	$\mathbf{A} \cdot \mathbf{B} = A_r B_r + A_\theta B_\theta + A_\phi B_\phi$
Cross product	$\mathbf{A} \times \mathbf{B} = (A_\theta B_\phi - A_\phi B_\theta)\mathbf{a}_r + (A_\phi B_r - A_r B_\phi)\mathbf{a}_\theta + (A_r B_\theta - A_\theta B_r)\mathbf{a}_\phi$
Path length	$d\mathbf{l} = dr\mathbf{a}_r + rd\theta\,\mathbf{a}_\theta + r\sin(\theta)d\phi\mathbf{a}_\phi$
Volume	$dv = r^2 \sin\theta\,drd\theta\,d\phi$
Surface areas	$d\mathbf{s}_r = r^2\sin\theta\,d\theta\,d\phi\,\mathbf{a}_r,\ d\mathbf{s}_\theta = r\sin\theta\,drd\phi\,\mathbf{a}_\theta,\ d\mathbf{s}_\phi = rdr\,d\theta\,\mathbf{a}_\phi$

---

6. **Line integral** $\int_a^b \mathbf{F} \cdot d\mathbf{l}$: sums the products of the differential path lengths and the components of vector field $\mathbf{F}$ that are *tangent* to the path. In the coordinate systems we have

$$\mathbf{F} \cdot d\mathbf{l} = F_x dx + F_y dy + F_z dz \quad \text{(rectangular coordinates)}$$
$$\mathbf{F} \cdot d\mathbf{l} = F_r dr + F_\phi rd\phi + F_z dz \quad \text{(cylindrical coordinates)}$$
$$\mathbf{F} \cdot d\mathbf{l} = F_r dr + F_\theta rd\theta + F_\phi r\sin\theta\,d\phi \quad \text{(spherical coordinates)}$$

7. **Surface integral** $\int_s \mathbf{F} \cdot d\mathbf{s}$: sums the products of the differential surface areas and the components of vector field $\mathbf{F}$ that are *perpendicular* to surface s. In the coordinate systems we have

$$\mathbf{F} \cdot d\mathbf{s} = F_x dydz + F_y dxdz + F_z dxdy \quad \text{(rectangular coordinate system)}$$
$$\mathbf{F} \cdot d\mathbf{s} = F_r rd\phi\,dz + F_\phi drdz + F_z rd\phi\,dr \quad \text{(cylindrical coordinate system)}$$
$$\mathbf{F} \cdot d\mathbf{s} = F_r r^2\sin\theta\,d\phi\,d\theta + F_\theta r\sin\theta\,dr\,d\phi + F_\phi r\,drd\theta \quad \text{(spherical coordinate system)}$$

▶ PROBLEMS

SECTION 2.1 VECTORS

2.1.1. A mass is suspended from a string whose length is 50 cm and is twirled about the vertical at an angular speed of 60 rpm (revolutions per minute). Determine the angle the string makes with the vertical. Gravitational acceleration is 9.78 m/s². [60.3°]

2.1.2. Consider an airplane which has to fly due north. The airspeed (groundspeed with no wind) is 200 mi/hr. During the flight there is a steady wind of 40 mi/hr from the northwest. Determine the direction the airplane must head in order to track north over the ground. Determine its groundspeed. Pilots of aircraft routinely use an E6B computer which plots the "wind triangle" and determine the heading correction and groundspeed.

2.1.3. A fisherman in a boat wishes to go directly across a river from the east side to the west side. The river flows from north to south at a speed of 3 mi/hr. The boat speed is 4 mi/hr. Determine the direction the boat must be headed in order to track across the river perpendicularly to both banks. [48.6° north of west]

2.1.4. A truck weighing 5000 lbs is traveling at a speed of 60 mi/hr and is rounding a curve whose radius is 1500 ft. Determine the angle the roadway must be banked in order that the truck successfully complete the turn with no friction between the tires and the roadway. Gravitational acceleration is 32 ft/s².

SECTION 2.3 THE DOT PRODUCT OF VECTORS

2.3.1. The law of cosines provides that for a triangle whose sides are of length A, B, and C $C^2 = A^2 + B^2 - 2AB \cos(\theta_{AB})$ where θ_{AB} is the angle between sides A and B. Prove this law using the dot product. Hint: Draw the triangle such that the sides are vectors and $\mathbf{C} = \mathbf{A} + \mathbf{B}$. Then take the dot product of $\mathbf{C}$ with itself to give its magnitude squared.

SECTION 2.4 THE CROSS PRODUCT OF VECTORS

2.4.1. The law of sines provides that for a triangle of side lengths A, B, and C that $A/\sin\theta_A = B/\sin\theta_B = C/\sin\theta_C$ where the angles θ_A, θ_B, and θ_C are the angles facing sides A, B, and C, respectively. Hint: Form the triangle of sides that are vectors such that $\mathbf{A} + \mathbf{B} + \mathbf{C} = 0$. Then successively take the cross product of each vector with this, for example, $\mathbf{A} \times (\mathbf{A} + \mathbf{B} + \mathbf{C}) = 0$. Remember that the angle involved in the cross product is the angle between the two vectors according to the right-hand rule.

2.4.2. Which of the following vector products do not make sense: (a) $(\mathbf{A} \cdot \mathbf{B}) \times \mathbf{C}$, (b) $(\mathbf{A} \times \mathbf{B}) \cdot \mathbf{C}$, (c) $(\mathbf{A} \cdot \mathbf{B}) \cdot \mathbf{C}$, (d) $\mathbf{A} + (\mathbf{B} \cdot \mathbf{C})$, (e) $\mathbf{A} + (\mathbf{B} \times \mathbf{C})$.

SECTION 2.5 THE RECTANGULAR COORDINATE SYSTEM

2.5.1. A vector $\mathbf{A}(x,y,z)$ is defined in a rectangular coordinate system as being directed from $(0,2,-4)$ to $(3,-4,5)$ where the units are in meters. Determine (a) an expression for $\mathbf{A}$, (b) the distance between the two points, and (c) a unit vector pointing in the direction of $\mathbf{A}$. [(a) $3\mathbf{a}_x - 6\mathbf{a}_y + 9\mathbf{a}_z$, (b) 11.22 m, (c) $0.27\mathbf{a}_x - 0.53\mathbf{a}_y + 0.8\mathbf{a}_z$]

2.5.2. Three vectors are given in a rectangular coordinate system as $\mathbf{A} = 2\mathbf{a}_x + 3\mathbf{a}_y - \mathbf{a}_z$, $\mathbf{B} = \mathbf{a}_x + \mathbf{a}_y - 2\mathbf{a}_z$, and $\mathbf{C} = 3\mathbf{a}_x - \mathbf{a}_y + \mathbf{a}_z$. Determine (a) $\mathbf{A} + \mathbf{B}$, (b) $\mathbf{B} - \mathbf{C}$, (c) $\mathbf{A} + 3\mathbf{B} - 2\mathbf{C}$, (d) $|\mathbf{A}|$, (e) $\mathbf{a}_B$, (f) $\mathbf{A} \cdot \mathbf{B}$, (g) $\mathbf{B} \cdot \mathbf{A}$, (h) $\mathbf{B} \times \mathbf{C}$, (i) $\mathbf{C} \times \mathbf{B}$, (j) $\mathbf{A} \cdot \mathbf{B} \times \mathbf{C}$.

2.5.3. If $\mathbf{A} = \mathbf{a}_x + 2\mathbf{a}_y - 3\mathbf{a}_z$ and $\mathbf{B} = 2\mathbf{a}_x - \mathbf{a}_y + \mathbf{a}_z$ determine (a) the magnitude of the projection of $\mathbf{B}$ onto $\mathbf{A}$, (b) the angle (smallest) between $\mathbf{A}$ and $\mathbf{B}$, (c) a unit vector perpendicular to the plane containing $\mathbf{A}$ and $\mathbf{B}$. [(a) 0.8, (b) 109.1°, (c) $-0.115\mathbf{a}_x - 0.808\mathbf{a}_y - 0.577\mathbf{a}_z$]

2.5.4. Two vectors are given by $\mathbf{A} = \mathbf{a}_x + 2\mathbf{a}_y - \mathbf{a}_z$ and $\mathbf{B} = \alpha\mathbf{a}_x + \mathbf{a}_y + 3\mathbf{a}_z$. Determine α such that the two vectors are perpendicular.

2.5.5. Two vectors are given by $\mathbf{A} = \mathbf{a}_x + 2\mathbf{a}_y + 3\mathbf{a}_z$ and $\mathbf{B} = \alpha\mathbf{a}_x + \beta\mathbf{a}_y - 9\mathbf{a}_z$. Determine α and β such that the two vectors are parallel.

2.5.6. Two vectors are given by $\mathbf{A} = 2\mathbf{a}_x - 3\mathbf{a}_y + \mathbf{a}_z$ and $\mathbf{B} = -\mathbf{a}_x + 2\mathbf{a}_y + 4\mathbf{a}_z$. Determine a vector $\mathbf{C}$ which is perpendicular to both $\mathbf{A}$ and $\mathbf{B}$ and has a length of 10. [$\mathbf{C} = -8.4\mathbf{a}_x - 5.4\mathbf{a}_y + 0.6\mathbf{a}_z$]

2.5.7. Prove the vector identity $\mathbf{A} \times (\mathbf{B} \times \mathbf{C}) = \mathbf{B}(\mathbf{A} \cdot \mathbf{C}) - \mathbf{C}(\mathbf{A} \cdot \mathbf{B})$.

2.5.8. Show that the cross product is not associative, that is, $\mathbf{A} \times (\mathbf{B} \times \mathbf{C}) \neq (\mathbf{A} \times \mathbf{B}) \times \mathbf{C}$.

2.5.9. Determine the distance between the two points $(-1,2,-4)$ and $(3,-1,5)$ directly and by integration.

2.5.10. Determine the area of a triangular surface formed between the three points $(1,1,1)$, $(1,3,1)$, and $(1,3,2)$ directly and by integration.

SECTION 2.6 THE CYLINDRICAL COORDINATE SYSTEM

2.6.1. Determine the relations between coordinates of a point in a cylindrical coordinate system and the point expressed in a rectangular system and vice versa. $[r = \sqrt{x^2 + y^2}, \phi = \tan^{-1} y/x, z = z$, and $x = r \cos\phi, y = r \sin\phi$, and $z = z]$

2.6.2. Determine (a) $\mathbf{a}_r \times \mathbf{a}_\phi$, (b) $\mathbf{a}_z \times \mathbf{a}_\phi$, (c) $\mathbf{a}_r \times \mathbf{a}_z$.

2.6.3. A vector is given in a cylindrical coordinate system as $\mathbf{A}(r,\phi,z) = A_r\mathbf{a}_r + A_\phi\mathbf{a}_\phi + A_z\mathbf{a}_z$. Determine the corresponding vector expressed in a rectangular coordinate system. $[A_x = A_r \cos\phi - A_\phi \sin\phi, A_y = A_r \sin\phi + A_\phi \cos\phi, A_z = A_z]$

2.6.4. Two vectors are given in a cylindrical coordinate system at a point $P(1, \pi/3, 2)$ as $\mathbf{A}(r,\phi,z) = 2\mathbf{a}_r - 3\mathbf{a}_\phi + \mathbf{a}_z$ and $\mathbf{B}(r,\phi,z) = 4\mathbf{a}_r + 6\mathbf{a}_\phi - 2\mathbf{a}_z$. Determine the dot product of these directly and by converting them to rectangular coordinates using the results of Problem 2.6.3.

2.6.5. Two points are defined in a cylindrical coordinate system as $P_1(2,\pi/2,1)$ and $P_2 = (3,\pi/3,-2)$. Determine the distance between the two points. Hint: convert the points to rectangular coordinates using the results of Problem 2.6.1. [3.407]

2.6.6. A surface is defined as $r = 2, 0 \le \phi \le \pi/3, 1 \le z \le 4$. Determine the area of the surface.

2.6.7. Determine the volume of a region defined by $1 \le r \le 2, 0 \le \phi \le \pi/3, 0 \le z \le 1$. $[\pi/2]$

SECTION 2.7 THE SPHERICAL COORDINATE SYSTEM

2.7.1. Determine the relations between coordinates of a point in a spherical coordinate system and the point expressed in a rectangular system and vice versa. $[r = \sqrt{x^2 + y^2 + z^2}, \theta = \tan^{-1}\sqrt{x^2 + y^2}/z, \phi = \tan^{-1} y/x$, and $x = r \sin\theta \cos\phi, y = r \sin\theta \sin\phi$, and $z = r \cos\theta]$

2.7.2. Determine (a) $\mathbf{a}_r \times \mathbf{a}_\phi$, (b) $\mathbf{a}_\theta \times \mathbf{a}_\phi$, (c) $\mathbf{a}_\theta \times \mathbf{a}_r$.

2.7.3. A vector is given in a spherical coordinate system as $\mathbf{A}(r,\theta,\phi) = A_r\mathbf{a}_r + A_\theta\mathbf{a}_\theta + A_\phi\mathbf{a}_\phi$. Determine the corresponding vector expressed in a rectangular coordinate system. $[A_x = A_r \sin\theta \cos\phi + A_\theta \cos\theta \cos\phi - A_\phi \sin\phi, A_y = A_r \sin\theta \sin\phi + A_\theta \cos\theta \sin\phi + A_\phi \cos\phi, A_z = A_r \cos\theta - A_\theta \sin\theta]$

2.7.4. Two vectors are given in a spherical coordinate system at a point $P(1,2\pi/3,\pi/3)$ as $\mathbf{A}(r,\theta,\phi) = 2\mathbf{a}_r + 3\mathbf{a}_\theta + \mathbf{a}_\phi$ and $\mathbf{B}(r,\theta,\phi) = 4\mathbf{a}_r + 2\mathbf{a}_\theta - 3\mathbf{a}_\phi$. Determine the dot product of these directly and by converting them to rectangular coordinates using the results of Problem 2.7.3.

2.7.5. Two points are defined in a cylindrical coordinate system as $P_1(2,\pi/2,2\pi/3)$ and $P_2 = (3,\pi/3,-\pi/6)$. Determine the distance between the two points. Hint: convert the points to rectangular coordinates using the results of Problem 2.7.1. [4.69]

2.7.6. A surface is defined as $r = 4, \pi/2 \le \theta \le \pi, \pi/2 \le \phi \le \pi$. Determine the area of the surface.

2.7.7. Determine the area of a surface defined by $r = 2, \pi/4 \le \theta \le \pi/3, 0 \le \phi \le 2\pi$. [5.21]

SECTION 2.8 THE LINE INTEGRAL

2.8.1. Determine the line integral of $\mathbf{F} = 2x\mathbf{a}_x + 4\mathbf{a}_y - y\mathbf{a}_z$ from $P_1(-1,3,-2)$ to $P_2(2,4,1)$ where the points are specified in a rectangular coordinate system. $[-7/2]$

2.8.2. If the force exerted on an object is given by $\mathbf{F} = 2x\mathbf{a}_x + 3z\mathbf{a}_y + 4\mathbf{a}_z$ N determine the work required to move the object in a straight line from $P_1(0,0,0)$ to $P_2(1,1,2)$ where the points are specified in a rectangular coordinate system in meters.

2.8.3. Determine the line integral of $\mathbf{F} = x\mathbf{a}_x + 2xy\mathbf{a}_y - y\mathbf{a}_z$ from $P_1(0,0,2)$ to $P_2(3,2,0)$ along paths consisting of (a) a straight-line path between the two points, and (b) a two segment path consisting of a straight-line path from $P_1(0,0,2)$ to the origin and a straight-line path from the origin to $P_2(3,2,0)$. The points are specified in a rectangular coordinate system. [(a) 29/2, (b) 25/2]

2.8.4. Determine the line integral of $\mathbf{F} = 2r\mathbf{a}_r + z\mathbf{a}_\phi + 4\mathbf{a}_z$ from $P_1(0,0,3)$ to $P_2(0,0,0)$ along two paths. The points are specified in a rectangular coordinate system. Along (a) a straight-line path from P_1 to P_2, and (b) a path consisting of straight-line segments from P_1 to $P_3(2,2,3)$ to $P_4(2,2,0)$ to P_2.

2.8.5. Determine the line integral of $\mathbf{F} = r\mathbf{a}_r + 2\mathbf{a}_\phi - z\mathbf{a}_z$ from $P_1(0,0,2)$ to $P_2(3,0,0)$ along two paths. The points are specified in a rectangular coordinate system. Along (a) straight-line paths from P_1 to the origin, $P_0(0,0,0)$, to P_2, and (b) straight-line paths from P_1 to $P_3(0,3,2)$ to $P_4(0,3,0)$ and a circular arc from P_4 to P_2. [(a) 13/2, (b) $13/2 - 3\pi$]

2.8.6. Determine the line integral of $\mathbf{F} = 2r\mathbf{a}_r + 3\mathbf{a}_\theta + 2\mathbf{a}_\phi$ from $P_1(0,0,3)$ to $P_2(3,0,0)$ along two paths. The points are specified in a rectangular coordinate system. Along (a) straight-line paths from P_1 to the origin, $P_0(0,0,0)$, to P_2, and (b) along circular curved paths from P_1 to $P_3(0,3,0)$ to P_2. *a.)=0   b.) $3\pi/2$*

2.8.7. Determine the line integral of $\mathbf{F} = r\mathbf{a}_r + 2r\mathbf{a}_\theta + 3\mathbf{a}_\phi$ from $P_1(0,0,2)$ to $P_0(0,0,0)$ along two paths. The points are specified in a rectangular coordinate system. Along (a) a circular curved path from P_1 to $P_3(0, \sqrt{2}, \sqrt{2})$ and along a straight-line path from P_3 to P_0, and (b) along a straight-line path from P_1 to the origin, $P_0(0,0,0)$. [(a) $2\pi - 2$, (b) -2]

SECTION 2.9 THE SURFACE INTEGRAL

2.9.1. Determine the flux of the vector field $\mathbf{F} = x\mathbf{a}_x + y\mathbf{a}_y + z\mathbf{a}_z$ out of the closed surface consisting of a square volume whose sides are of length 2 and is centered on the origin of a rectangular coordinate system. [24]

2.9.2. Determine the flux of the vector field $\mathbf{F} = xy\mathbf{a}_x + yz\mathbf{a}_y - xz\mathbf{a}_z$ out of a closed rectangular surface. The surface has corners at (1,0,0), (0,2,0), (0,0,3), (1,0,3), (0,0,0), (0,2,3), (1,2,3), and (1,2,0) which are given in rectangular coordinate system variables.

2.9.3. Determine the flux of the vector field $\mathbf{F} = 2r\mathbf{a}_r + 3\phi\mathbf{a}_\phi - 2\mathbf{a}_z$ out of the closed surface bounded by $0 \le r \le 2$, $0 \le \phi \le \pi/2$, $0 \le z \le 3$. [21π]

2.9.4. Determine the flux of the vector field $\mathbf{F} = 3r\mathbf{a}_r + \phi\mathbf{a}_\phi - z\mathbf{a}_z$ out of the closed surface bounded by $0 \le r \le 2$, $-\pi/2 \le \phi \le \pi/2$, $0 \le z \le 4$. *48π*

2.9.5. Determine the flux of the vector field $\mathbf{F} = r\mathbf{a}_r + 2\mathbf{a}_\theta - \phi\mathbf{a}_\phi$ out of the closed surface bounded by $0 \le r \le 2$, $0 \le \theta \le \pi/2$, $0 \le \phi \le \pi/2$. [$6\pi - \pi^2/2$]

2.9.6. Determine the flux of the vector field $\mathbf{F} = r\mathbf{a}_r - 2\mathbf{a}_\theta + 3\phi\mathbf{a}_\phi$ out of the closed surface bounded by $0 \le r \le 3$, $0 \le \theta \le \pi/2$, $-\pi/2 \le \phi \le \pi/2$.

CHAPTER 3

Static (DC) Electromagnetic Fields

We begin our study of electromagnetic fields by first examining the EM laws for static (dc) fields. Static distributions of charge (charge that is not moving) produce two of the four EM quantities, the *electric field intensity vector* denoted by **E** and the *electric flux density vector* denoted by **D**. Movement of charge constitutes a *current*, and a steady movement of charge is a dc current. Dc currents produce the remaining two EM quantities, the *magnetic field intensity vector* denoted by **H** and the *magnetic flux density vector* denoted by **B**.

In the remaining chapters of this text we will study the EM laws for *time-varying* movement of charge, that is, non-dc currents. The above four field vectors will remain our primary interest, *but* certain of the dc EM laws will need to be modified.

The dc field laws that we will study in this chapter are, technically, only valid for fixed charge distributions and/or steady (dc) currents. However, they can be applied, to a reasonable approximation, to charge whose movement is *slowly* varying with time. We refer to this as the quasi-static approximation. For example, at the commercial power frequency of 60 Hz, a wavelength is some 3000 mi (5000 km). Hence, for a circuit or device whose largest dimension is less than, say, 300 mi the structure is electrically small and the associated fields can be considered quasi static so that the dc field laws apply to a reasonable approximation. In ECE laboratories, we have frequently modeled electric circuits as lumped circuits without needing to use the complete EM field laws that we will study in subsequent chapters of this text. The reason we can do this is that the frequencies of the sources in that circuit are such that the largest circuit dimension is much less than a wavelength. For example, for a sinusoidal voltage or current source whose frequency is 1 MHz, a wavelength is 300 m. Hence circuits which contain this 1-MHz source and whose largest dimension is less than, say, 30 m are electrically small so that lumped-circuit models and Kirchhoffs laws are reasonable approximations and give reliable results.

Chapter Learning Objectives

After completing the contents of this chapter you should be able to

▶ use Coulomb's law to compute the force between charges,

▶ compute the electric field intensity vector **E** for various charge distributions,

▶ determine the effect of dielectric materials on the electric fields,

▶ use Gauss' law to compute the electric field for charge distributions that exhibit symmetry,

▶ compute voltages for various charge distributions,

▷ calculate the capacitance of various structures,

▷ determine the resistance of various structures using Ohm's law,

▷ use the Biot-Savart law to compute the magnetic flux density vector **B** for various current distributions,

▷ determine the effect of magnetic materials on the magnetic fields,

▷ use Ampere's law to compute the magnetic field for current distributions that exhibit symmetry,

▷ understand the meaning of Gauss' law for magnetic fields,

▷ calculate the inductance of various structures,

▷ use the Lorentz force law to determine the force due to charges and currents,

▷ cite examples of typical engineering applications of these EM principles.

▷ 3.1 CHARGE AND COULOMB'S LAW

We know that like charges repel and unlike charges attract. The force of attraction or repulsion is governed by *Coulomb's law*. Consider Fig. 3.1, where two point charges are separated in free space (approximately air) by a distance R. (We will reserve the symbol r for the radial distance variable in cylindrical and spherical coordinate systems. The symbol R will be used to denote the distance between two points.) Coulomb's law provides that the force of attraction or repulsion is given by

$$F = \frac{1}{4\pi\varepsilon_o} \frac{Q_1 Q_2}{R^2} \quad \text{N} \tag{3.1}$$

The force of attraction or repulsion is in newtons (N) and varies directly as the product of the charges and inversely as the *square of the separation distance*. Hence this is an inverse distance squared law like the law of gravity. The units of charge are coulombs (C), named after the discoverer of the law. The force is directed away from each charge for like charges and is along a line from one charge to the other. The quantity ε_o is the *permittivity* of the surrounding medium (here assumed to be free space or approximately air). The value is $\varepsilon_o = 8.8542 \cdots \times 10^{-12}$ F/m and the approximate value is

$$\varepsilon_o \cong \frac{1}{36\pi} \times 10^{-9} \quad \text{F/m} \tag{3.2}$$

The units are farads (F) per meter (m) or a capacitance per unit of distance. More will be said about these units later in this chapter. Hence the constant in Coulomb's law is

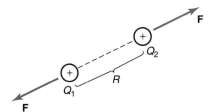

F

Figure 3.1 Illustration of Coulomb's law.

approximately

$$\frac{1}{4\pi\varepsilon_o} \cong 9 \times 10^9 \tag{3.3}$$

and Coulomb's law can be written (for charges in free space) as

$$F \cong 9 \times 10^9 \frac{Q_1 Q_2}{R^2} \tag{3.4}$$

For charges in material media other than free space, Coulomb's law will be the same except that the permittivity of free space, ε_o, will be replaced by the permittivity of that medium, ε, which will be discussed in Section 3.3.

▶ **EXAMPLE 3.1**

Three point charges are located in a rectangular coordinate system as shown in Fig. 3.2. Charge $Q_1 = 5 \, \mu C$ is located at $y = 3$ m, charge $Q_2 = -3 \, \mu C$ is located at $x = 2$ m, and charge $Q_3 = 1 \, \mu C$ is located at the origin. Determine the vector force exerted on Q_3 by the other two charges.

SOLUTION The vector force exerted on Q_3 by Q_1 is

$$\mathbf{F}_{13} = -9 \times 10^9 \frac{(5 \times 10^{-6})(1 \times 10^{-6})}{(3)^2} \mathbf{a}_y$$
$$= -5 \times 10^{-3} \mathbf{a}_y \quad N$$

The vector force exerted on Q_3 by Q_2 is

$$\mathbf{F}_{23} = 9 \times 10^9 \frac{(3 \times 10^{-6})(1 \times 10^{-6})}{(2)^2} \mathbf{a}_x$$
$$= 6.75 \times 10^{-3} \mathbf{a}_x \quad N$$

The force vector is

$$\mathbf{F} = \mathbf{F}_{13} + \mathbf{F}_{23}$$
$$= 6.75 \times 10^{-3} \mathbf{a}_x - 5 \times 10^{-3} \mathbf{a}_y$$

whose magnitude is $F = 8.4 \times 10^{-3}$ N.

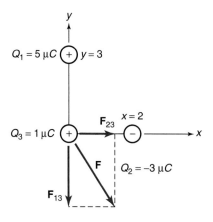

Figure 3.2 Example 3.1.

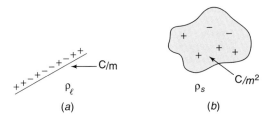

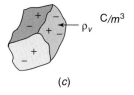

Figure 3.3 Illustration of various charge densities. (a) A line charge distribution. (b) A surface charge distribution. (c) A volume charge distribution.

▶ **QUICK REVIEW EXERCISE 3.1**

Determine the force exerted on each other by two 1-C point charges that are separated by 1 m.

ANSWER 9×10^9 N (1 million tons!). ◀

The point charge is a useful but somewhat artificial quantity. The charge of a point charge is considered to reside at an infinitesimally small point. Charge is usually distributed as a *line charge*, a *surface charge*, or a *volume charge* as shown in Fig. 3.3. Figure 3.3a shows a line charge distribution. The charge distribution is denoted as ρ_l C/m. This distribution may be uniform (uniformly distributed along the line) or nonuniform. The surface charge distribution shown in Fig. 3.3b is denoted as ρ_s and has the units of C/m^2, and the volume charge distribution shown in Fig. 3.3c is denoted as ρ_v and has the units of C/m^3. In the case of nonuniform distributions, an equation will describe that distribution.

▶ **EXAMPLE 3.2**

Charge is distributed nonuniformly over a disk of radius 2 m and has the following distribution: $\rho_s = 2 \times 10^{-6} r$ C/m^2. Determine the total charge contained on the disk.

SOLUTION The problem is sketched in Fig. 3.4. In cylindrical coordinates, an element of the surface is $ds = (rd\phi)(dr) = rdrd\phi$. The element of charge (which we will refer to as a "chunk

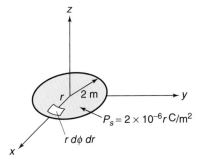

Figure 3.4 Example 3.2.

of charge") contained in that differential surface is

$$dQ = \rho_s ds$$
$$= 2 \times 10^{-6} r^2 dr d\phi$$

The total charge contained on the disk surface is

$$Q = \int_{\phi=0}^{2\pi} \int_{r=0}^{2} \underbrace{2 \times 10^{-6} r}_{\rho_s} \underbrace{r dr d\phi}_{ds}$$

$$= \frac{32\pi}{3} \times 10^{-6} \quad C$$

◀

QUICK REVIEW EXERCISE 3.2

Determine the charge contained in the volume charge distribution $\rho_v = 5 \times 10^{-4} r$ C/m^3 in a volume defined by $0 \leq r \leq 3, 0 \leq \phi \leq \pi/2, 0 \leq \theta \leq \pi/2$.

ANSWER 1.59×10^{-2} C.

◀

▶ 3.2 THE ELECTRIC FIELD INTENSITY VECTOR

Charges exert a force on other charges in their vicinity. Hence we may think of a *force field* existing around a point charge (and other charge distributions). This concept leads us to the first of the four fundamental EM quantities: the *electric field intensity vector* denoted as **E**. To illustrate the meaning of this quantity, consider a point charge Q shown in Fig. 3.5. If we introduce a positive test charge, q, a force will be exerted by Q on the test charge. *The electric field intensity vector is defined as the force per unit of positive test charge exerted on that charge*:

$$\mathbf{E} = \frac{\mathbf{F}}{q} \quad \text{V/m} \tag{3.5}$$

The units of electric field intensity are newtons per coulomb or, equivalently, volts per meter: a voltage per unit of distance. For the charge Q in Fig. 3.5, we select $Q_1 = Q$ and $Q_2 = q$ in (3.1) and obtain the electric field intensity of a point charge Q:

$$\boxed{\mathbf{E} = \frac{1}{4\pi\varepsilon_o} \frac{Q}{R^2} \mathbf{a}_R \quad \text{V/m}} \tag{3.6}$$

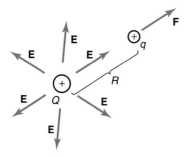

Figure 3.5 The electric field of a point charge.

This electric field intensity vector for a point charge is in the direction of the force exerted on the positive test charge and is directed radially away from Q as symbolized by the unit vector $\mathbf{a}_R$. The electric field intensity vector for a point charge decays inversely with distance squared.

3.2.1 Calculation of the Electric Field for Charge Distributions

The calculation of the electric field intensity vector (or simply the electric field) for other charge distributions such as line charge distributions, surface charge distributions, and volume charge distributions is very simple if we (1) divide the charge distribution into differential "chunks of charge," dQ, which are treated as point charges, (2) use *superposition* to add up the contributions from all these chunks of charge at the point where we are interested in computing $\mathbf{E}$, and (perhaps the most important step) (3) utilize symmetry to simplify the resulting integral.

This use of superposition to combine the contributions of all the differential chunks of charge and treat them as point charges is the heart of the computation of the electric field for other charge distributions. Once we have divided the charge distribution into chunks of charge, we treat them as point charges, dQ, and use the result for a point charge given in (3.6) to provide the differential contribution to the electric field at a point:

$$d\mathbf{E} = \frac{1}{4\pi\varepsilon_o} \frac{dQ}{R^2} \mathbf{a}_R \quad \text{V/m} \tag{3.7}$$

as shown in Fig. 3.6a. Then we sum (with an integral) the contributions to give the total electric field at the point. Observe that as we do this, the distance R between the chunk of charge, $dQ = \rho_v dv$, and the point where we are determining the electric field varies. Also, the differential contributions to the electric field vary in direction as shown in Fig. 3.6b, so that the integral summation of all these effects of the differential chunks of charge involves vectors. We will not be so ambitious as to calculate the electric field everywhere about the charge distribution but will only compute it at certain locations so that we can use symmetry to simplify the resulting integral. The following examples illustrate this method, which will be used virtually unchanged in solving magnetic field problems later in this chapter. Hence it is a very basic method and should be understood by the reader.

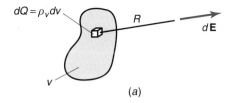

(a)

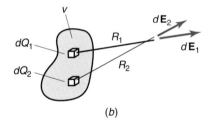

(b)

Figure 3.6 Illustration of determining the electric field intensity vector for a charge distribution. (a) Differential contributions to the vector from differential chunks of charge. (b) Addition of all the differential contributions involves the addition of vectors.

▶ **EXAMPLE 3.3**

Determine the electric field intensity vector away from a line charge distribution, ρ_l C/m, of length L where the charge is *uniformly distributed along the line*.

SOLUTION   First we align the line charge distribution along the z axis of a cylindrical coordinate system as shown in Fig. 3.7a. We center it at the origin of the coordinate system and place half of it on the $+z$ axis and the other half on the $-z$ axis to take advantage of symmetry. We divide the line charge into chunks of charge that are symmetrically disposed about the center. Each of

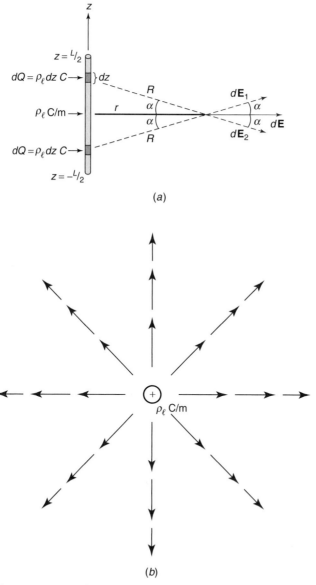

(a)

(b)

Figure 3.7   Example 3.3; determining the electric field intensity vector for a line charge. (a) Setting up the problem taking advantage of symmetry. (b) The electric field intensity vector for an infinite line charge is radially directed away from the line charge.

these chunks of charge is of value $dQ = \rho_l dz$ C and is treated as though it is a point charge. In order to simplify the resulting integral by taking advantage of symmetry, we will only compute the electric field at distances from its center and on a line perpendicular to its center. From the sketch we see an important observation from symmetry: the upper and lower chunks of charge will produce contributions such that their vertical (z-directed) components cancel, leaving only the radial component as the result. At other locations, the field will not be radially directed, so we restrict our calculation to only the points that are on a line perpendicular to the midpoint in order to simplify the calculation. This radial component (due to the upper and lower chunks of charge) is

$$dE = 2\frac{1}{4\pi\varepsilon_o}\frac{dQ}{R^2}\cos(\alpha)\,\mathbf{a}_r$$

and

$$\cos(\alpha) = \frac{r}{R}$$

where

$$R = \sqrt{r^2 + z^2}$$

Substituting and summing the contributions by integrating along the charge distribution, that is, with respect to z, gives

$$\mathbf{E} = 2\int_{z=0}^{L/2}\frac{1}{4\pi\varepsilon_o}\frac{\rho_l r}{(r^2 + z^2)^{3/2}}\,dz\,\mathbf{a}_r$$

The evaluation requires the integral

$$\int\frac{1}{(r^2 \pm z^2)^{3/2}}\,dz = \frac{z}{r^2\sqrt{r^2 \pm z^2}} \tag{3.8}$$

giving

$$\mathbf{E} = \frac{\rho_l}{2\pi\varepsilon_o r}\frac{L}{\sqrt{4r^2 + L^2}}\,\mathbf{a}_r \tag{3.9}$$

For an *infinite line charge* we let $L \to \infty$ and obtain

$$\boxed{\mathbf{E} = \frac{\rho_l}{2\pi\varepsilon_o r}\,\mathbf{a}_r \qquad L \to \infty} \tag{3.10}$$

This result for the electric field from an infinite line charge is a very important result that will be used on numerous occasions and should be committed to memory. With two or three integrals like (3.8), most such problems can be evaluated. Observe that the electric field for an infinite line charge is directed radially away from the line and decays directly with distance away from the line. A plot of this field is given in Fig. 3.7b. ◀

Many of the EM problems can be solved by using previously determined results. The following example illustrates the use of the result for an infinite line charge determined in the previous example.

▶ **EXAMPLE 3.4**

An infinite sheet of charge with uniform charge distribution ρ_s C/m^2 is located in the xz plane as shown in Fig. 3.8. Determine the electric field at some distance d away.

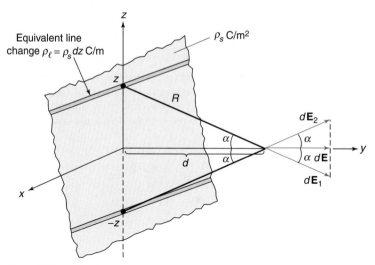

Figure 3.8 Example 3.4; determining the electric field intensity vector for a flat surface charge distribution by viewing the surface charge distribution as a sequence of infinite line charges and using symmetry and the result of Example 3.3.

SOLUTION Think of this charge distribution as being composed of line charge distributions in the x direction. The charge distribution of this line charge is

$$\rho_l = \rho_s dz \quad \text{C/m}$$

From Example 3.3, the electric field of an infinite line charge is radially directed away from the line charge and the magnitude is

$$dE = \frac{\rho_l}{2\pi\varepsilon_o R}$$
$$= \frac{\rho_s dz}{2\pi\varepsilon_o R}$$

Observe in Fig. 3.8 that the electric field of the symmetrically disposed line charges will cause the vertical components of the electric field to cancel, leaving only the horizontal components. Hence we obtain the total electric field as the horizontal components directed perpendicular to the sheet of charge:

$$d\mathbf{E} = 2\frac{\rho_s dz}{2\pi\varepsilon_o R}\cos(\alpha)\mathbf{a}_y$$

But

$$R = \sqrt{d^2 + z^2}$$

and

$$\cos(\alpha) = \frac{d}{R}$$

Substituting and summing all the contributions gives

$$\mathbf{E} = \int_{z=0}^{\infty} \frac{\rho_s d}{\pi\varepsilon_o R^2} dz\, \mathbf{a}_y$$
$$= \frac{\rho_s d}{\pi\varepsilon_o} \int_{z=0}^{\infty} \frac{1}{(d^2 + z^2)} dz\, \mathbf{a}_y$$

This requires the evaluation of the integral

$$\int \frac{1}{d^2 + z^2}\, dz = \frac{1}{d}\tan^{-1}\left(\frac{z}{d}\right) \tag{3.11}$$

The result is

$$\mathbf{E} = \frac{\rho_s}{2\varepsilon_o}\mathbf{a}_y \quad y > 0 \tag{3.12}$$

This also gives the field on the left side of the sheet, $y < 0$, if we reverse the sign since the field is directed in the $-y$ direction opposite to the unit vector $\mathbf{a}_y$. ◀

▷ QUICK REVIEW EXERCISE 3.3

Determine the electric field due to a sphere of charge of radius a for $r > a$ where the charge is uniformly distributed over the sphere surface with distribution ρ_s C/m². (Hint: Place the sphere at the origin of a spherical coordinate system.)

ANSWER $\mathbf{E} = (\rho_s a^2/\varepsilon_o r^2)\mathbf{a}_r$ for $r > a$. ◀

▷ QUICK REVIEW EXERCISE 3.4

Determine the electric field due to a sphere of charge of radius a for $r > a$ where the charge is uniformly distributed throughout the sphere with volume charge distribution ρ_v C/m³. (Hint: Place the sphere at the origin of a spherical coordinate system.)

ANSWER $\mathbf{E} = (\rho_v a^3/3\varepsilon_o r^2)\mathbf{a}_r$ for $r > a$. ◀

▷ 3.3 THE ELECTRIC FLUX DENSITY VECTOR AND DIELECTRIC MATERIALS

Up to this point we have been considering free space (approximately air) as the medium surrounding the charge distributions. We now examine the effect of other materials on those resulting electric fields. The important class of materials we will be considering are called *dielectrics*. Examples are quartz, Teflon, rubber, and glass. These materials are often also called *insulators* since there are not significant numbers of free charges present within them, unlike metals, where free electrons are available for current flow. The important distinguishing characteristic of dielectric materials is the presence of *bound charges*.

In order to illustrate the effect of bound charge, consider the *electric dipole* shown in Fig. 3.9a. The dipole consists of equal but opposite charges Q separated by a distance l. We determine the electric field off the ends of the dipole at $z = r$. We obtain

$$\mathbf{E} = \frac{Q}{4\pi\varepsilon_o\left(r - \frac{l}{2}\right)^2}\mathbf{a}_z - \frac{Q}{4\pi\varepsilon_o\left(r + \frac{l}{2}\right)^2}\mathbf{a}_z$$

$$= \frac{Q}{2\pi\varepsilon_o}\frac{lr}{\left(r^2 - \frac{l^2}{4}\right)^2}\mathbf{a}_z$$

Evaluating this for a distance much larger than the charge separation gives

$$\mathbf{E} = \frac{Ql}{2\pi\varepsilon_o r^3}\mathbf{a}_z \quad r \gg \frac{l}{2}, \quad \theta = 0° \tag{3.13a}$$

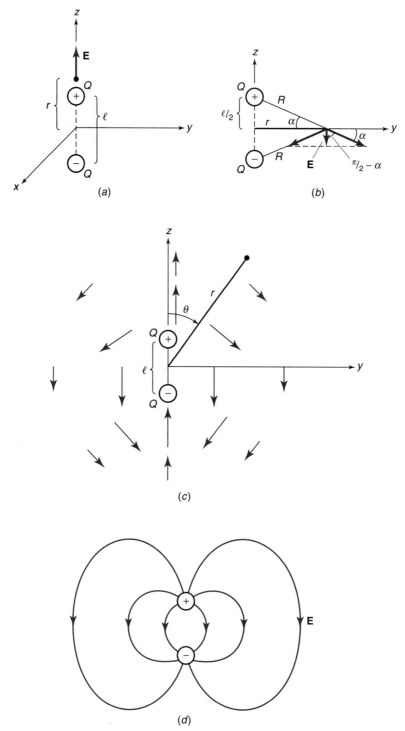

Figure 3.9 The electric dipole. (a) Determining the electric field off the ends of the dipole. (b) Determining the electric field for points perpendicular to the dipole. (c) Sketching the electric field as a quiver plot. (d) Sketching the field as a streamline plot.

Observe that the electric field depends on the product of the charge and the separation, Ql. It also decays inversely with distance cubed. Similarly we can determine the electric field broadside to the dipole perpendicular to and at a distance r from the line joining the charges as shown in Fig. 3.9b. Observe that the fields from the two charges of opposite sign are such that the horizontal (y-directed) components cancel, leaving only the z components (directed in the $-z$ direction). Combining the fields due to both charges gives

$$\mathbf{E} = -2\frac{1}{4\pi\varepsilon_o}\frac{Q}{R^2}\underbrace{\cos(90° - \alpha)}_{\sin(\alpha)}\mathbf{a}_z$$

where

$$R = \sqrt{r^2 + \left(\frac{l}{2}\right)^2}$$

and

$$\sin(\alpha) = \frac{l/2}{R}$$

Substituting these gives

$$\mathbf{E} = -\frac{1}{4\pi\varepsilon_o}\frac{Ql}{\left(r^2 + \left(\frac{l}{2}\right)^2\right)^{3/2}}\mathbf{a}_z$$

Evaluating this for distances much larger than the charge separation gives

$$\mathbf{E} = -\frac{1}{4\pi\varepsilon_o}\frac{Ql}{r^3}\mathbf{a}_z \qquad r \gg \frac{l}{2}, \quad \theta = 90° \tag{3.13b}$$

The derivation of the complete expression for the field at all points around the dipole is somewhat complicated. In terms of spherical coordinates as shown in Fig. 3.9c, the complete expression is

$$\mathbf{E} = \frac{Ql}{4\pi\varepsilon_o r^3}(2\cos\theta\,\mathbf{a}_r + \sin\theta\,\mathbf{a}_\theta) \tag{3.14}$$

This complete expression evaluated for $\theta = 0°$ gives the result in (3.13a). Broadside to the dipole, $\theta = 90°$, the result is directed in the θ or $-z$ direction and is parallel to the dipole and (3.14) gives (3.13b). The field is sketched using streamlines in Fig. 3.9d. Note that the electric field depends on the product of the charge and the separation distance. This is referred to as the *dipole moment*, which is denoted as $p = Ql$.

Electric dipoles consisting of a pair of charges of opposite polarity that are bound together in that they are not free to move independently of each other occur in dielectric materials at the atomic level. These are referred to as *bound charges*. Certain materials such as water contain microscopic dipoles which are randomly oriented in direction so that the electric fields of these dipoles tend to cancel. In other dielectric materials, the application of an external electric field causes the atomic centers of positive and negative charge to shift slightly, forming a dipole of charge. The application of an external electric field causes the dipoles of the dielectric to rotate so as to align with

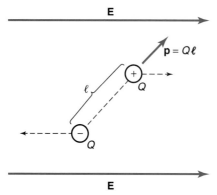

Figure 3.10 Illustration of dipole moment of bound charges in dielectrics.

the electric field as illustrated in Fig. 3.10. The force of rotation depends on the charge value and the separation distance, l, between them. This is contained in the *dipole moment*, which is expressed as a vector quantity

$$\mathbf{p} = Q\mathbf{l} \quad \text{Cm} \tag{3.15}$$

where the vector direction is directed *from* the negative charge *to* the positive charge. Note that the dipole moment is in the direction of the electric field off the ends of the dipole as obtained above and is defined as being *opposite to* the direction of force that one of the charges exerts on the other. In a volume containing a large number of these dipoles we define the *polarization vector* as the dipole moment per unit volume:

$$\mathbf{P} = \lim_{\Delta v \to 0} \frac{\sum_i \mathbf{p}_i}{\Delta v} \quad \text{C/m}^2 \tag{3.16}$$

In the absence of an electric field, the dipoles are randomly oriented so that their dipole moments cancel and the material is said to be *unpolarized*. When an external electric field is applied, the dipoles rotate so as to try to align with the field. Not all the dipoles completely align with the field, so that the net alignment is given by the polarization vector. This gives an indication of the net polarization of the dielectric.

   In order to illustrate the significance of this concept, consider a parallel-plate capacitor shown in Fig. 3.11a. A battery is attached to the plates and charge is transferred from the battery to the plates. This separation of charge creates an electric field between the plates. If a dielectric material is inserted between the plates as shown in Fig. 3.11b, the bound charge dipoles attempt to align with that field, creating a polarization vector in the dielectric. Observe another important result of this *polarization* of the dielectric: a surface charge is created on the two surfaces of the dielectric. Negative charge appears on the surface of the dielectric that is adjacent to the positively charged plate and positive charge appears on the surface of the dielectric that is adjacent to the negatively charged plate. Hence, *more free charge is drawn from the battery*. The capacitance of a capacitor is the ratio of the *free charge* stored on the plates, Q_f, to the voltage applied between them: $C = Q_f/V$. Since the voltage is unchanged, the increase in free charge stored on the plates

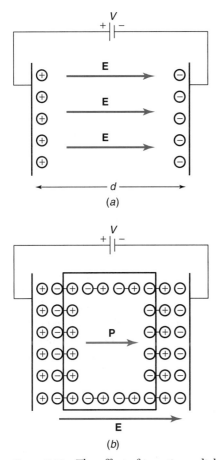

Figure 3.11   The effect of inserting a dielectric between the plates of a parallel-plate capacitor. (a) The electric field with the dielectric removed. (b) The electric field with the dielectric inserted, illustrating the polarization vector **P**; the bound charge on the surfaces of the dielectric; and the increased free charge on the capacitor plates.

resulting from the polarization of the dielectric results in increased capacitance of the structure.

There are two vector quantities involved in the capacitor, the applied electric field, **E**, and the polarization vector due to the polarization of the dielectric, **P**. Note in (3.16) that the units of **P** are C/m² or a *charge per unit of surface area*. It turns out the product of the free space permittivity and the electric field, $\varepsilon_o\mathbf{E}$, also has the units of C/m². Also note that **E** and **P** are in the same direction. Hence this suggests that we define the second of our fundamental EM quantities as the *electric flux density vector* since the units of **P** and $\varepsilon_o\mathbf{E}$ are both C/m². The electric flux density vector will be denoted as **D** and is defined as

$$\mathbf{D} = \varepsilon_o\mathbf{E} + \mathbf{P} \qquad \text{C/m}^2$$ (3.17)

The strength of the polarization vector depends on the strength of the applied electric field. This is related by the *electric susceptibility* of the material, χ_e, as

$$\mathbf{P} = \varepsilon_o\chi_e\mathbf{E}$$ (3.18)

Substituting (3.18) into (3.17) gives

$$\mathbf{D} = \varepsilon_o \underbrace{(1 + \chi_e)}_{\varepsilon_r}\mathbf{E} \tag{3.19}$$

The quantity $\varepsilon_r = (1 + \chi_e)$ is referred to as the *relative permittivity* or *dielectric constant* of the material. Some representative values are Teflon $\varepsilon_r = 2.1$, mica $\varepsilon_r = 5.4$, Styrofoam $\varepsilon_r = 1.03$, distilled water $\varepsilon_r = 80$. Hence if we insert a slab of Teflon between the plates of a capacitor, the capacitance essentially doubles.

Hence we can describe the effect of a dielectric by relating the electric flux density vector in the material to the electric field as

$$\boxed{\mathbf{D} = \varepsilon \mathbf{E} \qquad \text{C/m}^2} \tag{3.20}$$

where the *permittivity* of the material is

$$\boxed{\varepsilon = \varepsilon_r \varepsilon_o} \tag{3.21}$$

Equation (3.20) shows that we may freely interchange $\mathbf{D}$ and $\mathbf{E}$. We have been considering what are said to be *simple materials*. Simple materials are where the $\mathbf{D}$ and $\mathbf{E}$ vectors are parallel (isotropic materials) and the magnitude of $\mathbf{D}$ does not depend on the magnitude of $\mathbf{E}$ (linear materials). All dielectric materials considered in this text will be assumed to be linear and isotropic so that (3.20) relates $\mathbf{D}$ and $\mathbf{E}$ where ε is a scalar constant. It turns out that in some dielectric materials, the permittivity, ε, is dependent on the frequency of the electric field. This will be important when we later consider fields that are time varying instead of dc.

▶ 3.4 GAUSS' LAW FOR THE ELECTRIC FIELD

Gauss' law is an extraordinarily powerful computational tool. Until now the computation of the electric field due to various charge distributions, although straightforward when we utilized symmetry, required the evaluation of some moderately complicated integrals. Gauss' law will allow us to determine the electric field due to many of those charge distributions without the evaluation of any integrals and will, moreover, give us a great deal of insight into the problem that the direct evaluation did not provide.

Consider a closed surface containing some charge as shown in Fig. 3.12. We now know that electric field lines ($\mathbf{E}$ or $\mathbf{D}$ field lines) begin on positive charge and terminate on negative charge. If we count the number of $\mathbf{E}$ or $\mathbf{D}$ field lines exiting a closed surface, this should give us an indication of the *net positive charge enclosed by the surface*. Observe in Fig. 3.12 that field lines that begin and end on charge within the

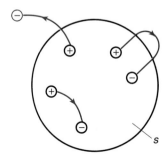

Figure 3.12 Illustration of Gauss' law; determining the net positive charge enclosed by a surface by determining the number of electric flux density lines exiting that surface.

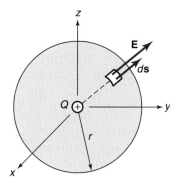

Figure 3.13 Illustration of using Gauss' law to determine the electric field of a point charge.

closed surface contribute nothing to the net field lines exiting that surface. However, the field lines that begin on positive charge within the surface but end on negative charge outside that surface contribute to field lines exiting that surface. This observation is *Gauss' law*, which is stated mathematically as

$$\oint_s \mathbf{D} \cdot d\mathbf{s} = Q_{\text{enclosed}} \tag{3.22}$$

Observe that the dot product in the integral requires that we obtain *the component of* **D** *that is perpendicular to the surface*. This is logical since the component of **D** that is parallel to the surface contributes nothing to the number of field lines exiting the surface.

In order to illustrate this important law, consider a single point charge as shown in Fig. 3.13. In order to compute Gauss' law we place the point charge at the origin. Next we decide on the coordinate system and the surface shape to be used. This is the critical aspect of Gauss' law: *we are allowed to select the surface and should choose one that simplifies the computation*. Since the **E** and **D** field lines from a point charge emanate radially away from it, we will choose a spherical coordinate system and let the surface be a sphere of radius r. Why did we make this choice? The answer is that this will simplify the evaluation of Gauss' law more than any other choice *for this particular problem*. Since the **E** and **D** field lines emanate radially away from the point charge, *they will be perpendicular to the chosen spherical surface*. Hence the dot product in (3.22) is not needed and Gauss' law reduces to

$$\oint_s \mathbf{D} \cdot d\mathbf{s} = \oint_s D\,ds$$

Furthermore, the **D** field has the same value at a fixed distance from the charge, that is, over this surface. Hence, the magnitude of **D**, D, can be removed from the integral and Gauss' law simplifies, *for this judicious choice of the surface*, to

$$D \underbrace{\oint_s ds}_{4\pi r^2} = Q$$

Solving this gives

$$\mathbf{D} = \frac{Q}{4\pi r^2}\mathbf{a}_r \tag{3.23}$$

and we have made the result into a vector quantity, since the field lines emanate radially away from the charge. Substituting $\mathbf{D} = \varepsilon\,\mathbf{E}$ gives

$$\mathbf{E} = \frac{Q}{4\pi\varepsilon r^2}\mathbf{a}_r \tag{3.24}$$

which is the result derived earlier for a point charge in free space (ε here replaced by ε_o) and given in (3.6). Gauss' law provides that the flux of $\mathbf{D}$ through the surface equals the net positive charge enclosed. Hence the name for $\mathbf{D}$: the *electric flux density vector*. We view lines of electric field as flux much in the same way as we visualize light flux through an opening.

Although Gauss' law was demonstrated for a point charge, it holds for any type of charge distribution. The remarkable thing about this is that it essentially relates back to the inverse square dependence on distance in Coulomb's law. If Coulomb's law had depended on distance to any other power other than $2.000\cdots$, Gauss' law would not have occurred. For example, let us redo the previous problem but substitute (3.6) to give

$$\oint_s \mathbf{D}\cdot d\mathbf{s} = \oint_s \underset{D}{\underbrace{\frac{Q}{4\pi r^2}}}r^2\underset{ds}{\underbrace{\sin(\theta)d\theta d\phi}}$$

$$= \frac{Q}{4\pi}\int_{\phi=0}^{2\pi}\int_{\theta=0}^{\pi}\sin(\theta)d\theta d\phi$$

$$= Q$$

Observe in this evaluation that the r^2 in the expression for the D field and the r^2 in the expression for ds in spherical coordinates cancel. If the electric field had not been an inverse square law, this would not result. So the inverse square dependence is a fundamental dependence in electromagnetics (and many other physical laws).

3.4.1 Calculation of the Electric Field for Charge Distributions

The power of Gauss' law lies in the fact that we can choose the surface over which it is evaluated. In all such problems we should attempt to choose the surface such that

1. the D and E field lines are perpendicular to the chosen surface, and
2. the D and E field lines are constant over that surface.

The first condition removes the dot product from the integral, leaving

$$\oint_s D\,ds = Q_{\text{enclosed}}$$

and the second condition means that D is independent of the position on the surface and hence can be removed from the integral, leaving

$$D\oint_s ds = Q_{\text{enclosed}}$$

The first condition, choosing the surface so that the D and E field lines are everywhere on the surface perpendicular to it, is a necessary first step. If this cannot be done, then

Gauss' law will not provide a *computational* simplification. The second condition, choosing the surface so that D is independent of position on the surface, may not be possible in all cases but when it is it should be taken advantage of. Once again, the key to making EM problems simple is to *visualize* the fields.

▷ **EXAMPLE 3.5**

Determine the electric field about an infinite line charge bearing a uniform line charge distribution ρ_l C/m using Gauss' law.

SOLUTION This problem was solved in Example 3.3 by direct integration and superposition of differential chunks of charge. An infinite line charge bearing a charge distribution that is uniformly distributed along it will have, by symmetry, an electric field that is directed radially away from the line as illustrated in Fig. 3.14a. Note that this would not be the case if (1) the line is not infinite in length, and/or (2) the charge is not *uniformly* distributed along it. In order to take advantage of this symmetry observation we choose the Gaussian surface as a cylinder of radius r centered on the line as shown in Fig. 3.14b. The length of this cylinder will be chosen as l. Over the ends of the cylinder, the field is parallel to these surfaces since the line is infinite in length and contributes nothing to Gauss' law. Over the sides of the cylinder, the field is perpendicular to it and hence the dot product in Gauss' law may be removed. Further, over this constant radius surface, the field is constant, so that Gauss' law reduces to

$$\oint_s \mathbf{D} \cdot d\mathbf{s} = D \oint_s ds$$

$$= D2\pi rl$$

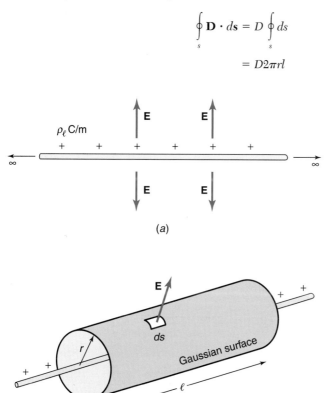

(a)

(b)

Figure 3.14 Example 3.5; determining the electric field of an infinite line charge. (a) Using symmetry to determine that the electric field intensity vector must be radially directed away from the line charge. (b) Surrounding the line charge with an appropriate Gaussian surface (a cylinder) for determining the electric field intensity vector via Gauss' law.

The total charge enclosed by the surface is

$$Q_{\text{enclosed}} = \rho_l l$$

Hence we obtain

$$\mathbf{D} = \frac{\rho_l}{2\pi r}\mathbf{a}_r \tag{3.25}$$

or, substituting (3.20),

$$\mathbf{E} = \frac{\rho_l}{2\pi \varepsilon r}\mathbf{a}_r \tag{3.26}$$

where we assume that the material surrounding the line charge has permittivity ε. This agrees with (3.10) obtained in Example 3.3. ◀

▶ EXAMPLE 3.6

Determine the electric field outside and within an infinitely long cylinder of radius a that bears a surface charge distribution ρ_s C/m^2 that is uniformly distributed both along the cylinder length *and* around the cylinder periphery.

SOLUTION Again we observe that because of symmetry (the cylinder is infinitely long and the charge is uniformly distributed along its length and around its periphery) the electric field is radially directed. Hence we choose the Gaussian surface to be a cylinder of length l and radius r and center it on the axis of the charged cylinder as illustrated in Fig. 3.15. Again, the field is parallel to the ends of the cylinder and perpendicular to its side surface. Hence Gauss' law reduces to

$$\oint_s \mathbf{D} \cdot d\mathbf{s} = D \oint_s ds$$

$$= D2\pi rl$$

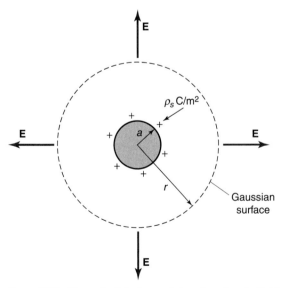

Figure 3.15 Example 3.6; determining the electric field about a cylinder of charge.

and the charge enclosed is

$$Q_{\text{enclosed}} = \rho_s 2\pi a l$$

Thus

$$\mathbf{D} = \frac{\rho_s a}{r}\mathbf{a}_r \quad r > a \tag{3.27}$$

or

$$\mathbf{E} = \frac{\rho_s a}{\varepsilon r}\mathbf{a}_r \quad r > a \tag{3.28}$$

For points interior to the cylinder of charge, no charge is enclosed by the Gaussian surface and hence the field is zero:

$$\mathbf{E} = 0 \quad r < a$$

We could relate this result to that of an infinite line charge solved in the previous example. The charge density per unit of cylinder length is $\rho_l = \rho_s 2\pi a$. Rewriting the above result in terms of this equivalent line charge gives

$$\mathbf{E} = \frac{\rho_l}{2\pi\varepsilon r}\mathbf{a}_r \quad r > a \tag{3.29}$$

which is the same as for an equivalent line charge located on its axis. Hence, at points away from the cylinder of charge, the electric field is the same as if the surface charge were concentrated as a line charge at the axis of the cylinder. ◀

▶ EXAMPLE 3.7

A coaxial transmission line consists of two concentric cylinders shown in Fig. 3.16a. This is a common type of structure used to guide EM waves from one point to another which will be considered

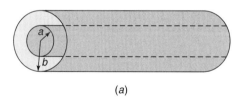

(a)

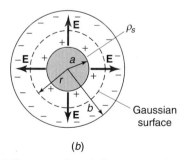

(b)

Figure 3.16 Example 3.7; determining the electric field for a coaxial cable. (a) The cable dimensions. (b) Surrounding the inner cylinder with an appropriate Gaussian surface (a cylinder) for determining the electric field intensity vector via Gauss' law.

in Chapter 6. The inner cylinder has radius a and the outer cylinder has radius b. Determine the electric field between the two cylinders.

SOLUTION   Suppose the inner cylinder has a surface charge distribution of ρ_s C/m^2 that is uniformly distributed along its length and around its periphery. The outer cylinder has the same *total* charge as the inner cylinder distributed over its inner surface but is of opposite polarity. (The distributions are different but both are uniform around the inner and outer peripheries.) Both cylinders are considered infinite in length. Because of the uniform charge distributions and the infinite lengths of the cylinders, the electric field will be radially directed from the inner cylinder toward the outer cylinder. We enclose the inner cylinder with a cylindrical Gaussian surface of radius r as shown in Fig. 3.16b. The electric field will be perpendicular to the sides of the Gaussian surface and parallel to the end surfaces. Because the Gaussian surface is the same as in the previous example, the field in the space between the two cylinders is given by (3.28):

$$\mathbf{E} = \frac{\rho_s a}{\varepsilon r}\mathbf{a}_r \quad a < r < b \tag{3.30}$$

Observe that this is independent of the presence of the outer cylinder. In terms of an equivalent line charge distribution, $\rho_l = \rho_s 2\pi a$, this becomes

$$\mathbf{E} = \frac{\rho_l}{2\pi\varepsilon r}\mathbf{a}_r \quad a < r < b \tag{3.31}$$

The electric field inside the inner cylinder is zero, since a cylindrical Gaussian surface inside that cylinder will contain no charge:

$$\mathbf{E} = 0 \quad r < a$$

Observe that the electric field outside the cable is also zero, since the total charge enclosed by a cylindrical Gaussian surface surrounding *both* cylinders is zero:

$$\mathbf{E} = 0 \quad r > b$$

This is the reason for using such a cable; the exterior is "shielded" from the fields interior to the cable. The coaxial cable is therefore often referred to as a "shielded cable." ◀

▶ EXAMPLE 3.8

Determine the electric field of an infinite plane of charge that is uniformly distributed over its surface using Gauss' law.

SOLUTION   This problem was solved by direct integration in Example 3.4. We place the plane of charge in the xz plane. Once again, we observe that because of the infinite extent of the plane and the uniform distribution of charge over it, the electric field will be perpendicular to its surface. Suppose that charge distribution is ρ_s C/m^2. Since the field will be perpendicular to the surface, an appropriate choice of the Gaussian surface is a rectangular surface extending to the right and to the left of the plane as shown in Fig. 3.17. The electric field will be parallel to the sides of this rectangular Gaussian surface and contribute nothing there but will be perpendicular to the front and back surfaces. Hence Gauss' law becomes

$$\oint_s \mathbf{D} \cdot d\mathbf{s} = D \oint_s ds$$
$$= 2DA$$

where the front and back surfaces have area A. The total charge enclosed by the Gaussian surface is

$$Q_{\text{enclosed}} = \rho_s A$$

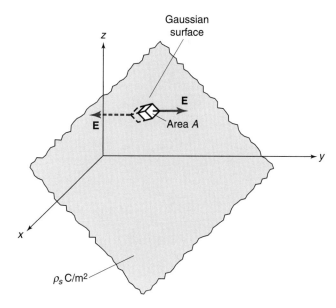

Figure 3.17 Example 3.8; determining the electric field of a flat surface charge distribution via Gauss' law by generating an appropriate Gaussian surface (a rectangular box extending on both sides).

Hence the electric field becomes

$$\mathbf{E} = \begin{cases} \dfrac{\rho_s}{2\varepsilon}\,\mathbf{a}_y & y > 0 \\[2ex] -\dfrac{\rho_s}{2\varepsilon}\,\mathbf{a}_y & y < 0 \end{cases} \tag{3.32}$$

as determined in Example 3.4.   ◀

▷ QUICK REVIEW EXERCISE 3.5

Determine, using Gauss' law, the electric field due to a sphere of charge of radius a where the charge is uniformly distributed over the sphere surface with distribution ρ_s C/m^2.

ANSWER $\mathbf{E} = (\rho_s a^2 / \varepsilon r^2)\mathbf{a}_r$ for $r > a$ and $\mathbf{E} = 0$ for $r < a$.   ◀

▷ QUICK REVIEW EXERCISE 3.6

Determine, using Gauss' law, the electric field due to a sphere of charge of radius a where the charge is uniformly distributed throughout the sphere with volume charge distribution ρ_v C/m^3.

ANSWER $\mathbf{E} = (\rho_v a^3 / 3\varepsilon r^2)\,\mathbf{a}_r$ for $r > a$ and $\mathbf{E} = (\rho_v r / 3\varepsilon)\,\mathbf{a}_r$ for $r < a$.   ◀

▷ 3.5 VOLTAGE

Moving a charge in an electric field requires energy expenditure. This brings us to the concept of *voltage*. Consider Fig. 3.18a, where we wish to move a positive charge q from point a to point b in the presence of an electric field $\mathbf{E}$. The force exerted on the charge by the field is $\mathbf{F} = q\mathbf{E}$. The component of this force along the path is $\mathbf{F} \cdot d\mathbf{l}$. Along

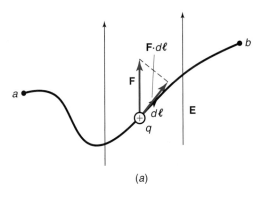

(a)

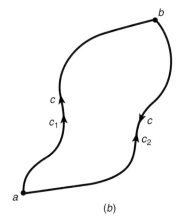

(b)

Figure 3.18 Illustration of determining the voltage between two points. (a) Determining the work required to move a positive charge between two points in the presence of an electric field. (b) Illustration of the conservative nature of the static electric field and the unique determination of voltage between two points independent of the path taken.

certain portions of the path work is required on our part to move the charge while along other portions, the electric field provides the energy to move the charge. The net energy expended in moving the charge from a to b is

$$
\begin{aligned}
W_{ba} &= -\int_a^b \mathbf{F} \cdot d\mathbf{l} \\
&= -q \int_a^b \mathbf{E} \cdot d\mathbf{l}
\end{aligned}
$$

(3.33)

A negative sign is required since we are asking for the work required by us to move the charge against the field. The *voltage* or *potential difference* between points a and b with *point b at the assumed higher potential* is the work per unit of charge required to move the charge from a to b:

$$
\begin{aligned}
V_{ba} &= \frac{W_{ba}}{q} \\
&= -\int_a^b \mathbf{E} \cdot d\mathbf{l} \quad \text{V}
\end{aligned}
$$

(3.34)

The unit of voltage is the volt (V), whose underlying units are joules (J) per coulomb (C).

If we move the charge from a to b and return to a as shown in Fig. 3.18b, the net work done is zero. Hence we obtain the first of the important EM laws signifying that the static (dc) electric field is *conservative*:

$$\oint_c \mathbf{E} \cdot d\mathbf{l} = 0 \qquad (3.35)$$

Hence *the line integral of the electric field around a closed path c yields a result of zero.* We will find in the next chapter that for a time-varying field this is no longer true. However, for static fields the result shows that *voltage between two points is unique regardless of the path taken between those two points.* This is because we can break (3.35) into

$$\underbrace{\int_a^b \mathbf{E} \cdot d\mathbf{l}}_{\text{along } c_1} - \underbrace{\int_a^b \mathbf{E} \cdot d\mathbf{l}}_{\text{along } c_2} = 0$$

where c_2 is opposite the direction of the closed path c along this portion. Hence

$$
\begin{aligned}
V_{ba} &= -\underbrace{\int_a^b \mathbf{E} \cdot d\mathbf{l}}_{\text{along } c_1} \\
&= -\underbrace{\int_a^b \mathbf{E} \cdot d\mathbf{l}}_{\text{along } c_2}
\end{aligned}
\qquad (3.36)
$$

Equation (3.35) is essentially a statement of Kirchhoff's voltage law (KVL), where contour c is a lumped-circuit loop.

We will first determine a fundamental result for the voltage due to a point charge and then compute the voltage due to other charge distributions. Consider Fig. 3.19, where we place the charge at the origin of a spherical coordinate system. We want to determine the voltage between two points that are at radial distances r_a and r_b from point charge Q with point b at radius r_b at an assumed positive or higher voltage than

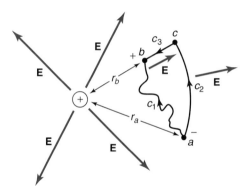

Figure 3.19 Determination of the voltage between two points produced by a point charge, and the appropriate choice of path to simplify its determination.

point a at radius r_a. We previously determined that the electric field of a point charge is radially directed away from the charge and is given by

$$\mathbf{E} = \frac{Q}{4\pi\varepsilon r^2}\mathbf{a}_r$$

Since the path does not matter for static fields, we have several choices. We could choose a rather circuitous path c_1 for which the evaluation of (3.34) would be very difficult or we could choose a path to simplify the evaluation. For example, suppose we choose the combination of a path c_2 of constant radius r_a from the assumed lower potential point a to point c which is along a radial line through point b, and a path c_3 which is directed radially inward to the assumed higher potential point b at radius r_b. The voltage between the two points with point b at the assumed higher potential is

$$V_{ba} = -\int_{c_2} \mathbf{E} \cdot d\mathbf{l} - \int_{c_3} \mathbf{E} \cdot d\mathbf{l}$$

Along the constant radius path c_2 the electric field is perpendicular to the path and hence contributes nothing to the integral:

$$\int_{c_2} \mathbf{E} \cdot d\mathbf{l} = 0$$

Thus

$$V_{ba} = -\int_{c_3} \mathbf{E} \cdot d\mathbf{l}$$

Along c_3 the electric field is parallel to the path and hence the dot product is removed, leaving

$$V_{ba} = -\int_{c_3} \mathbf{E} \cdot d\mathbf{l}$$

$$= -\int_{r_a}^{r_b} \frac{Q}{4\pi\varepsilon} \frac{1}{r^2} dr$$

Evaluating this gives

$$\boxed{V_{ba} = \frac{Q}{4\pi\varepsilon}\left(\frac{1}{r_b} - \frac{1}{r_a}\right)} \tag{3.37}$$

This result for the voltage between two points due to a point charge is a very important result and will be used on numerous occasions. Hence it should be committed to memory. It is important to "sanity check" all such results where possible. For example, if point a is farther from the charge than point b, $r_a > r_b$, then (3.37) indicates that V_{ba} is positive. This is a sensible result because moving a positive test charge from a to b, which is at a closer distance to the point charge, will require energy expenditure on our part in overcoming the repulsive force of Q. Observe that the constant voltage or equipotential surfaces about a point charge are spheres as illustrated in Fig. 3.20. Movement

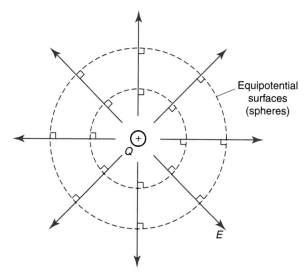

Figure 3.20 Illustration of the electric field and equipotential surfaces (spheres) about a point charge.

of a charge on an equipotential surface requires no energy expenditure. Also observe that *the electric field is perpendicular to the equipotential surfaces*.

3.5.1 Calculation of Voltage for Charge Distributions

Next we will compute the voltage due to various charge distributions considered earlier.

▶ **EXAMPLE 3.9**

Determine the voltage between two points at radial distances r_a and r_b away from an infinite line charge bearing a charge distribution ρ_l C/m that is uniformly distributed along it.

SOLUTION   The electric field about the line charge was determined earlier as

$$\mathbf{E} = \frac{\rho_l}{2\pi\varepsilon r}\mathbf{a}_r$$

and is directed radially away from the line. Hence the voltage between two points is

$$V_{ba} = -\int_{r_a}^{r_b} \frac{\rho_l}{2\pi\varepsilon r}dr$$

Evaluating this gives

$$V_{ba} = \frac{\rho_l}{2\pi\varepsilon}\ln\left(\frac{r_a}{r_b}\right) \qquad (3.38)$$

This result for the voltage between two points due to an infinite line charge is a very important result that will be used on numerous occasions. Hence it should be committed to memory. Again it is important to sanity check the result. If $r_a > r_b$, then the voltage will be positive (point b at distance

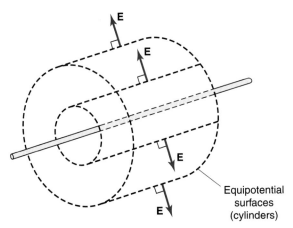

Figure 3.21 Example 3.9; illustration of the electric field and equipotential surfaces (cylinders) about an infinite line charge.

r_b is the assumed higher potential or positive voltage point). Again this is sensible since for $r_a > r_b$ moving a positive test charge from a to b (closer to the positive line charge) will require energy expenditure on our part in order to overcome the repulsive force of the line charge. Observe that the constant voltage or equipotential contours about an infinite line charge are cylinders. Movement of charge on one of these cylinders will require no work. The equipotential surfaces (cylinders) are perpendicular to the electric field at every point as shown in Fig. 3.21. ◀

▶ **EXAMPLE 3.10**

Determine the voltage between two points away from an infinite sheet of charge.

SOLUTION The electric field away from an infinite sheet of charge bearing a distribution ρ_s C/m^2 was determined previously as

$$
\mathbf{E} = \begin{cases} \dfrac{\rho_s}{2\varepsilon}\mathbf{a}_y & y > 0 \\[2mm] -\dfrac{\rho_s}{2\varepsilon}\mathbf{a}_y & y < 0 \end{cases}
$$

where the sheet is placed in the xz plane as shown in Fig. 3.17. Observe that the field is independent of distance away from the sheet and is directed perpendicular to the sheet. Hence the voltage is evaluated using (3.34) to give

$$
V_{ba} = \frac{\rho_s}{2\varepsilon}(d_a - d_b) \tag{3.39}
$$

Observe that if the distance from the sheet to point a (the assumed lower voltage point), d_a, is greater than the distance from the sheet to point b (the assumed higher or positive voltage point), d_b, $d_a > d_b$, then the voltage V_{ba} with point b at the assumed higher potential is positive. Again this makes sense because movement of a positive test charge from a to point b, which is closer to the sheet, will require work on our part to overcome the repulsive force of the sheet of charge. The equipotential surfaces are planes parallel to the sheet. Movement of charge on these planes requires no energy expenditure. Once again, the electric field is perpendicular to these equipotential surfaces as shown in Fig. 3.22. ◀

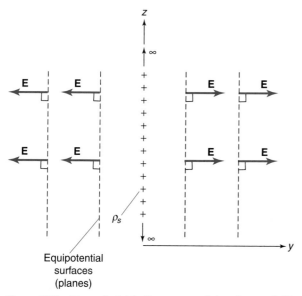

Figure 3.22 Example 3.10; illustration of the electric field and equipotential surfaces (planes) about an infinite sheet of charge.

▶ **EXAMPLE 3.11**

Determine the voltage between the inner and outer cylinders of a coaxial cable.

SOLUTION The problem is depicted in Fig. 3.16 and the electric field was determined in Example 3.7 to be

$$\mathbf{E} = \frac{\rho_s a}{\varepsilon r}\mathbf{a}_r \quad a < r < b$$

Hence the voltage of the inner cylinder of radius a with respect to the outer cylinder whose radius is b with the inner cylinder at the assumed higher potential is

$$V_{ab} = -\int_b^a \frac{\rho_s a}{\varepsilon r}dr$$

Evaluating this gives

$$V_{ab} = \frac{\rho_s a}{\varepsilon}\ln\left(\frac{b}{a}\right)$$
$$= \frac{\rho_l}{2\pi\varepsilon}\ln\left(\frac{b}{a}\right) \tag{3.40}$$

and we have written this result in terms of the equivalent line charge, $\rho_l = \rho_s 2\pi a$. Since $b > a$, the voltage of the inner cylinder is positive with respect to the outer cylinder. Again this makes sense because the electric field is directed from the inner cylinder toward the outer cylinder and hence work would be required by us to move a positive test charge from the outer cylinder to the inner cylinder. Again, the equipotential surfaces are cylinders surrounding the inner cylinder. The electric field vector is radially directed and perpendicular to these equipotential surfaces (see Fig. 3.21).

◀

▶ **QUICK REVIEW EXERCISE 3.7**

Determine the voltage between two concentric spheres of radii a and b with $a < b$. The inner sphere supports a total charge Q that is uniformly distributed over its surface, and the outer sphere supports the same total charge but of opposite polarity over its inner surface.

ANSWER $V_{ab} = \dfrac{Q}{4\pi\varepsilon}\left(\dfrac{1}{a} - \dfrac{1}{b}\right).$ ◀

In the previous examples we computed the voltage from the electric field via (3.34). This required that we first determine an expression for the electric field vector. It is possible to bypass this calculation of the electric field and to compute voltage directly from the charge distribution. Voltage is defined as being between two points. We can talk about the voltage of a point if we refer that voltage to some reference point. If we choose this reference point as being at infinity, then the voltage between the point and infinity is referred to as the *absolute potential* of a point and is defined as

$$V = -\int_{\infty}^{r} \mathbf{E} \cdot d\mathbf{l} \qquad (3.41)$$

Here the voltage at a point r is the work required to move a charge from infinity to the point per unit of the charge moved. The voltage between two points due to a point charge is given in (3.37). The absolute potential of a point at a distance r from the point charge can be obtained by letting $r_a \to \infty$ and setting $r_b = r$ in that expression, giving

$$V = \frac{Q}{4\pi\varepsilon r} \qquad (3.42)$$

We can similarly determine the absolute potential of a point due to a distribution of charge by summing the contributions to differential chunks of charge, dQ. Treating these chunks of charge as point charges and using (3.42), we obtain the differential contribution—the absolute potential of a point due to this chunk of charge as shown in Fig. 3.23:

$$dV = \frac{dQ}{4\pi\varepsilon R} \qquad (3.43)$$

where the differential chunk of charge is $dQ = \rho_v dv$. The total absolute potential of a point due to the charge distribution can be obtained in the usual fashion by summing (with an integral) the contributions from all the chunks of charge:

$$V = \int_{v} \frac{\rho_v dv}{4\pi\varepsilon R} \qquad (3.44)$$

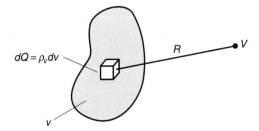

Figure 3.23 Determining the absolute potential of a point due to a distribution of charge by superimposing the contributions due to differential chunks of charge.

This is very similar to the method for determining the electric field due to a charge distribution as shown in Fig. 3.6. However, in the case of voltage there is a significant difference. In the case of the electric field, the integral summation involved vectors, whereas the integral summation in determining voltage does not. Hence, the advantage in determining the absolute potential via this method is that vector quantities are not dealt with and the integrals are much simpler than the direct method in terms of the electric field as in (3.34). This method is only applicable to charge distributions that are finite. If the charge distribution extends to infinity, such as the infinite line charge, absolute potential cannot be defined.

▷ **EXAMPLE 3.12**

Determine the absolute potential of an electric dipole.

SOLUTION   The dipole consists of a pair of point charges of opposite polarity separated by a distance l as shown in Fig. 3.24. We choose a spherical coordinate system and place the charges on the z axis symmetrically located about the origin. We desire the absolute potential at a point that is a distance r from the origin of the coordinate system and at an angle θ from the positive z axis. Summing the contributions due to each charge gives

$$V = \frac{Q}{4\pi\varepsilon R^+} - \frac{Q}{4\pi\varepsilon R^-} \tag{3.45}$$

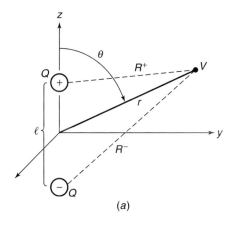

(a)

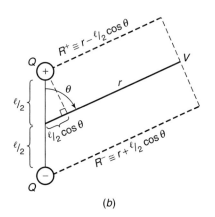

(b)

Figure 3.24   Example 3.12; determining the potential distribution about an electric dipole of charge. (a) Setting up the problem in a spherical coordinate system. (b) Using the parallel line approximation to simplify the determination of the potential assuming that the distance to the point is much greater than the charge separation.

Let us assume that the point where we desire to determine the absolute potential is at a very large distance r from the point charges compared to their spacing, l, that is, $r \gg l$. In this case, the distance lines from each charge to the point, R^+ and R^-, are approximately parallel as shown in Fig. 3.24b. Hence these distances can be written as

$$R^+ \cong r - \frac{l}{2}\cos\theta$$

and

$$R^- \cong r + \frac{l}{2}\cos\theta$$

Hence (3.45) becomes

$$V = \frac{Q}{4\pi\varepsilon}\left[\frac{l\cos\theta}{r^2 - \frac{l^2}{4}\cos^2\theta}\right]$$

Evaluating this for $r \gg l$ gives

$$V \cong \frac{Ql}{4\pi\varepsilon r^2}\cos\theta \qquad r \gg l \tag{3.46}$$

3.5.2 Determining the Electric Field from the Potential Distribution

For those problems having a charge distribution of finite extent such that the absolute potential can be defined and determined via (3.44), it is possible to determine the electric field from that potential. Consider the definition of the voltage in terms of the electric field as

$$V = -\int \mathbf{E} \cdot d\mathbf{l}$$

Taking the total derivative of both sides in rectangular coordinates gives

$$dV = \frac{\partial V}{\partial x}dx + \frac{\partial V}{\partial y}dy + \frac{\partial V}{\partial z}dz$$
$$= -\mathbf{E} \cdot d\mathbf{l}$$

But this can be written as

$$\underbrace{\left(\frac{\partial V}{\partial x}\mathbf{a}_x + \frac{\partial V}{\partial y}\mathbf{a}_y + \frac{\partial V}{\partial z}\mathbf{a}_z\right)}_{\text{gradient of V}} \cdot d\mathbf{l} = -\mathbf{E} \cdot d\mathbf{l}$$

Comparing both sides yields

$$\boxed{\mathbf{E} = -\text{gradient } V} \tag{3.47}$$

where in rectangular coordinates

$$\boxed{\text{gradient } V = \left(\frac{\partial V}{\partial x}\mathbf{a}_x + \frac{\partial V}{\partial y}\mathbf{a}_y + \frac{\partial V}{\partial z}\mathbf{a}_z\right)} \tag{3.48a}$$

Hence the electric field can be determined in terms of the *gradient* of the potential. In cylindrical and spherical coordinate systems the gradient operation becomes

$$\text{gradient } V = \left(\frac{\partial V}{\partial r} \mathbf{a}_r + \frac{1}{r} \frac{\partial V}{\partial \phi} \mathbf{a}_\phi + \frac{\partial V}{\partial z} \mathbf{a}_z \right) \quad \text{cylindrical} \tag{3.48b}$$

$$\text{gradient } V = \left(\frac{\partial V}{\partial r} \mathbf{a}_r + \frac{1}{r} \frac{\partial V}{\partial \theta} \mathbf{a}_\theta + \frac{1}{r \sin\theta} \frac{\partial V}{\partial \phi} \mathbf{a}_\phi \right) \quad \text{spherical} \tag{3.48c}$$

The term *gradient* refers to the fact that as we move perpendicular to the equipotential surfaces, the associated electric field either increases or decreases at a maximum rate with respect to distance moved. So in order to move in a direction such that the electric field increases at its maximum rate with respect to distance, we choose a path that is perpendicular to the equipotential surfaces.

▷ EXAMPLE 3.13

Determine the electric field of the electric dipole from the potential obtained in Example 3.12.

SOLUTION We will use a spherical coordinate system. The potential at a large distance away from the dipole is given in (3.46) as

$$V \cong \frac{Ql}{4\pi\varepsilon r^2} \cos\theta$$

Substituting this into (3.47) and using the gradient expression in spherical coordinates given in (3.48c) yields

$$\mathbf{E} = -\text{gradient } V$$
$$= -\frac{\partial}{\partial r} \left(\frac{Ql}{4\pi\varepsilon r^2} \cos\theta \right) \mathbf{a}_r - \frac{1}{r} \frac{\partial}{\partial \theta} \left(\frac{Ql}{4\pi\varepsilon r^2} \cos\theta \right) \mathbf{a}_\theta - 0 \ \mathbf{a}_\phi$$

giving

$$\mathbf{E} = \left(\frac{Ql}{2\pi\varepsilon r^3} \cos\theta \right) \mathbf{a}_r + \left(\frac{Ql}{4\pi\varepsilon r^3} \sin\theta \right) \mathbf{a}_\theta \tag{3.49}$$

which agrees with (3.14). ◀

▷ 3.6 CAPACITANCE

The concept of capacitance was introduced in earlier electric circuits courses. We now investigate the computation of the capacitance of various structures. If a battery is applied to two metallic conductors, charge will be transferred from the battery to the conductors: a positive charge resides on one conductor and an equal but opposite charge resides on the other. The charge transferred is *free charge*, such as electrons that flow through the metallic connection wires. For a voltage V applied between the two conductors, resulting in a total free charge, Q_f, deposited or stored on the conductors of the capacitor, the capacitance of the structure is defined as

$$C = \frac{Q_f}{V} \quad \text{F} \tag{3.50}$$

The units of capacitance are coulombs per volt, which is given the name of farads (F). This is named in honor of Michael Faraday, who made significant contributions to electromagnetics. His celebrated law will be studied in the next chapter.

Perhaps the simplest capacitance structure is the parallel-plate capacitor consisting of two parallel conducting plates of area A and separation d shown in Fig. 3.25a. A cross section is shown in Fig. 3.25b. The free charge deposited on the plates causes an electric field to be directed from one plate to the other. The majority of the field lines are perpendicular to the plates, but some of that electric field at the plate edges "bends" which is called the fringing field. We will assume that the plate separation is small enough compared to the plate area such that we can ignore the fringing field. Hence, the electric field is perpendicular to the plates and is

$$E = \frac{V}{d}$$

From our earlier results for the electric field due to an infinite sheet of charge, the electric field is also [see (3.12)]

$$E = \frac{\rho_s}{2\varepsilon}$$

$$= \frac{Q_f}{2\varepsilon A}$$

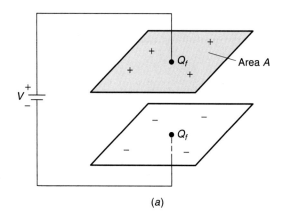

(a)

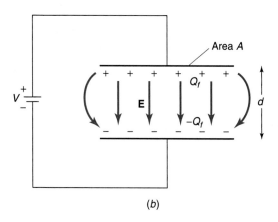

(b)

Figure 3.25 Determining the capacitance of a parallel-plate capacitor. (a) Attaching a battery supplies charge Q_f, which is stored on the plates. (b) The electric field between the plates illustrating fringing of the field at the edges of the plates. If the plate area is much larger than the plate separation, this fringing can be neglected and the electric field lines are perpendicular to the plates.

where the surface charge density on each plate is assumed to be uniformly distributed over that plate, so that $Q_f = \rho_s A$. But this was for one plate, so that the field for both plates is

$$E = \frac{Q_f}{\varepsilon A}$$

Solving these yields

$$\boxed{\begin{aligned} C &= \frac{Q_f}{V} \\ &= \frac{\varepsilon A}{d} \end{aligned}}$$

(3.51)

We have assumed that the intervening medium between the two plates has permittivity ε.

3.6.1 Calculation of the Capacitance of Structures

Calculations of the capacitance for other structures follows a similar path. We deposit a certain charge (positive on one conductor and an equal but opposite charge on the other conductor). Then we compute the resulting electric field and hence the resulting voltage between the two conductors and obtain the capacitance from (3.50). The following examples illustrate this method.

▶ **EXAMPLE 3.14**

Determine the capacitance per unit of length for a two-wire transmission line.

SOLUTION The problem consists of two parallel conductors of radius a that are separated a distance s as shown in Fig. 3.26. The conductors are infinite in length and have a charge distribution ρ_s C/m^2 that is uniformly distributed along their length and around their peripheries with a positive charge on one and an equal but negative charge on the other. The assumption that the charge is uniformly distributed around the peripheries is an approximation. As the conductors are brought closer together, the charge distributions will tend to concentrate on the facing surfaces and will no longer be uniform. However, this complicates the calculation, so we will assume that the ratio of the separation to conductor radius, s/a, is large enough (usually a ratio of 5–10 is sufficient and is typical for transmission lines) so that the charge is uniformly distributed around the peripheries. With this assumption we can again utilize a previous result. We

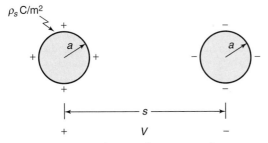

Figure 3.26 Example 3.14; determining the per-unit-length capacitance of a two-wire transmission line.

obtained, in Example 3.9, the voltage between two points at radial distances r_a and r_b from an infinite line charge bearing a charge distribution of ρ_l C/m as

$$V_{ba} = \frac{\rho_l}{2\pi\varepsilon} \ln\left(\frac{r_a}{r_b}\right)$$

The total voltage between the two cylinders in Fig. 3.26 is, by superposition, twice this result, with $r_a = s$ and $r_b = a$ ($\rho_l = \rho_s 2\pi a$):

$$V = 2\frac{\rho_l}{2\pi\varepsilon} \ln\left(\frac{s}{a}\right)$$

The charge per unit of length is ρ_l, so that the capacitance per unit length (denoted as lowercase c) is

$$c = \frac{\rho_l}{V}$$

$$= \frac{\pi\varepsilon}{\ln\left(\dfrac{s}{a}\right)} \quad \text{F/m} \tag{3.52}$$

A typical transmission line used to carry a rooftop television antenna signal into a household is referred to as "twin lead" and consists of two parallel wires of radius of $a = 0.5$ mm and separation of $s = 0.6$ cm, giving a per-unit-length capacitance of 11 pF/m. ◀

▶ EXAMPLE 3.15

Determine the per-unit-length capacitance of a coaxial cable.

SOLUTION The dimensions are shown in Fig. 3.16. The voltage between the inner and outer cylinders was determined in (3.40) as

$$V = \frac{\rho_s a}{\varepsilon} \ln\left(\frac{b}{a}\right)$$

The per-unit-length charge distribution is $\rho_l = \rho_s 2\pi a$, so that the voltage becomes

$$V = \frac{\rho_l}{2\pi\varepsilon} \ln\left(\frac{b}{a}\right)$$

The charge per unit of length is ρ_l, so that the capacitance per unit length (denoted as lowercase c) is

$$c = \frac{\rho_l}{V}$$

$$= \frac{2\pi\varepsilon}{\ln\left(\dfrac{b}{a}\right)} \quad \text{F/m} \tag{3.53}$$

A common coaxial cable is designated as RG-58U whose interior cylinder is a wire of radius 0.406 mm and whose outer "shield" has an interior radius of 1.47 mm. The dielectric within the cable is polyethylene having $\varepsilon_r = 2.3$. This gives a per-unit-length capacitance of about 100 pF/m. ◀

▶ QUICK REVIEW EXERCISE 3.8

Determine the capacitance of two concentric spheres with the inner sphere of radius a and the outer sphere of radius b.

ANSWER $C = 4\pi\varepsilon/(1/a - 1/b)$. ◀

▶ 3.7 CURRENT AND THE MAGNETIC FLUX DENSITY VECTOR

In the previous sections we found that static (fixed) distributions of charge produce two of the four fundamental EM vectors, the electric field intensity vector **E** and the electric flux density vector **D**. In the remaining sections we will find that a steady movement of charge, a dc current, produces the remaining two fundamental EM vectors, the magnetic field intensity vector **H** and the magnetic flux density vector **B**. Hence static charge distributions produce electric fields, and dc currents produce magnetic fields. In the remaining chapters of this text we will find that for time-varying currents, all four EM vectors are interrelated.

Current is the flow of *free charge*. In a conducting material such as a metal, the outer electrons of the atoms are free to move under the influence of an electric field. In certain other materials, the charge flow is a combination of positive and negative charge, such as in a plasma and in semiconductor materials. We will describe this current with the *current density vector* **J**, which has the units of A/m². The net current flowing through an open surface, *s*, as shown in Fig. 3.27 is obtained with a surface integral:

$$I = \int_s \mathbf{J} \cdot d\mathbf{s} \tag{3.54}$$

This is logical since the component of **J** parallel to the surface contributes nothing to the flux of **J** through the surface, and only the component of **J** that is perpendicular to the surface provides a contribution. In many conducting materials, the current density is proportional to the electric field that is providing the force to move the charge:

$$\mathbf{J} = \sigma \mathbf{E} \tag{3.55}$$

This is referred to as Ohm's law. The quantity σ is the *conductivity* of the material whose units are S/m where S is the unit of conductivity in siemens. Note that the product of the units of σ and the units of **E** gives VS/m² or A/m². Copper is a common metal used to fabricate wires and has a conductivity of $\sigma = 5.8 \times 10^7$ S/m. For the *simple* materials that we will consider, the directions of **J** and **E** are the same and the material is said to be *isotropic*. Also in these simple materials, the conductivity σ does not depend on the magnitude of **E** and the materials are said to be *linear*.

The *resistance* of a block of conducting material is the ratio of the voltage drop between the two end contacts and the total current through the material: $R = V/I$.

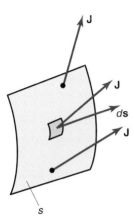

Figure 3.27 Illustration of determining the total current through a surface by adding the components of the current density vector **J** that are perpendicular to the surface.

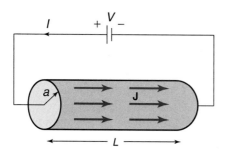

Figure 3.28 Determination of the resistance of a wire.

Consider a resistor consisting of a cylindrical block of conducting material such as copper of radius a as shown in Fig. 3.28. A battery of voltage V is applied across the ends and a current I passes through the cylinder. We assume that the current is uniformly distributed across the cross section of the cylinder (which is true for a dc current). Hence the current density is

$$J = \frac{I}{\pi a^2}$$

The voltage applied across a length L of the material develops an electric field through it of

$$E = \frac{V}{L}$$

Substituting these into Ohm's law in (3.55) yields

$$\frac{I}{\pi a^2} = \sigma \frac{V}{L}$$

Solving yields

$$V = \left(\frac{L}{\pi a^2 \sigma}\right) I$$

from which we identify the resistance as

$$R = \frac{L}{\pi a^2 \sigma} \quad \Omega \tag{3.56}$$

Charge can be neither created nor destroyed. Hence if we consider a closed surface, s, and examine the net outflow of charge from that surface, we obtain the mathematical statement of *conservation of charge* as

$$\oint_s \mathbf{J} \cdot d\mathbf{s} = 0 \tag{3.57}$$

This is valid for currents that are not time varying (dc). It is essentially Kirchhoff's current law if the closed surface s is considered to be a node of a lumped circuit.

3.7.1 The Biot-Savart Law

Currents must necessarily form closed paths. However, we can examine the contribution to the magnetic field due to a differential length of that current. Consider Fig. 3.29,

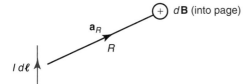

Figure 3.29 Illustration of the Biot-Savart law for determining the magnetic flux density vector **B** due to a segment of current.

where we have shown a differential current, $Id\mathbf{l}$. At a distance R from that differential current, a differential contribution to the *magnetic flux density* vector, $d\mathbf{B}$, is given by the *Biot-Savart law*:

$$d\mathbf{B} = \frac{\mu_o I}{4\pi R^2} d\mathbf{l} \times \mathbf{a}_R \quad \text{Wb/m}^2 \qquad (3.58)$$

where $\mathbf{a}_R$ is a unit vector pointing from the current element to the point at which we are determining the magnetic field. The vector **B** is the magnetic flux density vector whose units are webers (Wb) per square meter. These units are equivalently given in terms of tesla (T) where $1\text{ T} = 1\text{ Wb/m}^2$ and are named for another pioneer in the field of EM. The constant μ_o is the *permeability* of free space and is

$$\mu_o = 4\pi \times 10^{-7} \quad \text{H/m} \qquad (3.59)$$

The units are henries (H) per meter, which is an inductance per unit of distance. More will be said about inductance later in this chapter. Other magnetic media that we will consider will be characterized by a permeability μ. Observe that the magnetic flux density vector is proportional to the product of the length of the current element and the current and is inversely proportional to the distance from the element to the point *squared*. Hence this is an inverse distance squared law like Coulomb's law and bears many similarities to that law. The main difference is that *the direction of the resulting magnetic flux density vector is perpendicular to the plane containing the current element and the direction vector to the point according to the right-hand-rule*. Since the cross product is proportional to the sine of the angle between the current element and the distance vector, the magnetic flux density vector is zero off the ends of the current element and a maximum at points perpendicular to the current element.

3.7.2 Calculation of the Magnetic Field for Current Distributions

The Biot-Savart law can be used to determine the total magnetic flux density vector due to a closed loop of current by summing the contributions from all the differential current elements that make up that current loop using *superposition*. This is very similar to the method used to determine the electric field due to distributions of charge considered previously. The main difference here is that the vector direction of the result is a bit more involved and must be determined with the right-hand-rule. The following examples illustrate this process for various current paths.

▷ **EXAMPLE 3.16**

Detemine the magnetic field due to a current I of length L.

SOLUTION To take advantage of symmetry, we place the current along the z axis of a cylindrical coordinate system, with half of it along the $+z$ axis and the other half along the $-z$ axis as

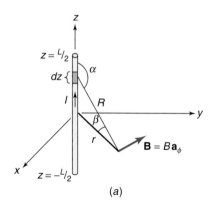

(a)

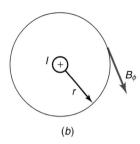

(b)

Figure 3.30 Example 3.16; determining the magnetic field due to a line current. (a) Setting up the problem to take advantage of symmetry. (b) Illustration of the fact that the magnetic field about an infinite line current is circumferentially directed.

shown in Fig. 3.30a. We will compute the magnetic field only at distances r from the midpoint of and perpendicular to the current in order to simplify the integral. By the right-hand-rule and symmetry, the magnetic flux density vector is directed in the ϕ direction in a cylindrical coordinate system around the current. The differential contribution to the field due to a current element is

$$d\mathbf{B} = \frac{\mu_o I}{4\pi} \frac{dz}{R^2} \sin(\alpha)\mathbf{a}_\phi$$

The distance from the current element to the point is

$$R = \sqrt{r^2 + z^2}$$

and the sine of the angle involved in the cross product is

$$\sin(\alpha) = \sin\left(\beta + \frac{\pi}{2}\right)$$
$$= \cos(\beta)$$
$$= \frac{r}{R}$$

Hence, to obtain the total field we sum the contributions from all current elements to give

$$\mathbf{B} = \int_{-L/2}^{L/2} \frac{\mu_o I}{4\pi} \frac{r\,dz}{(r^2 + z^2)^{3/2}}\mathbf{a}_\phi$$

In order to evaluate this integral we use (3.8) to give

$$\mathbf{B} = \frac{\mu_o I r}{4\pi}\left[\frac{z}{r^2\sqrt{r^2 + z^2}}\right]_{-L/2}^{L/2}\mathbf{a}_\phi$$

resulting in

$$\mathbf{B} = \frac{\mu_o I}{2\pi r} \frac{L/2}{\sqrt{r^2 + L^2/4}} \mathbf{a}_\phi \tag{3.60}$$

For an infinite current filament we let $L \to \infty$, giving

$$\boxed{\mathbf{B} = \frac{\mu_o I}{2\pi r} \mathbf{a}_\phi} \tag{3.61}$$

This latter result for the magnetic flux density vector due to an infinite current is an important result that will be used on numerous occasions. Hence it should be committed to memory. It will be derived more simply later using Ampere's law. Observe that the magnetic field around an infinite current is (1) circumferentially directed according to the right-hand-rule and (2) is constant along circles of constant radii as shown in Fig. 3.30b. ◄

► **EXAMPLE 3.17**

Determine the magnetic flux density due to a circular loop of current of radius a at a distance d from the midpoint and on a line perpendicular to the loop.

SOLUTION The problem is sketched in Fig. 3.31. From the sketch we see that the horizontal components cancel when the contributions from all differential current elements are summed, leaving only the z-directed vertical contributions. Hence the differential contribution to this total is

$$d\mathbf{B} = \frac{\mu_o I}{4\pi} \frac{a\, d\phi}{R^2} \sin(\alpha)\cos(\beta)\mathbf{a}_z \quad z > 0$$

Here the angle α involved in the cross product is 90° and hence $\sin(\alpha) = 1$. But $\cos(\beta)$, which is required to give the vertical z-directed component, is

$$\cos(\beta) = \cos\left(\frac{\pi}{2} - \psi\right)$$
$$= \sin(\psi)$$
$$= \frac{a}{R}$$

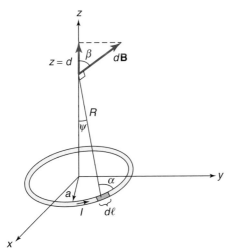

Figure 3.31 Example 3.17; determination of the magnetic field for a current loop (magnetic dipole).

and

$$R = \sqrt{a^2 + d^2}$$

Hence

$$\mathbf{B} = \int\limits_{\phi = 0}^{2\pi} \frac{\mu_o I}{4\pi} \frac{a^2}{(a^2 + d^2)^{3/2}} d\phi \, \mathbf{a}_z \quad z > 0$$

But this is a simple integral since the integrand does not depend on the variable of integration, ϕ. Hence we obtain

$$\mathbf{B} = \frac{\mu_o I}{2} \frac{a^2}{(a^2 + d^2)^{3/2}} \mathbf{a}_z \quad z > 0 \tag{3.62}$$

At the center of the loop

$$\mathbf{B} = \frac{\mu_o I}{2a} \mathbf{a}_z \quad d = 0 \tag{3.63}$$

At a very large distance from the loop compared to its radius, $d \gg a$, we obtain

$$\mathbf{B} = \frac{\mu_o I a^2}{2d^3} \mathbf{a}_z \quad d \gg a \tag{3.64}$$

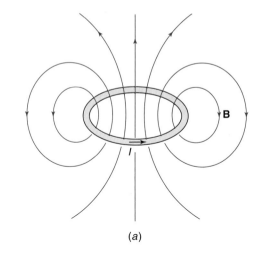

(a)

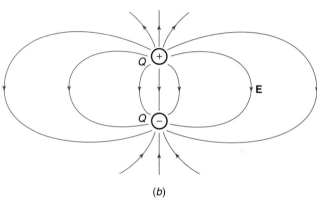

(b)

Figure 3.32 Illustration of the similarity of (a) the magnetic field about a current loop (a magnetic dipole) and (b) the electric field about a pair of electric charges (an electric dipole).

The result for a large distance from the loop in (3.64) indicates that the magnetic field from a loop of current falls off in magnitude with the cube of the distance.

We could similarly determine the magnetic field at points other than perpendicular to the loop and on a line from its center by summing the contributions of the differential current elements using the Biot-Savart law. But the resulting integrals would be difficult to evaluate. The complete result for large distances from the loop, $r \gg a$, in a spherical coordinate system is

$$\mathbf{B} = \frac{\mu_o m}{4\pi r^3}(2\cos\theta\,\mathbf{a}_r + \sin\theta\,\mathbf{a}_\theta) \quad r \gg a \qquad (3.65)$$

where the *magnetic dipole moment* is

$$m = \pi a^2 I \qquad (3.66)$$

The dipole moment is the product of the area of the loop and the current. Along the z axis, $\theta = 0°$, (3.65) reduces to the earlier result for $d \gg a$ in (3.64). Broadside to the loop, $\theta = 90°$, (3.65) shows that the field is directed in the $-z$ direction. Figure 3.32a shows the field distribution around the loop. It is rotationally symmetric about the z axis. Compare this to the plot of the electric field for the electric dipole shown in Fig. 3.32b. The circular loop of current is said to be a *magnetic dipole* and will be instrumental in explaining how magnetic materials are magnetizable. ◄

► EXAMPLE 3.18

Determine the magnetic flux density for an infinite sheet of current carrying a surface current density K A/m that is directed along one axis of the sheet. A surface current density can be visualized as an infinite set of line currents of infinite length that are distributed over a surface. Hence one dimension (along the current) is removed, and the distribution referred to is with respect to the dimension perpendicular to the current.

SOLUTION We place the sheet of current in the xz plane of a rectangular coordinate system as shown in Fig. 3.33. Next we utilize a previous result for an infinite current filament given in (3.61). To do so we divide the current density into filaments of current $I = Kdz$ A directed in the

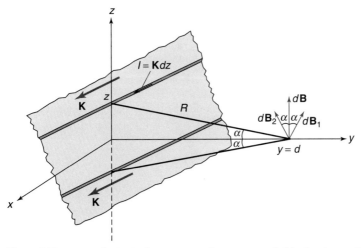

Figure 3.33 Example 3.18; determining the magnetic field of a sheet of current by viewing it as a sequence of line currents and superimposing those fields.

direction of the current density vector: the x direction. The differential contribution to the magnetic field at a point a distance $y = d$ away from the sheet is, according to (3.61),

$$dB_1 = \frac{\mu_o K dz}{2\pi R}$$

The differential contribution to the magnetic field from a symmetrically disposed current is the same in magnitude

$$dB_2 = \frac{\mu_o K dz}{2\pi R}$$

The difference between the two is in their directions. However, we observe that the horizontal, y-directed components cancel, leaving the total as sum of the z-directed components.

$$dB = dB_1 \cos(\alpha) + dB_2 \cos(\alpha)$$
$$= \frac{\mu_o K dz}{\pi R} \cos(\alpha)$$

which is directed in the $+z$ direction and

$$\cos(\alpha) = \frac{d}{R}$$

and

$$R = \sqrt{d^2 + z^2}$$

Hence we obtain the total magnetic field by summing the contributions from these symmetrically disposed currents to give

$$\mathbf{B} = \int\limits_{z=0}^{\infty} \frac{\mu_o K d}{\pi(d^2 + z^2)} dz \, \mathbf{a}_z$$

Evaluation of this integral requires use of the integral

$$\int \frac{1}{d^2 + z^2} dz = \frac{1}{d} \tan^{-1}\left(\frac{z}{d}\right)$$

giving

$$\mathbf{B} = \frac{\mu_o K}{2} \mathbf{a}_z$$
$$= \frac{\mu_o}{2} \mathbf{K} \times \mathbf{a}_n$$

(3.67)

where $\mathbf{a}_n$ is a unit vector perpendicular or normal to the sheet. This result in terms of the unit normal applies to the field on either side of the sheet. ◀

▶ QUICK REVIEW EXERCISE 3.9

Determine the magnetic flux density between two large sheets of current of surface current density K A/m that are directed in opposite directions.

ANSWER $B = \mu_o K$ and is directed perpendicular to the current direction according to (3.67), that is, if the current in the top sheet is right to left, then the magnetic field is perpendicular to and out of the page. ◀

▶ 3.8 THE MAGNETIC FIELD INTENSITY VECTOR AND MAGNETIC MATERIALS

Magnetic materials are, in a general sense, similar to dielectric materials in that dipoles are either available or created which are able to align with an applied field (an electric field in the case of dielectrics and a magnetic field in the case of magnetic materials). Magnetic materials are a bit more diverse in their properties than are dielectric materials. The key to understanding how certain materials are magnetizable is to view the atoms in the Bohr sense as electrons orbiting a nucleus. These orbiting electrons constitute a current loop (whose direction is opposite to the direction of electron movement). In Example 3.17 we found that a current loop of radius a produced a magnetic flux density at a distance d from the center and on a line perpendicular to the loop of

$$B = \frac{\mu_o I a^2}{2d^3} \quad d \gg a$$

In order to put this into some meaningful form, let us again define the *magnetic dipole moment* of the loop as the product of the loop current and the area of the loop:

$$\mathbf{m} = \pi a^2 I \mathbf{a}_n \quad \text{Am}^2 \tag{3.68}$$

where $\mathbf{a}_n$ is a unit vector perpendicular or normal to the loop. The units of magnetic dipole moment are amperes times meters squared. Substituting this into the previous result yields the magnetic flux density due to and perpendicular to the plane of a current loop as

$$\mathbf{B} = \frac{\mu_o \mathbf{m}}{2\pi d^3} \tag{3.69}$$

The orbiting electrons of an atom can be viewed as having a magnetic dipole moment and generate a magnetic flux density according to (3.69) and shown in Fig. 3.34a.

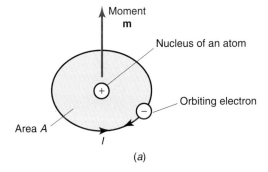

(a)

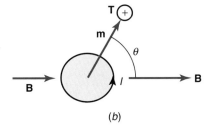

(b)

Figure 3.34 Illustration of the magnetic dipole in magnetic materials. (a) The dipole formed by orbiting charges of the atoms. (b) Rotation of the magnetic dipole to align with an external magnetic field.

If another external magnetic field, **B**, is applied, the two magnetic fields interact to produce a torque on this magnetic dipole given by

$$\mathbf{T} = \mathbf{m} \times \mathbf{B} \tag{3.70}$$

as shown in Fig. 3.34b. This torque tends to rotate the magnetic dipole so as to align with the applied field. In a material containing a large number of such microscopic magnetic dipoles, the alignment with an applied field causes the dipoles to align, increasing the magnetic field as illustrated in Fig. 3.35. The net magnetic field is the sum of the applied field and the field produced by the alignment of the magnetic dipoles. Hence the alignment increases the total magnetic field, resulting in *magnetization* of the material.

In addition to the orbital magnetization discussed above, the electrons themselves possess dipole moments due to their spin. For many types of materials the orbital and spin magnetic moments approximately cancel, and these materials are called *diamagnetic*. Typical such materials are copper, lead, and silver. In diamagnetic materials there is a very small net magnetic moment. The internal magnetic field of diamagnetic materials is slightly less than but approximately equal to the applied field. In other materials where the orbital and spin moments do not cancel, the atom has a small magnetic dipole moment. However, the random orientation of these dipoles results in only a very small net dipole moment and the application of an external magnetic field can cause the moments to align. The total magnetic field is slightly larger than the applied field. These materials are called *paramagnetic*. Typical such materials are aluminum and air.

There are four final classes of materials that exhibit strong magnetic moments. These are *ferromagnetic, antiferromagnetic, ferrimagnetic, and superparamagnetic* materials.

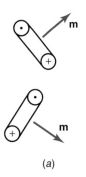

(a)

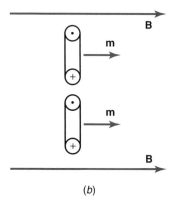

(b)

Figure 3.35 Illustration of the magnetic polarization by dipole alignment. (a) The random orientaion of the dipole moments in the absence of an external magnetic field, and (b) the alignment of the dipoles due to an external magnetic field.

Ferromagnetic materials have strong dipole moments. In segments of the material called domains, these moments line up, creating a large magnetic dipole moment. In some such materials when the applied field is removed, the dipole moments do not return to random cancellation and a residual magnetic field is created. This phenomenon is called *hysteresis* and is common in materials such as iron, cobalt, and nickel. In other materials, called antiferromagnetic, the magnetic moments of adjacent atoms align opposing each other, giving a net magnetic moment of zero. The ferrimagnetic materials also exhibit opposing alignments of the atomic dipole moments but they are not equal, giving them a large response to an applied magnetic field. Ferrites are an important class of ferrimagnetic materials and have found increasing use in electronic devices and in preventing interference in electronic systems. They tend to have high resistivity and losses that increase with increasing frequency of the applied field. Examples of ferrites are nickel-zinc and manganese-zinc. Superparamagnetic materials consist of ferromagnetic particles held together by a nonferromagnetic dielectric. These can be used to be selectively magnetized in very short distances and are typical of materials used in magnetic recording tapes.

We saw that the dipoles of dielectric materials produce an internal electric field and associated polarization vector, **P**, which adds to the applied electric field, $\varepsilon_o\mathbf{E}$, creating a net electric field that is their sum according to (3.17). Similarly, the alignment of magnetic dipoles in magnetic materials produces an internal magnetic field which adds to the applied magnetic field, creating a net magnetic field that is greater than the applied magnetic field. In order to characterize such materials, we define the *magnetization vector* as the *dipole moment per unit volume* [compare to (3.16)]:

$$\mathbf{M} = \lim_{\Delta v \to 0} \frac{\sum_i \mathbf{m}_i}{\Delta v} \quad \text{A/m} \tag{3.71}$$

The units of the individual dipole moments in (3.68) are Am^2. Hence the units of the magnetization vector are A/m. The magnetization vector and the applied magnetic flux density vector are parallel.

In a fashion similar to electric fields and dielectric materials, we are led to define the last of our important EM vector quantities, the *magnetic field intensity vector*, **H**. In free space, the magnetic flux density vector **B** and this magnetic field intensity vector **H** are related by the permeability of free space, μ_o, as

$$\boxed{\mathbf{B} = \mu_o\mathbf{H} \quad \text{free space}} \tag{3.72}$$

The units of **H** are A/m. When a magnetic material is introduced into this field, the magnetic dipoles align with the applied field and add to it. Hence the total magnetic flux density vector is

$$\boxed{\mathbf{B} = \mu_o\mathbf{H} + \mu_o\mathbf{M}} \tag{3.73}$$

Recall that the units of the magnetization vector, **M**, are A/m, which are the same as **H**. The magnetization vector **M** is related to the magnetic field intensity vector as

$$\mathbf{M} = \chi_m\mathbf{H} \tag{3.74}$$

where χ_m is the *magnetic susceptibility* of the material. Substituting (3.74) into (3.73) and rearranging gives

$$\mathbf{B} = \mu_o\underbrace{(1 + \chi_m)}_{\mu_r}\mathbf{H} \tag{3.75}$$

The quantity $(1 + \chi_m)$ is called the *relative permeability* of the material:

$$\mu_r = (1 + \chi_m) \tag{3.76}$$

Hence the magnetic flux density vector and the magnetic field intensity vector can be related as

$$\boxed{\mathbf{B} = \mu\mathbf{H}} \tag{3.77}$$

where the *permeability* of the material is

$$\boxed{\mu = \mu_r\mu_o \quad \text{H/m}} \tag{3.78}$$

Diamagnetic materials have a relative permeability slightly less than unity and paramagnetic materials have a relative permeability slightly greater than unity. For example, silver has $\mu_r = 0.99999981$ and aluminum has $\mu_r = 1.00000065$. Ferromagnetic materials have very large relative permeabilities. For example, at dc, sheet steel has $\mu_r = 2000$ and mumetal has $\mu_r = 30,000$.

An *isotropic* material is one in which **B** and **H** are in the same direction. A *linear* material is one in which the permeability does not depend on **H**. There are important nonlinear materials such as iron which are ferromagnetic. These ferromagnetic materials exhibit a nonlinear relation between **B** and **H** via a *hysteresis curve* shown in Fig. 3.36a. Not only is the relation nonlinear but it also shows that as the material is magnetized by increasing H, when H is returned to zero, there is a remaining B or residual

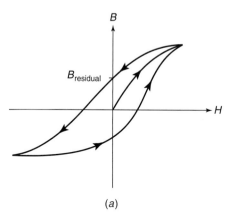

(a)

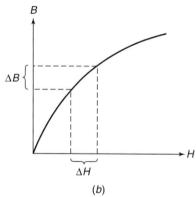

(b)

Figure 3.36 Illustration of the nonlinearity of the relation between the B and H fields in a ferromagnetic material. (a) The hysteresis curve. (b) Illustration of saturation effect in a ferromagnetic material such as iron.

magnetization of the material. Furthermore, the permeability depends on the level of the applied field as shown in Fig. 3.36b. The incremental permeability is the slope of the B–H curve or the ratio

$$\mu = \frac{\Delta B}{\Delta H}$$

For increasingly larger values of H, the slope of the B–H curve decreases, thereby giving smaller values of effective permeability. The largest slope and hence effective permeability occurs for small values of H. The decrease in effective permeability for increasing applied magnetic field levels is referred to as *saturation*. We will not consider nonlinear or anisotropic materials in this text, so that **B** and **H** can be freely interchanged with (3.77). This linear relation holds, as a reasonable approximation, for many magnetic materials so long as the applied magnetic field is sufficiently small to prevent saturation.

In addition to the phenomenon of saturation, the relative permeability of ferromagnetic materials often depends on frequency. For example, at 1 kHz the relative permeability of mumetal is on the order of $\mu_r = 30{,}000$, yet at 20 kHz, its relative permeability drops to $\mu_r = 2000$. Sheet steel, on the other hand, has a relative permeability of $\mu_r = 2000$ from dc to well over 100 kHz.

▶ 3.9 AMPERE'S LAW

In Section 3.7.2 we showed how to calculate the magnetic flux density vector using superposition of current elements and the Biot-Savart law. This required the evaluation of some moderately complicated integrals for many problems. Ampere's law will allow the direct solution of many of those problems without the evaluation of any integrals. Ampere's law is stated as

$$\oint_c \mathbf{H} \cdot d\mathbf{l} = I_{\text{enclosed}} \tag{3.79}$$

where

$$I_{\text{enclosed}} = \int_s \mathbf{J} \cdot d\mathbf{s} \tag{3.80}$$

It essentially provides that *the sum of the products of the components of* **H** *that are tangent to a closed contour c around that contour and the differential lengths of the contour will yield the net current passing through the surface enclosed by that contour.* This is illustrated in Fig. 3.37. This is somewhat similar to Gauss' law for the electric field in (3.22), where the sum of the products of the components of the electric flux density vector that are perpendicular to a closed surface and the differential surfaces yielded the *net* positive charge enclosed by that surface.

3.9.1 Calculation of the Magnetic Field for Current Distributions

The power of Ampere's law is that *we are allowed to choose the contour c*. A judicious choice of that contour that exploits symmetry will simplify the computation considerably. Furthermore, considerably more insight will be gained than with the superposition and

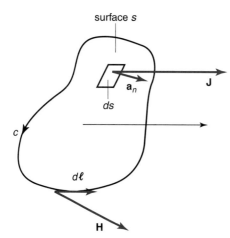

Figure 3.37 Illustration of Ampere's law relating the magnetic field to the conduction current and displacement current.

direct integration method using the Biot-Savart law. As with Gauss' law for the electric field, we should attempt to choose the contour such that (1) the **H** field is everywhere tangent to that contour, resulting in the removal of the dot product from the integral:

$$\oint_c \mathbf{H} \cdot d\mathbf{l} = \oint_c H \, dl$$

and (2) the magnitude of **H** is independent of position on that contour, so that it may be removed from the integral, resulting further in

$$\oint_c \mathbf{H} \cdot d\mathbf{l} = H \underbrace{\oint_c dl}_{\substack{\text{length of} \\ \text{contour}}}$$

The following examples illustrate its use and computational power.

▶ **EXAMPLE 3.19**

Determine the magnetic field due to an infinite current.

SOLUTION   We place the current along the z axis of a cylindrical coordinate system as shown in Fig. 3.38. By symmetry we see that (1) the magnetic field is directed in the ϕ direction everywhere and (2) it is constant for a fixed radius away from the line. Hence we choose the contour c as a circle of radius r parallel to the xy plane. The first property allows us to remove the dot product from Ampere's law, and the second property allows us to remove H from the integral, leaving

$$H \underbrace{\oint_c dl}_{2\pi r} = I$$

Hence we easily obtain the magnetic field intensity vector as

$$\mathbf{H} = \frac{I}{2\pi r}\mathbf{a}_\phi \tag{3.81}$$

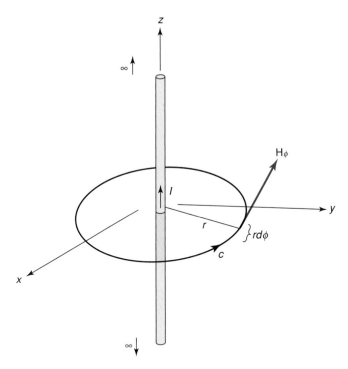

Figure 3.38 Example 3.19; determining the magnetic field intensity about an infinite line current with Ampere's law.

Substituting (3.77) gives the magnetic flux density vector as

$$\mathbf{B} = \frac{\mu I}{2\pi r}\mathbf{a}_\phi \tag{3.82}$$

This is the same result obtained in Example 3.16, Equation (3.61), by direct integration using the Biot-Savart law but was achieved with considerably less effort by using Ampere's law. ◄

► **EXAMPLE 3.20**

Determine the magnetic field of a coaxial cable having interior conductor of radius a and an outer cylinder of interior radius b.

SOLUTION The problem is sketched in Fig. 3.39. We place a current I on the inner cylinder and an equal but oppositely directed current on the interior of the outer cylinder. By symmetry, the currents will be uniformly distributed around the peripheries of the cylinders and the resulting magnetic field will be circumferentially directed about the inner cylinder. So we choose contour c to be a circle of radius r around the inner cylinder. As in the previous problem, the magnetic field is constant on this contour and is tangent to it. Hence

$$\oint_c \mathbf{H} \cdot d\mathbf{l} = H \oint_c dl$$
$$= I_{\text{enclosed}}$$
$$= I$$

and we obtain the result in the previous example:

$$\mathbf{H} = \frac{I}{2\pi r}\mathbf{a}_\phi \quad a < r < b \tag{3.83}$$

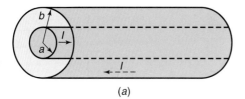

(a)

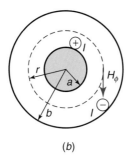

(b)

Figure 3.39 Example 3.20; determining the magnetic field of a coaxial cable. (a) The physical dimensions. (b) Utilizing symmetry to determine the magnetic field with Ampere's law.

and

$$\mathbf{B} = \frac{\mu I}{2\pi r}\mathbf{a}_\phi \quad a < r < b \tag{3.84}$$

Observe that interior to the inner cylinder and exterior to the outer cylinder, the net current enclosed is zero. Because the current is uniformly distributed around the peripheries of each cylinder, H can be removed from the integral of Ampere's law, giving

$$\oint_c \mathbf{H} \cdot d\mathbf{l} = H \oint_c dl$$
$$= I_{\text{enclosed}}$$
$$= 0$$

for $r < a$ and $r > b$. Hence the magnetic field is also zero inside the inner cylinder and outside the outer cylinder:

$$\mathbf{H}, \mathbf{B} = 0 \begin{cases} r < a \\ r > b \end{cases} \tag{3.85}$$

As in the electric field problem for this coaxial cable it is said to be *shielded* since no fields exist outside it.   ◀

▶ EXAMPLE 3.21

Determine the magnetic field of an infinite sheet of current carrying a surface current density of K A/m using Ampere's law.

SOLUTION   This problem was solved by direct integration in Example 3.18. In using Ampere's law to solve it, we need to observe that the magnetic field will, by symmetry, be directed parallel to the surface of the sheet. We can see this by visualizing the current sheet as consisting of filaments of current whose magnetic fields are circumferentially directed about them. The components of these fields that are perpendicular to the surface cancel, leaving only the parallel

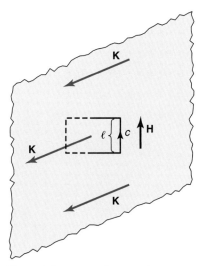

Figure 3.40 Example 3.21; determining the magnetic field about a plane surface current density with Ampere's law.

components. Hence we choose a rectangular contour shown in Fig. 3.40 that is perpendicular to the sheet. Along the components of the path that are perpendicular to the surface the field is perpendicular to the path and contributes nothing to (3.79). Along the portions of the contour that are parallel to the sheet (front and back), and of length l, the magnetic field is parallel to the path. Hence Ampere's law reduces to

$$\oint_c \mathbf{H} \cdot d\mathbf{l} = H(2l)$$

and

$$I_{\text{enclosed}} = Kl$$

This gives the result obtained in Example 3.18:

$$\mathbf{B} = \frac{\mu}{2} \mathbf{K} \times \mathbf{a}_n \qquad (3.86)$$

where $\mathbf{a}_n$ is a unit vector perpendicular to the sheet.   ◄

► 3.10 GAUSS' LAW FOR THE MAGNETIC FIELD

Unlike point charges, there are no known *isolated* magnetic sources. For example, if we cut a permanent magnet into two pieces, the north and south poles will occur on the ends of each section as illustrated in Fig. 3.41a. Gauss' law for the magnetic field mathematically states this important fact as

$$\oint_s \mathbf{B} \cdot d\mathbf{s} = 0 \qquad (3.87)$$

which essentially provides that *the net flux of the magnetic field out of a closed surface is zero* as illustrated in Fig. 3.41b. Therefore magnetic field lines must form *closed paths*, unlike point charges whose electric field lines begin and end on charge.

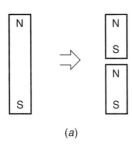

(a)

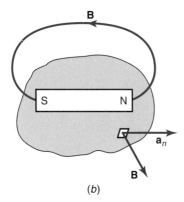

(b)

Figure 3.41   Illustration of Gauss' law for the magnetic field. (a) Cutting a permanent magnet creates additional north and south poles. (b) Illustration of Gauss' law that the net magnetic flux lines passing through a closed surface are zero, illustrating that the magnetic field has no isolated sources or sinks.

▶ EXAMPLE 3.22

Show that Gauss' law for the magnetic field is satisfied for the coaxial cable shown in Fig. 3.39.

SOLUTION   The magnetic flux density in the space between the inner and outer cylinders was determined in Example 3.20 to be

$$\mathbf{B} = \frac{\mu I}{2\pi r}\mathbf{a}_\phi \quad a < r < b$$

Choosing the closed surface as a cylinder of radius r and length l encircling the inner cylinder, Gauss' law becomes

$$\oint_s \mathbf{B} \cdot d\mathbf{s} = \underset{\text{sides}}{0} + \underset{\text{end caps}}{0}$$

Since the magnetic field is circumferentially directed in the space between the two cylinders, it is parallel to the side of the cylinder and parallel to the end caps, giving zero contribution to the surface integral in both cases. Hence Gauss' law is satisfied. We could have chosen any other closed surface and obtained the same result, but the choice of a cylindrical surface simplified the integration. ◀

▶ QUICK REVIEW EXERCISE 3.10

Verify Gauss' law for an infinite sheet of current. (Hint: Observe the direction of the magnetic field found in Examples 3.18 and 3.21 and choose an appropriate shape for the closed surface.) ◀

► 3.11 INDUCTANCE

We have examined the concept of inductance in our earlier electric circuits courses. The definition of inductance is *the ratio of the magnetic flux through a loop to the current of that loop that produces it*:

$$L = \frac{\psi}{I} \qquad (3.88)$$

where the flux is

$$\psi = \int_s \mathbf{B} \cdot d\mathbf{s} \qquad (3.89)$$

as illustrated in Fig. 3.42a. If N turns are tightly wound so that none of the field leaks out between the coils as shown in Fig. 3.42b, the inductance is *the ratio of the flux linkages linking the current*:

$$L = \frac{\Lambda}{I} \qquad (3.90)$$

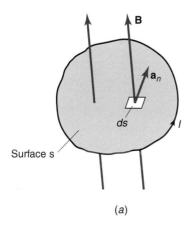

(a)

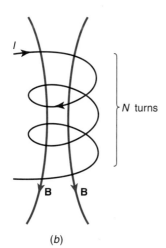

(b)

Figure 3.42 Illustration of inductance. (a) The magnetic flux through a loop produced by the current of that loop. (b) Magnetic flux linkages for a multiturn coil.

where the *flux linkages* are

$$\Lambda = N\psi$$

(3.91)

3.11.1 Calculation of the Inductance of Structures

The following examples illustrate the computation of inductance for some common structures.

▷ **EXAMPLE 3.23**

A toroidal inductor shown in Fig. 3.43a consists of N turns of wire carrying current I that are tightly wound on a core of rectangular cross section. The core is constructed of a material having permeability μ and has an inner radius a, an outer radius b, and a thickness t. Determine the inductance of the toroid.

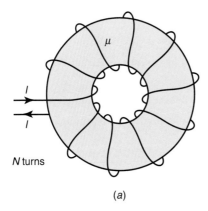

(a)

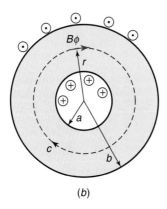

(b)

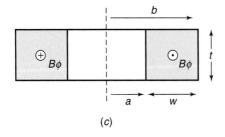

(c)

Figure 3.43 Example 3.23; inductance of a toroid. (a) The physical dimensions. (b) Illustration of the magnetic field in the toroid. (c) A cross-sectional view for computing the flux linkages.

SOLUTION   Because the turns of wire are tightly wound on the core, the magnetic flux density is circumferentially directed in the core as shown in Fig. 3.43b. This suggests choosing the contour c in Ampere's law to be a circle of radius r. The magnetic field is tangent to this contour and is constant at all points on it. Hence Ampere's law yields

$$\underset{c}{\oint} \mathbf{H} \cdot d\mathbf{l} = H_\phi \underset{\underset{2\pi r}{\underbrace{c}}}{\oint} dl$$

$$= NI$$

Hence the magnetic flux density in the core is circumferentially directed and is given by

$$\mathbf{B} = \frac{\mu NI}{2\pi r}\mathbf{a}_\phi$$

A cross-sectional view is shown in Fig. 3.43c. The total flux through the cross section is

$$\psi = \int_s \mathbf{B} \cdot d\mathbf{s}$$

$$= \int_{z=0}^{t} \int_{r=a}^{b} \frac{\mu NI}{2\pi r} dr dz$$

$$= \frac{\mu NIt}{2\pi} \ln\left[\frac{b}{a}\right]$$

and the inductance becomes

$$L = \frac{N\psi}{I}$$

$$= \frac{\mu N^2 t}{2\pi} \ln\left[\frac{b}{a}\right] \tag{3.92}$$

This expression may be simplified for cores where the width, $w = b - a$, is much less than the inner radius, $w \ll a$, by using the expansion of the logarithm:

$$\ln\left(\frac{b}{a}\right) = \ln\left(\frac{w}{a} + 1\right)$$

$$\cong \frac{w}{a} \quad w \ll a$$

Evaluating (3.92) using this yields

$$L \cong \frac{\mu N^2 tw}{2\pi a} \quad a/w \gg 1$$

But the cross-sectional area of the core is $A = tw$. Hence, the inductance of a toroid of rectangular cross section of cross-sectional area A and radius a is approximately

$$L \cong \frac{\mu N^2 A}{2\pi a} \tag{3.93}$$

▷ **QUICK REVIEW EXERCISE 3.11**

Determine the inductance of a core constructed of iron ($\mu_r = 5000$) having an inner radius of 4 cm, a width and thickness of $w = t = 2$ mm, and containing 100 turns.

ANSWER $L = 1$ mH.

▶ **EXAMPLE 3.24**

Determine the per-unit-length inductance of a coaxial cable having an inner conductor of radius a and an inner radius of the outer conductor of b.

SOLUTION   The problem is sketched, in Fig. 3.44. The magnetic flux density between the two cylinders was determined in Example 3.20 to be circumferentially directed and given by

$$\mathbf{B} = \frac{\mu I}{2\pi r}\mathbf{a}_\phi$$

Since the field is circumferentially directed, we choose the surface s through which we evaluate the flux via (3.89) to be a flat surface extending radially from $r = a$ to $r = b$ and along the cable a length d. The flux through this surface becomes

$$\psi = \int_s \mathbf{B} \cdot d\mathbf{s}$$

$$= \int_{z=0}^{d} \int_{r=a}^{b} \frac{\mu I}{2\pi r} \underbrace{drdz}_{ds}$$

$$= \frac{\mu d I}{2\pi} \ln\!\left(\frac{b}{a}\right)$$

Observe that the magnetic flux density is perpendicular to this surface, so the dot product may be removed. However, B varies with r over this surface and so cannot be removed from the integral. Hence the inductance becomes

$$L = \frac{\psi}{I}$$

$$= \frac{\mu d}{2\pi} \ln\!\left(\frac{b}{a}\right) \quad \text{H}$$

Hence, the *per-unit-length inductance* is

$$l = \frac{L}{d}$$

$$= \frac{\mu}{2\pi} \ln\!\left(\frac{b}{a}\right) \quad \text{H/m} \tag{3.94}$$

For transmission lines, unlike the toroid of the previous problem, the medium between the two cylinders is typically a dielectric which is nonmagnetic, so that $\mu = \mu_o$.   ◀

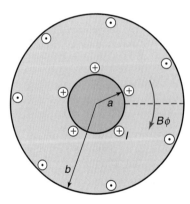

Figure 3.44   Example 3.24; computation of the per-unit-length inductance of a coaxial cable.

▶ **QUICK REVIEW EXERCISE 3.12**

Determine the per-unit-length inductance of a RG-58U coaxial cable, which has an inner conductor of radius 0.406 mm and an outer conductor of inner radius 1.47 mm. The medium is Teflon ($\mu_r = 1$).

ANSWER 257 nH/m. ◀

▶ **EXAMPLE 3.25**

Determine the per-unit-length inductance of a transmission line consisting of two parallel cylinders (wires) of radii a and separation s. The surrounding medium is air.

SOLUTION The problem is sketched in Fig. 3.45. In order to simplify the problem, we will assume that the conductors are separated sufficiently with respect to their radii so that the current I on each conductor is uniformly distributed around the conductor peripheries but oppositely directed along the conductors. This is usually a good approximation for $s/a > 5$. The magnetic flux density for each of the conductors individually was determined in Example 3.19 to be

$$\mathbf{B} = \frac{\mu I}{2\pi r}\mathbf{a}_\phi$$

Since the flux is circumferentially directed, we choose the surface through which we evaluate the magnetic flux to be a flat surface between the two facing sides and a distance d down the line. Combining the flux due to both currents gives

$$\psi = 2 \int_{z=0}^{d} \int_{r=a}^{s-a \cong s} \frac{\mu I}{2\pi r}\underbrace{drdz}_{ds}$$

Evaluating this gives

$$\psi = \frac{\mu I d}{\pi} \ln\left(\frac{s}{a}\right)$$

Hence the inductance of a section of the line of length d is

$$L = \frac{\psi}{I}$$

$$= \frac{\mu d}{\pi} \ln\left(\frac{s}{a}\right) \quad \text{H}$$

Thus the per-unit-length inductance becomes

$$l = \frac{L}{d}$$

$$= \frac{\mu}{\pi} \ln\left(\frac{s}{a}\right) \quad \text{H/m} \tag{3.95}$$

◀

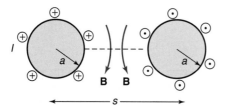

Figure 3.45 Example 3.25; computation of the per-unit-length inductance of a two-wire transmission line.

▶ **QUICK REVIEW EXERCISE 3.13**

Determine the per-unit-length inductance of a transmission line consisting of two parallel wires of radii 0.406 mm and separation 1 cm. The surrounding medium is air

ANSWER   1.28 μH/m.   ◀

▶ 3.12 FORCES PRODUCED BY CHARGE AND CURRENT

An electric field exerts a force on a point charge Q of $\mathbf{F} = Q\mathbf{E}$ as shown in Fig. 3.46a. Similarly, a point charge moving with velocity v in a magnetic field $\mathbf{B}$ will have a force exerted on it of $\mathbf{F} = Q\mathbf{v} \times \mathbf{B}$ as shown in Fig. 3.46b. This latter force acts perpendicular to the plane containing the velocity vector $\mathbf{v}$ and the magnetic flux vector $\mathbf{B}$ according to the right-hand-rule. The combination of these two forces is referred to as the *Lorentz force equation*:

$$\boxed{\mathbf{F} = Q\mathbf{E} + Q\mathbf{v} \times \mathbf{B}} \tag{3.96}$$

Consider a wire moving with velocity v in a magnetic field shown in Fig. 3.47. The wire "cuts" the magnetic field lines. The free charges in the wire (electrons) will have a force exerted on them (downward on the free electrons or, equivalently, upward on positive charge) that will move positive charge to the top of the wire and negative charge to the bottom. This separation of charge will produce an electric field that is directed from the positive accumulated charge on the top end toward the negative accumulated charge on the bottom end. Hence an electric field will be developed along the wire. This electric field is effectively

$$\mathbf{E}' = -\frac{\mathbf{F} = Q\mathbf{v} \times \mathbf{B}}{Q}$$
$$= -\mathbf{v} \times \mathbf{B}$$

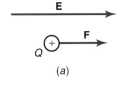

(a)

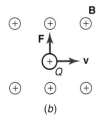

(b)

Figure 3.46 Illustration of the Lorentz force equation. (a) Force on a point charge due to an electric field. (b) Force on a moving point charge due to a magnetic field.

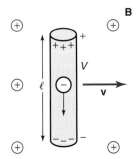

Figure 3.47 Illustration of the production of a voltage between the two ends of a wire when that wire "cuts" a magnetic field.

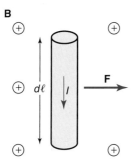

Figure 3.48 Illustration of the force exerted on a current-carrying wire by a magnetic field.

Hence the voltage developed between the two ends of the wire (positive at the top) is

$$
\begin{aligned}
V &= -\int \mathbf{E}' \cdot d\mathbf{l} \\
&= \int (\mathbf{v} \times \mathbf{B}) \cdot d\mathbf{l} \\
&= vBl \quad \text{V}
\end{aligned}
\tag{3.97}
$$

where the integration is performed along the wire. This result assumes that the wire axis is perpendicular to the plane containing the velocity vector and the magnetic flux density vector, **B**, so that the cross product can be removed from (3.97). Hence a conductor that is moving in a magnetic field and "cuts" those magnetic field lines will have a voltage induced between its two end points. This is often referred to as the *generator equation* since it explains how a coil of wire rotating in a magnetic field will generate a voltage as in an electric generator. (See Engineering Applications Section 3.13.6.)

A similar relation can be obtained to explain the force exerted on a wire carrying a current I in a magnetic field. Once again consider a segment of a wire of differential length dl carrying a current I that is immersed in a magnetic field as shown in Fig. 3.48. The force exerted on the charges constituting the current will be subject to a force given by $d\mathbf{F} = dQ\mathbf{v} \times \mathbf{B}$. But this may be rewritten in terms of the current by noting that $I d\mathbf{l} = (dQ/dt)\, d\mathbf{l} = dQ\,(d\mathbf{l}/dt) = dQ\mathbf{v}$. Hence we may write

$$
d\mathbf{F} = I d\mathbf{l} \times \mathbf{B}
\tag{3.98}
$$

The force vector is perpendicular to a plane containing the current and the magnetic field. This is often referred to as the BIL law in reference to the three items in it. This is also referred to as the *motor equation* since it explains how a current-carrying coil of wire in a magnetic field is caused to rotate, producing an electric motor. (See Engineering Applications Section 3.13.6.)

▷ QUICK REVIEW EXERCISE 3.14

A 1-cm wire is moving at a velocity of 1 cm/s in a magnetic field of 10 Wb/m². Determine the voltage induced between the two ends of the wire.

ANSWER 1 mV.   ◁

▶ **QUICK REVIEW EXERCISE 3.15**

A 5-cm wire carrying a current of 100 mA is immersed in a magnetic field of 5 Wb/m². Determine the force exerted on the wire if the magnetic field is perpendicular to the wire.

ANSWER   25 mN.   ◀

▶ 3.13 ENGINEERING APPLICATIONS

In this final section we will discuss several additional practical applications of the principles studied in this chapter.

3.13.1   High-Voltage Power Transmission Lines

A three-phase, high-voltage transmission line consists of three conductors carrying the three-phase voltages as shown in Fig. 3.49. A typical line-to-line voltage between the two conductors is V_{LL} = 765 kV. The line is supplying 1000 MVA power. Typical heights of the lines above ground are h = 12 m and the separations are s = 10 m. The conductors, which are actually bundled rather than solid, have effective radii of approximately 15 cm. We will calculate the electric field intensity beneath the center conductor at a height of 2 m, the height of a person standing beneath the line. Although the frequency is 60 Hz and not dc, we can approximately use the results of this chapter since the electrical dimensions are small (a wavelength at 60 Hz is 3107 mi). The electric field intensity about an infinite line charge was determined in (3.10) as

$$E = \frac{\rho_l}{2\pi\varepsilon_o r}$$

The voltage between each pair of conductors is 765 kV. Hence, using (3.38) for the voltage between two points due to a line charge,

$$V = 765 \text{ kV}$$
$$= \frac{\rho_l}{2\pi\varepsilon_o} \ln\left(\frac{10}{0.15}\right)$$

gives the line charge distribution as ρ_l = 10.12 μC/m. Therefore, the electric field at the person's head beneath the center conductor is the superposition of the fields due

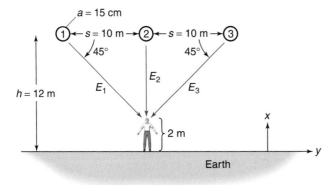

Figure 3.49   Computing the electric and magnetic fields of a high-voltage power transmission line.

to all three conductors:

$$E_1 = \frac{\rho_l}{2\pi\varepsilon_o \sqrt{(10)^2 + (10)^2}}\left[-\sin(45°)\mathbf{a}_x + \cos(45°)\mathbf{a}_y\right]$$

$$= -9.1\frac{\text{kV}}{\text{m}}\mathbf{a}_x + 9.1\frac{\text{kV}}{\text{m}}\mathbf{a}_y$$

$$E_2 = -\frac{\rho_l}{2\pi\varepsilon_o(10)}\mathbf{a}_x$$

$$= -18.22\frac{\text{kV}}{\text{m}}\mathbf{a}_x$$

$$E_3 = \frac{\rho_l}{2\pi\varepsilon_o \sqrt{(10)^2 + (10)^2}}\left[-\sin(45°)\mathbf{a}_x - \cos(45°)\mathbf{a}_y\right]$$

$$= -9.1 \text{ kV/m } \mathbf{a}_x - 9.1 \text{ kV/m } \mathbf{a}_y$$

The voltages and resulting electric fields are 120° out of phase. Hence the total electric field is

$$E = E_1\angle 0° + E_2\angle 120° + E_3\angle -120°$$

whose magnitude is $|E| = 18.2$ kV/m. This would be sufficient to cause a fluorescent bulb to glow if held by the person.

Next we compute the magnetic field intensity at the person's head. The power transmitted, 1000 MVA, is related to the line current and the line-to-line voltage as

$$P = \sqrt{3}V_{LL}I_L$$

Solving gives a line current of 755 A. The magnetic field due to an infinite current is obtained in (3.61) in Example 3.16 as

$$H = \frac{I}{2\pi r}$$

The magnetic field intensity due to the currents in the three wires are

$$H_1 = \frac{I}{2\pi \sqrt{(10)^2 + (10)^2}}\left[\cos(45°)\mathbf{a}_x + \sin(45°)\mathbf{a}_y\right]$$

$$= 6\mathbf{a}_x + 6\mathbf{a}_y \quad \text{A/m}$$

$$H_2 = \frac{I}{2\pi 10}\mathbf{a}_y$$

$$= 12\mathbf{a}_y \quad \text{A/m}$$

$$H_3 = \frac{I}{2\pi \sqrt{(10)^2 + (10)^2}}\left[-\cos(45°)\mathbf{a}_x + \sin(45°)\mathbf{a}_y\right]$$

$$= -6\mathbf{a}_x + 6\mathbf{a}_y \quad \text{A/m}$$

The currents and resulting magnetic fields are 120° out of phase. Hence the total magnetic field intensity is

$$H = H_1\angle 0° + H_2\angle 120° + H_3\angle -120°$$

whose magnitude is $|H| = 17$ A/m. The magnetic flux density is

$$B = \mu_o H$$

$$= 21.4 \ \mu\text{T}$$

It should be noted that the earth's magnetic field has a value of approximately 50 μT over the middle latitudes of the United States.

In recent years there have been numerous studies as to the health effects due to high-voltage power transmission lines and systems. Some studies seem to provide a weak link to causing cancer in humans while studies of others do not. Most recommendations are to avoid exposure to these fields where possible. For example, use of electric blankets is generally discouraged.

3.13.2 Electrostatic Discharge (ESD)

Most of the readers have personally experienced the phenomenon of electrostatic discharge or ESD. Walking across a nylon rug and bringing one's hand close to a doorknob may produce a discharge between the body and the doorknob. This miniature lightning bolt can result from voltages of some 10 kV. (Discharges may occur for voltages less than 3 kV but will usually not be felt.) This is the phenomenon of ESD, and the discharge can damage digital as well as analog electronic equipment. Today's closely packed digital electronics contain miniature semiconductor devices that can either be upset or permanently damaged by ESD. ESD is becoming of increasing concern in the reliable operation of digital electronics.

What causes this intense discharge? The answer is *separation of charge*. When two initially neutral materials are placed in contact, charge may be transferred from one material to the other. When the materials are separated, they may become charged: one negatively and the other positively. The degree to which charge is transferred depends on many factors. One factor is determined by the *triboelectric series* given in Table 3.1. The triboelectric series indicates which materials tend to give up electrons easily and become positively charged (those toward the top, or positive, end of the chart) and which tend to accept electrons and become negatively charged (those toward the bottom, or negative, end of the chart). For example, rubbing nylon against Teflon can cause electrons to be transferred from the surface of the nylon to the surface of the Teflon. Therefore the nylon may acquire a positive charge, whereas the Teflon may acquire a negative charge. Other examples are combing cat's fur with an acrylic comb and causing the comb to become charged thereby attacting other materials in the chart, and nylon apparel being attracted to the body. The degree to which this charge transfer takes place depends on a number of factors, and the triboelectric series is only a rough indicator of this. The order of the two materials

TABLE 3.1 The Triboelectric Series

POSITIVE	Aluminum	Polystyrene foam
Air	Paper	Acrylic
Human skin	Cotton	Polyester
Asbestos	Wood	Celluloid
Glass	Steel	Orlon
Mica	Sealing wax	Polyurethane foam
Human hair	Hard rubber	Polyethylene
Nylon	Mylar	Polypropylene
Wool	Epoxy-glass	Polyvinyl chloride (PVC)
Fur	Nickel, copper	Silicon
Lead	Brass, silver	Teflon
Silk	Gold, platinum	NEGATIVE

in the triboelectric series is an important factor, but it does not completely determine the degree of charge separation. Other factors such as the smoothness of the surface, surface cleanliness, contact surface area, contact pressure, degree of rubbing, and the speed of separation are more important. Charge may also be separated when two like materials are in contact, as with the opening of a plastic bag used to carry produce in a grocery store.

The key here is the *separation of charge*. This is particularly troublesome for insulating materials whose surface resistivity (inverse of conductivity) is very large. The transferred charge tends to accumulate in regions, and, because of the high surface resistivity, it remains there and cannot flow to other spots on the material to be neutralized. There are antistatic agents that can be sprayed on the surface to lower the surface resistivity and allow the charge to move about and become neutralized. Pink polyethylene plastic bags are routinely used to transport sensitive electronic components and protect them from ESD. The plastic is coated with a substance much like laundry detergent that greatly reduces the surface resistivity, thereby allowing any accumulated charge to move about and neutralize. High humidity causes a similar ability of the charge to move about on the surface and neutralize. Dry climates therefore tend to be more prone to ESD events than humid ones. Charge can also be transferred to a conductor (such as the moist human skin) by contact and then separation.

We now know that separation of charge creates an electric field. It is important to understand that air, normally nonconductive, has a certain electric field level at which it breaks down, liberating free charge. We see this in a lightning bolt and in the blue corona around a high-voltage power line on a damp day. The breakdown electric field for air is around 3×10^6 V/m or 3 MV/m. Over a shorter distance, a lower voltage difference between the two objects is required to break down the air. For example, the breakdown field strength of air of 3 MV/m is equivalent to 3000 V/mm. When charge is transferred to two materials by contact and the materials are separated, a voltage is generated between them due to the capacitance between the materials. Since $Q = CV$, as the materials are separated, the capacitance between them decreases and hence the voltage between them increases. For example, for a capacitance of 100 pF and a separated charge of 1 μC, the voltage between the two objects is 10 kV. Hence separated charge can create intense electric fields that break down the air, generating a miniature lightning bolt and intense currents. This phenomenon is called electrostatic discharge or ESD.

Since the surface resistivity of insulating materials is very large, any accumulated charge cannot easily move around, much less rapidly discharge to a nearby conductor. The conductivity of conducting materials is so large (resistivity so small) that charges can easily move about on their surface as well as discharge to other nearby conductors. An electrostatic discharge generally results from (1) charging an insulator, (2) transferring that charge to a conductor (such as the human skin) either by contact or induction, and (3) that charged conductor being brought close to another conductor, resulting in an electrostatic discharge. Charging a conductor by placing a charged insulator close to it is called *induction*. For example, suppose a charged material (perhaps an insulating material charged by contact earlier with another insulating material at the opposite end of the triboelectric chart) approaches a (neutral) conductor but does not physically touch it as shown in Fig. 3.50. Suppose the charge on the insulator

Figure 3.50 Illustration of electrostatic discharge (ESD) by the process of induction.

is negative. This will cause the charge in the neutral conductor to be separated; positive charge will be attracted to or induced on the face of the conductor facing the insulator. If this conductor momentarily touches another conductor, the negative charge on the opposite face of the conductor will be removed, thereby leaving a positively charged conductor.

This process of induction occurs frequently when one walks across a nylon carpet with rubber-soled shoes. Electrons are transferred from the carpet to the rubber shoes, leaving a positively charged footprint on the carpet and a negative charge on the soles of the shoes. This induces a charge separation (by induction) on the body (which is a conductor). Positive charge is induced on the soles of the feet in response to the negative charge on the soles of the shoes, and negative charge is induced on the upper parts of the body such as the hand. When the negatively charged finger approaches, for example, a digital computer keyboard, induction again induces charge (positive) on the keyboard and an intense electric field is created between the finger and the keyboard. The closer the finger to the keyboard, the smaller the required voltage difference to cause breakdown of the air. Once the discharge takes place, it creates an intense electromagnetic field (that is no longer static but varies with time) which can destroy sensitive semiconductor components in the computer or simply cause logic problems, such as resetting the computer and loss of memory. ESD is becoming one of the most important design obstacles in digital electronics.

We can roughly estimate the ability of the body to store charge. In Quick Review Exercise Problem 3.8 it is determined that the capacitance between two concentric spheres of radii $a < b$ is

$$C = \frac{4\pi\varepsilon_o}{\left(\dfrac{1}{a} - \dfrac{1}{b}\right)}$$

If we let the outer sphere recede to infinity, $b \rightarrow \infty$, then the capacitance of a sphere of radius a is

$$C = 4\pi\varepsilon_o a$$
$$\cong 111a \quad \text{pF}$$

If we model the body as a sphere of radius 1 m, this gives its capacitance of 111 pF. If a charge of 1 μC is deposited on the body, the body is raised to a potential of 9 kV. Admittedly, this is a rather crude calculation, but it does illustrate that relatively small amounts of charge can create very large voltages capable of breaking down air and creating an ESD discharge.

Manufacturers of modern digital electronics routinely test their products for susceptibility to ESD by simulating the ESD event. ESD "guns" are brought close to digital computer keyboards and other parts of the computer and allowed to discharge to the device. The levels of these discharges are set by regulatory agencies (governmental bodies) and are meant to simulate what happens when, for example, a person walks across a carpet, thereby acquiring charge, and then approaches or touches a keyboard of a digital device, creating an intense discharge to that device. By these tests, the manufacturers want to ensure that their products will be immune to the inevitable ESD discharges that they will encounter in a normal environment such as an office.

3.13.3 Interference

In addition to ESD, discussed in the previous section, there are numerous other instances where intense electromagnetic fields (even dc ones) can upset electronic devices (digital as well as analog). In this section we will investigate how magnetic fields can create interference in an electronic device.

Video display units are an integral part of computers, allowing a visual display of the information being processed. A video display unit consists of a glass tube which has the air evacuated from it as shown in Fig. 3.51. A filament has current passed through it, causing heat (as in an incandescent light bulb) which heats a cathode. This causes electrons to be "boiled off" the surface of the cathode. The interior of the front face is lined with a metal, constituting an anode, and a large dc voltage is applied between the anode and the cathode. This creates a large horizontal electric field that accelerates the electrons according to $\mathbf{F} = q\mathbf{E}$. When these accelerated electrons strike the inner face of the tube, the phosphorescent coating on that face glows, creating a spot. In order to create words, images, etc., on that face, the electron beam is swept across the face in horizontal lines. The beam is swept across the face using either electrostatic or magnetostatic deflection. The electrostatic deflection of the beam is accomplished with pairs of deflection plates, one pair mounted vertically about the beam and the other pair mounted horizontally. Large voltages are applied to the plates, creating an electric field between each pair. The electron beam is deflected in the vertical and horizontal directions by the force of these electric fields as $\mathbf{F} = q\mathbf{E}$. The majority of the displays seem to use magnetic deflection coils to create magnetic fields that move the electron beam across the face. Two coils are mounted on the top and bottom of the neck of the tube and two more coils are mounted on the sides as illustrated in Fig. 3.51. These create a magnetic field intensity between them that is vertically and horizontally directed. Each magnetic field exerts a force $\mathbf{F} = q\mathbf{v} \times \mathbf{B}$ on the electron stream, hence deflecting it and causing it to sweep across the face of the unit. External magnetic fields can affect the electron beam for either the electrostatic or the magnetostatic deflection scheme. Manufacturers of video displays routinely calibrate their displays before shipment to compensate for the earth's magnetic field, which is on the order of 50 μT in the middle latitudes of the United States. This understanding of the operation of the video display unit explains why the display can create magnetic fields that may interfere with magnetic recording devices such as disk drives and also why magnetic fields external to the display can interfere with it.

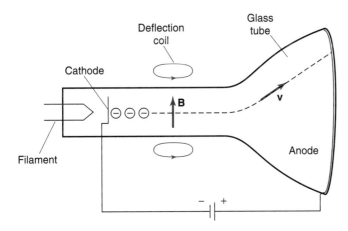

Figure 3.51 A video display tube using magnetic deflection of the beam.

For example, each deflection coil of a magnetostatic deflection scheme can be visualized as a loop consisting of a large number of turns of wire. The magnetic fields of these loops of wire can be obtained by using the result in (3.62). Considering a deflection coil as a loop of N turns of wire where the loop has radius a, the magnetic flux density vector a distance d away and broadside to the loop is given by

$$B = \frac{\mu_o IN}{2} \frac{a^2}{(a^2 + d^2)^{3/2}}$$

Consider a coil of radius 5 cm carrying 0.6 A and having 1000 turns. At a distance of 10 cm, the magnetic field is 674 μT. This is 14 times the magnetic field of the earth and can create errors in magnetic floppy disk drives and hard drives. Some 20 years ago, a manufacturer of personal computers produced a computer that had a restriction stated in the user's manual that the video display unit must not be placed on top of the computer. Why was this restriction imposed? It turns out that the magnetic fields of the video display unit were so intense that they caused read–write errors in the floppy disk drive that was located at the top of the computer and directly below the video display unit. Inserting a sheet of high-permeability metal such as mumetal ($\mu_r = 30{,}000$) on top of the computer diverted the magnetic field lines of the video display unit and stopped the interference with the floppy disk drive.

The author is aware of another similar incident of interference with a video display unit. In a certain newly constructed office complex, computer operators complained of "wavy lines" on their monitors. It turns out that the phase and neutral wires in the power distribution wiring for that room were not run together but instead one wire was routed over the top of the room and the other, return, wire was routed below the room. This created a wire loop carrying some 20 A of current that had dimensions of 2.5 m × 5 m. Approximating this as a circular loop and using the result for the magnetic field at the center of a loop given in (3.63) of Example 3.17 yields (approximating the loop as a circle with area $\pi a^2 = 2.5 \times 5$ or an effective radius of 2.0 m)

$$B = \frac{\mu_o I}{2a}$$
$$= 6.3 \text{ μT}$$

This level was evidently sufficient to cause movement of the electron beam in the video display units, creating wavy lines on the face of those units.

Submarines typically use an ac generator to generate the 400-Hz ac power that is routed around the submarine. The fields of the currents in these power lines sometimes cause interference with the sonar system. Some modern submarines send power around the sub that is dc in order to reduce this interference with the sonar. At points where it is needed, dc–dc converters convert the dc power distribution level to the lower or higher level needed by the supplied devices.

These are some of the numerous examples of interference caused by dc or low-frequency magnetic fields. We will find in the next chapter that currents that are not dc but have high-frequency variation will induce more significant signals in electronic circuits, causing even more interference problems.

3.13.4 Parasitic Effects in Components

In modern electronic devices such as digital computers there are numerous *parasitic* capacitances and inductances. These are unintended and are not shown on the schematic. This is referred to as the "hidden schematic." At the high frequencies of

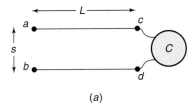

(a)

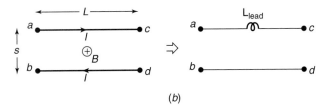

(b)

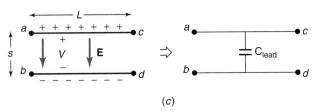

(c)

Figure 3.52 Illustration of the parasitic inductance and capacitance of the connection leads of a lumped-circuit component such as a capacitor. (a) The physical dimensions of the connection leads. (b) Inductance of the connection leads. (c) Capacitance of the connection leads.

today's modern electronic devices, these parasitic elements must be considered by designers; otherwise, the device will not operate as intended.

An example of this is the effect of the component's connection leads that connect the component to the electronics board. For example, consider a capacitor shown in Fig. 3.52a. The connection leads are of length L and separation s. Current flowing along these connection leads produces a magnetic field that threads the enclosed loop. Hence the connection leads form an inductance as shown in Fig. 3.52b. The inductance of this loop can be obtained using a result derived earlier for a pair of parallel wires in Example 3.25. The per-unit-length inductance was obtained in (3.95):

$$l = \frac{\mu_o}{\pi} \ln\left(\frac{s}{a}\right) \qquad \text{H/m}$$

where a is the wire radius. Multiplying this by the total length of the wires, L, gives the lead inductance:

$$L_{\text{lead}} = \frac{\mu_o L}{\pi} \ln\left(\frac{s}{a}\right) \qquad \text{H} \qquad (3.99)$$

Similarly, the voltage between the two connection leads causes charge to be deposited on those leads as shown in Fig. 3.52c and hence the connection leads form a capacitance. The lead capacitance can be obtained using the per-unit-length capacitance of a pair of parallel wires obtained in Example 3.14 Equation (3.52):

$$c = \frac{\pi \varepsilon_o}{\ln\left(\dfrac{s}{a}\right)} \qquad \text{F/m}$$

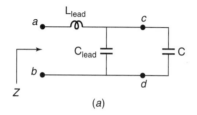

(a)

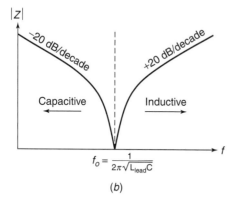

$$f_o = \frac{1}{2\pi\sqrt{L_{lead}C}}$$

(b)

Figure 3.53 Illustration of the effect of connection leads on the high-frequency performance of a capacitor. (a) Lumping the lead capacitance and inductance. (b) The frequency response of the input impedance of the leads showing that above the resonant frequency, the lumped capacitor appears to be an inductor!

Multiplying this by the length of the leads gives the total lead capacitance:

$$C_{lead} = \frac{\pi\varepsilon_o L}{\ln\left(\dfrac{s}{a}\right)} \qquad F \tag{3.100}$$

Hence an equivalent circuit of the capacitor and its connection leads is shown in Fig. 3.53a. The lead capacitance is in parallel with the component capacitance, which is usually much larger than the lead capacitance. Hence, the lead capacitance may be neglected. The input impedance to the complete structure is

$$Z = j\omega L_{lead} - j\frac{1}{\omega C}$$

This is a series LC circuit which has a resonant frequency of

$$f_o = \frac{1}{2\pi \sqrt{L_{lead}C}} \qquad Hz \tag{3.101}$$

At this resonant frequency the impedance seen looking into the leads is zero. The magnitude of the input impedance is sketched in Fig. 3.53b. For frequencies below this resonant frequency, the magnitude of the input impedance approximates to

$$|Z| \cong \frac{1}{\omega C} \qquad f < f_o$$

whereas at frequencies above this resonant frequency, the magnitude of the input impedance approximates to

$$|Z| \cong \omega L_{lead} \qquad f > f_o$$

Hence, above the resonant frequency, the capacitor appears to be an inductor!

Consider a typical capacitance of $C = 1000\,\text{pF}$ and leads whose separation is $s = \frac{1}{4}$ in. (.635 cm) and length is $L = \frac{1}{2}$ in. (1.27 cm). The connection leads are nominally #20 gauge wire whose radius is 16 mils (.406 mm). The lead inductance becomes, by substituting these into (3.99), $L_{\text{lead}} = 14\,\text{nH}$. The lead capacitance is obtained by substitution into (3.100), giving $C_{\text{lead}} = 0.128\,\text{pF}$. Hence the lead capacitance is much less than the component capacitance and can be neglected. The resonant frequency in (3.101) becomes $f_o = 42.6\,\text{MHz}$. Hence, above 42.6 MHz the capacitor appears to be an inductor! See C.R. Paul, *Introduction to Electromagnetic Compatibility*, Chapter 6, John Wiley Interscience, 1992, for experimental data that confirm this model.

Nonideal parasitic components such as these can cause high-frequency operation of electronic devices to appear "mysterious" and can drastically alter their desired behavior. Design of modern high-frequency devices such as digital computers demand that these nonideal effects be considered in the design; otherwise, the devices will not operate properly. As frequencies of operation of modern electronic devices increase, this is one aspect of the field of electromagnetic compatibility (EMC) that is rapidly becoming a significant part of an ECE's knowledge base and must be considered in design of those devices.

3.13.5 Electrostatic Shielding

Metallic enclosures are frequently used to *shield* sensitive electronics from being interfered with by sources that are outside the enclosure. These take the form of rooms of either conducting metal or wire screen. Smaller-scale enclosures such as the metal case of a digital computer contain electronic systems for the purpose of preventing EM signals inside the computer from interfering with electronics such as TVs outside it. We will discuss shielding enclosures for protection against static (dc) fields in this section. In Chapter 5 we will discuss the use of shielded enclosures for protection against high-frequency fields.

The principle of shielding against static electric fields (*electrostatic shielding*) is

> Inside a cavity formed by a closed, conducting surface of arbitrary shape, the electric field is zero. This is true so long as no charge has been introduced into the cavity and the fields outside are static (dc).

Since the electric field is zero in the cavity, *the potential at any point in the cavity is the same as that of the conductor, that is, there is no potential difference or voltage between two points in the cavity.* As an example, consider a spherical conducting shell of radius a where a charge Q is uniformly distributed over the surface of the sphere as shown in Fig. 3.54. By symmetry, the electric field is directed radially and is constant over spheres of radius r. Applying Gauss' law gives

$$\oint_s \varepsilon_o \mathbf{E} \cdot d\mathbf{s} = Q$$

$$= \varepsilon_o E_r 4\pi r^2$$

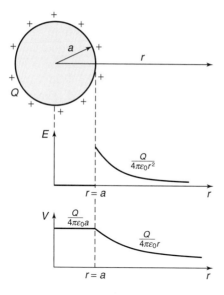

Figure 3.54 Illustration of electrostatic shielding with a charge-bearing sphere.

where we have chosen the Gaussian surface to be another sphere of radius r. Hence we obtain

$$E_r = \frac{Q}{4\pi\varepsilon_o r^2} \qquad r > a$$

$$= 0 \qquad r < a$$

The result that the field is zero for $r < a$ is due to the fact that the charge enclosed by the Gaussian surface (a sphere of radius r) is zero for $r < a$. The potential (absolute) is

$$V = -\int_{\infty}^{r} \mathbf{E} \cdot d\mathbf{l}$$

$$= \frac{Q}{4\pi\varepsilon_o r} \qquad r \geq a$$

$$= \frac{Q}{4\pi\varepsilon_o a} \qquad r \leq a$$

The result that the potential of points inside the sphere is the same as the potential of the sphere surface is a result of the fact that the electric field inside the sphere is zero. If we move a unit positive charge from infinity toward the sphere, work will be required to overcome the repulsion of the radially directed electric field. Inside the sphere, no *additional* work will be required to move the charge because the electric field is zero.

Although the principle of an electrostatic shield was demonstrated in Fig. 3.54 using a spherical conductor, it also applies to conducting enclosures of arbitrary shape. Such shielded enclosures are often referred to as Faraday shields after a Scottish mathematician, Michael Faraday, who conducted experiments in the early 1800s to confirm this. (Benjamin Franklin made a similar discovery in the mid-1700s in the United States.) In Chapter 4 we will study the famous Faraday's law that was also discovered by this gifted scientist. The Faraday shield is sometimes called a Faraday "cage" or screened enclosure since these enclosures are frequently constructed of metallic screen wire rather than a contiguous metal. Screened rooms are used to perform electromagnetic compatibility (EMC) testing. In order to prevent interference, governmental regulatory agencies limit the amount of electric field produced by digital electronic devices such as computers. Testing of the device inside a screened room or Faraday shield confirms whether the device conforms to those regulations. The Faraday screen prevents outside fields from interfering with the test. There are numerous other applications of the Faraday shield. Modern aircraft frequently fly through bad weather containing intense lightning and thunderstorms. These lightning discharges result from the large-scale separation of charge between a cloud and the earth as well as between two clouds. Hence large electric fields exist in the vicinity of these storms. The passengers inside the metallic fuselage of an aircraft are unharmed since the fuselage serves as a Faraday cage so that there can be no electric field and hence no potential difference inside it. The necessary windows, doors, and other penetrations of the fuselage make it a less than perfect Faraday cage but it nevertheless serves its purpose rather well. The metal frame of an automobile can also serve a similar purpose.

Another aspect of electrostatic shielding involves the inevitable capacitances between metallic objects. Consider two metallic objects shown in Fig. 3.55a. Capacitance

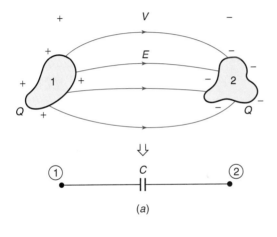

(a)

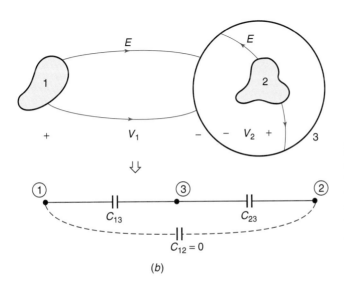

(b)

Figure 3.55 Illustration of electrostatic shielding. (a) Electric field lines between two bodies determine the capacitance between them. (b) Surrounding a conductor with another conductor provides an electrostatic shield.

between the objects, C, can be demonstrated by either placing charge Q on one and $-Q$ on the other and computing the voltage, V, that occurs between them or applying a voltage between them and computing the charge on the bodies. In either case the capacitance between the two bodies is

$$C = \frac{Q}{V}$$

Observe that in either case electric field lines begin on one conductor and end on the other. The number of electric field lines determines the capacitance; the more that originate on one conductor and terminate on the other, the larger the capacitance. In order to *shield* one conductor from another, we essentially want to *divert* electric field lines so that they do not terminate on the conductor which is to be shielded. Suppose we surround conductor 2 with a conducting enclosure (conductor 3) as shown in Fig. 3.55b. The electric field lines from conductor 1 must terminate on conductor 3 and cannot terminate on conductor 2 as before. Hence there are capacitances between conductors 1 and 3, C_{13}, and between conductors 2 and 3, C_{23}. Observe that there is no capacitance directly between conductors 1 and 2, $C_{12} = 0$. This represents electrostatic shielding in a similar way to the Faraday shield. Observe here that conductors 1 and 2 are not completely isolated since they are connected through the series combination of C_{13} and C_{23}.

Consider three metallic objects, one of which is enclosed by one of the other objects as shown in Fig. 3.56. Suppose all three objects are above a large metallic plane, which will be referred to as "ground." The word "ground" is perhaps the most

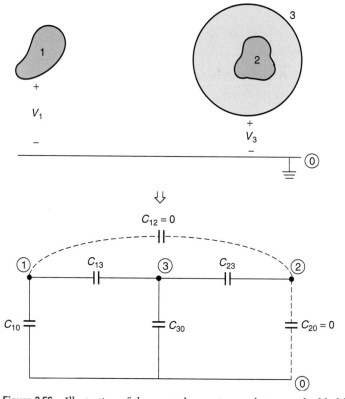

Figure 3.56 Illustration of the mutual capacitances between shielded bodies.

misunderstood and misused term in electrical engineering. In the commercial power area, engineers routinely mean "earth" when they speak of "ground." In high-frequency electronic devices there are several types of ground: safety ground for shock protection, electronic ground to return high-frequency currents to their source, etc. Here we use the term ground to mean a point of referencing voltages. We will designate this fourth or ground conductor as the *zeroth* conductor. If we apply (dc) voltages V_i between conductor i and ground, we will find that these are related to the charges on the conductors by the capacitances of the structure as shown in Fig. 3.56. These can be written as

$$Q_1 = C_{10}V_1 + C_{13}(V_1 - V_3)$$
$$Q_3 = C_{30}V_3 + C_{13}(V_3 - V_1) + C_{23}(V_3 - V_2)$$
$$Q_2 = C_{23}(V_2 - V_3)$$

The items C_{i0} are the self-capacitances between the ith conductor and ground, whereas, C_{ij} is said to be the mutual capacitance between the ith and jth conductors. Observe that some of the mutual and self-capacitances are absent from the equivalent circuit. For example, there is no self-capacitance between conductor 2 and ground, $C_{20} = 0$. Similarly, there is no mutual capacitance between conductor 1 and conductor 2, $C_{12} = 0$. These are a result of the fact that conductor 3 surrounds conductor 2 and hence electric field lines originating on 1 cannot terminate on 2 and vice versa. However, note that conductor 3 is "floating," that is, its voltage with respect to ground, V_3, is not zero. If we apply a time-varying voltage to conductor 1, V_1, this can cause, through the capacitive network, a voltage of conductor 2 (with respect to ground). However, suppose we "ground" conductor 3 to the ground plane. This sets $V_3 = 0$ and hence isolates conductors 1 and 2. Hence by "grounding" the enclosure, we can eliminate any inducement of a voltage on items inside a shield due to sources outside it. We will see this principle of grounding shields occur in many instances later in the text.

3.13.6 The Electric Generator and Motor

Consider a rectangular loop of wire that is rotating in a magnetic field at a speed of ω rad/s as shown in Fig. 3.57a. We determined in Section 3.12 that a straight wire of length l that cuts across a magnetic field B with a velocity v will have a voltage $V = Blv$ between its two endpoints. (See Fig. 3.47.) This principle is used to construct electric generators.

Consider a loop of wire rotating in a magnetic field as shown in Fig. 3.57a and shown in cross section in Fig. 3.57b. A voltage will be induced in the top and bottom wires of length l according to (3.97), and none will be induced in the sides since they are parallel to the magnetic field. In terms of the angle θ in Fig. 3.57b, the voltage induced will be

$$V = 2vBl \sin(\theta)$$

But we can write the angle in terms of the rotation speed as

$$\theta = \omega t$$

The sides of length l move an incremental arc length

$$dr = \frac{w}{2}d\theta$$

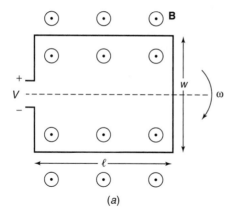

(a)

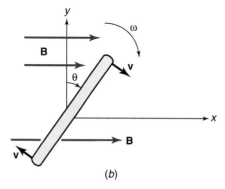

(b)

Figure 3.57 The electric generator. (a) Physical dimensions of the rotating loop. (b) Side view of the rotating loop.

Hence the velocity is

$$v = \frac{dr}{dt}$$

$$= \frac{w}{2}\omega$$

and the induced voltage is

$$V = wBl\omega \sin(\omega t)$$

In terms of the area of the loop, $A = lw$, this becomes

$$V = \omega BA \sin(\omega t)$$

This essentially illustrates the construction of an electric generator. A large number of coils of wire are rotated by a "prime mover" such as a steam turbine, an electric motor, or a water wheel. The voltages of all these coils add to produce a sufficiently large output voltage. This induced voltage depends, according to the above result, on (1) the speed of rotation, (2) the number of coils (all are connected in series), and (3) the area of the coils. Hence we have converted mechanical energy to electrical energy.

We also determined in Section 3.12 that a wire of length l that is carrying a current I will have a force exerted on it by a magnetic field B. That force is $F = BIl$ and is perpendicular to the plane containing B and I as shown in Fig. 3.48. This principle

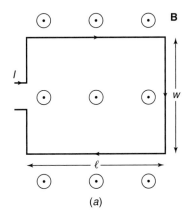

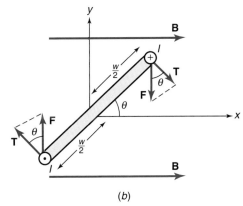

Figure 3.58 The electric motor. (a) Physical dimensions of the rotating loop. (b) Side view of the rotating loop and the forces and torque exerted on it.

is used to construct electric motors. Consider a loop of wire that has a current applied to it as shown in Fig. 3.58a. The cross-sectional view is shown in Fig. 3.58b. A force is exerted on the horizontal wires according to (3.98) of

$$F = IlB$$

The component of this force acting perpendicular to the loop to turn it is $F = IlB \cos(\theta)$, giving a torque to produce rotation of

$$
\begin{aligned}
T &= F \frac{w}{2} \cos(\theta) \\
&= \frac{wIIB}{2} \cos(\theta) \\
&= \frac{IBA}{2} \cos(\theta)
\end{aligned}
$$

and we have written the result in terms of the area of the coil, $A = wl$. This essentially illustrates the construction of an electric motor. A large number of coils are wound into a multiturn loop and a current is applied. An external magnetic field is obtained by a similar process. The interaction between the current-carrying wires and the magnetic field causes the multiturn loop to rotate, thus converting electric energy to mechanical energy.

▶ SUMMARY OF IMPORTANT CONCEPTS AND FORMULAE

1. **Coulomb's law:** gives the force exerted on one point charge by another point charge: $F = Q_1Q_2/4\pi\varepsilon R^2$. The force is directed on a line between the two charges.

2. **Permittivity of free space (essentially air):** $\varepsilon_o \cong 1/36\pi \times 10^{-9}$ F/m.

3. **Electric field intensity vector:** is denoted by $\mathbf{E}$ and is the ratio of the force exerted on a charge q by another charge or charge distribution per unit of charge: $\mathbf{E} = \mathbf{F}/q$ V/m.

4. **Electric field intensity due to a point charge:** $\mathbf{E} = Q/4\pi\varepsilon r^2 \mathbf{a}_r$ which is directed radially away from the charge.

5. **Electric field intensity due to distributions of charge:** can be determined by superimposing the effects of chunks of the charge distribution dQ.

6. **Electric field intensity due to an infinite line charge:** $\mathbf{E} = \rho_l/2\pi\varepsilon r\, \mathbf{a}_r$ which is directed radially away from the line charge.

7. **Dielectric materials:** consist of dipoles of charge which when subjected to an external electric field are caused to align. This creates an additional *polarization* field that increases the total field. The electric flux density vector $\mathbf{D}$ whose units are C/m^2 is the sum of these effects: $\mathbf{D} = \varepsilon_o\mathbf{E} + \mathbf{P}$. This allows us to freely interchange $\mathbf{E}$ and $\mathbf{D}$ via the permittivity of the material as $\mathbf{D} = \varepsilon\mathbf{E}$, where $\varepsilon = \varepsilon_r\varepsilon_o$ and ε_r is the relative permittivity of the dielectric material.

8. **Gauss' law for the electric field:** $\oint_s \mathbf{D} \cdot ds = Q_{\text{enclosed}}$, which plainly states that if we add the products of the differential surface areas and the components of $\mathbf{D}$ that are perpendicular to the closed surface s, we will obtain the *net positive charge* enclosed by that surface. In order for Gauss' law to be useful in computation, the charge distribution must be such that we can choose a Gaussian surface s such that (1) $\mathbf{D}$ is everywhere perpendicular to it and (2) $\mathbf{D}$ is constant over the surface.

9. **Voltage:** between two points is the work required to move a charge from the assumed negative voltage point to the assumed positive voltage point per unit of charge moved. In terms of the electric field $V_{ba} = W_{ba}/q = -\int_a^b \mathbf{E} \cdot dl$, that is, if we add the products of the differential path lengths and the components of $\mathbf{E}$ that are parallel to the path, we will obtain the voltage between the two points. In addition, $\oint_c \mathbf{E} \cdot dl = 0$, which states that the static electric field is conservative, so that voltage between two points is independent of path taken. We will later find that for time-varying fields, voltage between two points is no longer independent of path.

10. **Voltage between two points due to a point charge:** $V_{ba} = Q/4\pi\varepsilon(1/r_b - 1/r_a)$.

11. **Voltage between two points due to an infinite line charge:** $V_{ba} = \rho_l/2\pi\varepsilon \ln(r_a/r_b)$.

12. **Capacitance between two bodies:** is the ratio of free charge (equal but opposite on the bodies) to the voltage between the two bodies: $C = Q_f/V$. To compute this, place equal and opposite charges on the two bodies and compute the resulting voltage between them.

13. **Current:** is the flow of charge. Current density $\mathbf{J}$(A/m^2) and electric field $\mathbf{E}$(V/m) that produced it are related by Ohm's law, $\mathbf{J} = \sigma\mathbf{E}$, where σ is the conductivity of the material. For static (dc) currents, conservation of charge dictates that $\oint_s \mathbf{J} \cdot ds = 0$.

14. **Biot-Savart law:** gives the magnetic flux density vector $\mathbf{B}$ whose units are Wb/m^2 or tesla (T) due to a segment of current as $d\mathbf{B} = (\mu I/4\pi R^2)\, dl \times \mathbf{a}_R$. Observe that the $\mathbf{B}$ field is perpendicular to the plane containing the current and the radial distance vector from the current to the desired point according to the right-hand-rule. The magnetic flux density vector for current loops or long segments can be obtained by summing (with an integral) the $\mathbf{B}$ fields of these chunks of current.

15. **Permeability of free space (essentially air):** $\mu_o = 4\pi \times 10^{-7}$ H/m.

16. **Magnetic flux density due to an infinite line current:** $\mathbf{B} = \mu I/2\pi r\, \mathbf{a}_r$ which is circumferentially directed about the current according to the right-hand-rule.

17. **Magnetic materials:** consist of magnetic dipoles that will tend to align with an external magnetic field. This creates an additional *magnetization* field that increases the total field. The

magnetic flux density vector **B** is the sum of these effects: $\mathbf{B} = \mu_o\mathbf{H} + \mu_o\mathbf{M}$. The magnetic field intensity vector is denoted as **H** and has units of A/m. This allows us to freely interchange **B** and **H** via the permeability of the material as $\mathbf{B} = \mu\mathbf{H}$, where $\mu = \mu_r\mu_o$ and μ_r is the relative permeability of the magnetic material.

18. **Ampere's law:** $\oint_c \mathbf{H} \cdot d\mathbf{l} = I_{enclosed}$, which states that if we add the products of the differential path lengths and the components of **H** that are tangent to the path around a closed contour c, we will obtain the current enclosed by that contour. For currents that exhibit symmetry, this allows a considerably simplified calculation of the **H** field. But the contour must be chosen such that (1) **H** is tangent to the contour at all points and (2) the **H** field is the same at all points on that contour.

19. **Gauss' law for the magnetic field:** $\oint_s \mathbf{B} \cdot d\mathbf{s} = 0$, which states that if we add the products of the differential surface areas of a closed surface s and the components of **B** that are perpendicular to the surface, we will always obtain a result of zero. Essentially this states that, unlike the electric field, the magnetic field has no isolated sources and the magnetic field lines must form closed paths with no beginning or end.

20. **Inductance:** inductance of a closed loop is the ratio of the magnetic flux threading the loop to the current that produced the flux: $L = \psi/I$. In order to compute this, we apply a current I and compute the magnetic flux ψ resulting from that current that threads the loop.

21. **Lorentz force equation:** the force exerted on a charge Q by an electric field **E** and a magnetic field **B**, where the charge is moving with respect to the magnetic field with velocity **v**, is $\mathbf{F} = Q\mathbf{E} + Q\mathbf{v} \times \mathbf{B}$. A wire of length l moving perpendicular to a magnetic field B with velocity v has a voltage generated between its endpoints of $V = vBl$, where the wire, the B field, and the velocity are all mutually orthogonal. A wire of length l carrying a current I in a magnetic field B has a force $F = BIl$ exerted on it. The wire and the magnetic field are orthogonal and the force is exerted perpendicular to the plane containing the current and the magnetic field.

▶ PROBLEMS

SECTION 3.1 CHARGE AND COULOMB'S LAW

3.1.1. A volume charge distribution $\rho_v = 2z$ C/m^3 is contained in a region defined in cylindrical coordinates as $0 \le z \le 2$ m, $0 \le r \le 1$ m, $45° \le \phi \le 90°$. Determine the total charge contained in the region.

3.1.2. A surface charge distribution is contained in a flat, wedge-shaped surface whose corners are defined in a rectangular coordinate system by (2,1,2) m, (1,1,2) m, and (1,3,2) m. The charge distribution is given by $\rho_s = 3xyz$ C/m^2. Determine the total charge contained on the surface. [13 C].

3.1.3. A 100-μC point charge is located in a rectangular coordinate system at (1,1,1) m, and another point charge of 50 μC is at (−1,0,−2) m. Determine the vector force on the first charge.

3.1.4. Four 100-μC point charges are located on the corners of a square that are defined in a rectangular coordinate system by (1,0,0) m, (0,1,0) m, (−1,0,0) m, and (0,−1,0) m. Determine the vector force exerted on another 100-μC charge that is located at (a) (0,0,0) m, (b) (0,0,1) m, (c) (1,1,0) m. [(a) 0, (b) 127.28 $\mathbf{a}_z$ N, (c) 114.15$\mathbf{a}_x$ + 114.15$\mathbf{a}_y$ N]

3.1.5. Two point charges, $Q_1 = 18$ μC and $Q_2 = 72$ μC, are separated by 3 cm. A third charge, $Q_3 = -8$ μC, is introduced between the other two. Determine the distance between Q_1 and Q_3 such that Q_3 will not move. [1 cm]

3.1.6. Two equal and opposite-sign 35-μC charges are separated in air by a distance of 20 cm, and an electron is located midway between these charges. Determine the force experienced by the electron. [1×10^{-11} N toward the positive charge]

3.1.7. Two point charges Q are suspended with identical, massless strings of length l from a common point. If each point charge has a mass m, determine an expression for charge in terms of the angle between the vertical and the strings and the gravitational constant g.

SECTION 3.2 THE ELECTRIC FIELD INTENSITY VECTOR

3.2.1. A positive point charge, $Q_1 = 5$ μC, is located at $(0,0,2$ m$)$ in a rectangular coordinate system, and a negative point charge, $Q_2 = -10$ μC, is located at $(0,4$ m,$0)$. Determine the electric field intensity vector at $(0,4$ m,2 m$)$.

3.2.2. A positive 50-μC point charge and two negative 50-μC charges are placed on the corners of an equilateral triangle whose sides are of length 5 m. Determine the magnitude of the electric field intensity vector at the center of the triangle. [108 kV/m]

3.2.3. Two point charges Q of opposite sign are separated by a distance l and placed along the z axis of a rectangular coordinate system. The positive charge is placed at $(0,0,l/2)$ and the negative charge is placed at $(0,0,-l/2)$. Determine an expression for the electric field intensity vector (a) along the z axis and (b) along the y axis. [(a) $Ql/2\pi\varepsilon_o \, z/(z^2 - l^2/4)^2 \, \mathbf{a}_z$, (b) $-Ql/4\pi\varepsilon_o \, 1/(y^2 + l^2/4)^{3/2} \, \mathbf{a}_z$]

3.2.4. A ring of charge of radius a has a linear charge density of ρ_l C/m that is uniformly distributed around the ring. Determine the electric field at a distance d along a line perpendicular to the ring and centered on it. Evaluate this result for a distance much greater than the radius of the ring. This result at a large distance away should reduce to that for a point charge. Why?

3.2.5. Determine the electric field of a disk of charge of radius a having a uniform charge distribution of ρ_s C/m^2 over its surface at a distance d away from its center and on a line through its axis. You will need the integral $\int r/(d^2 + r^2)^{3/2} dr = -1/\sqrt{d^2 + r^2}$. Evaluate this expression at a distance much larger than the radius of the disk. This result at a large distance should reduce to that for a point charge. Why? [$\mathbf{E} = \rho_s/2\varepsilon_0[1 - d/\sqrt{d^2 + a^2}]\mathbf{a}_z$, $\mathbf{E} = \pi a^2\rho_s/4\pi\varepsilon_o d^2\mathbf{a}_z$ $d \gg a$, At large distances away, the disk of charge can be thought of as a point charge equal to the total charge contained in the disk: $Q = \pi a^2\rho_s$.]

3.2.6. Two infinite line charges of opposite polarity, ρ_l C/m, and $-\rho_l$ C/m, are directed in the z directions in a rectangular coordinate system. The positive line charge is placed at $y = l/2$ and the negative line charge is placed at $y = -l/2$. Determine the electric field intensity vector (a) along the y axis and (b) along the x axis. [(a) $\mathbf{E} = \rho_l/2\pi\varepsilon_o \, l/(y^2 - l^2/4)\mathbf{a}_y$ (b) $\mathbf{E} = -\rho_l/2\pi\varepsilon_o \, l/(x^2 + l^2/4)\mathbf{a}_y$]

3.2.7. Charge is uniformly distributed over a strip which is infinite in length and which has a width W with surface charge distribution ρ_s C/m^2. Determine the electric field intensity vector at a point that is perpendicular to the strip and located a distance d from its center.

SECTION 3.3 THE ELECTRIC FLUX DENSITY VECTOR AND DIELECTRIC MATERIALS

3.3.1. A parallel-plate capacitor has plates of very large extent such that the electric field is perpendicular to the plates. The plates are separated by a distance of 1 mm, and a voltage of 100 V is applied to them as shown in Fig. 3.11. A slab of mica ($\varepsilon_r = 5.4$) is inserted between the plates. Determine the magnitude of the polarization vector in the mica. [3.89 μC/m^2]

SECTION 3.4 GAUSS' LAW FOR THE ELECTRIC FIELD

3.4.1. A spherical volume charge distribution, $\rho_v = kr$ C/m^3, is contained in a spherical volume of radius a, and the medium is free space. Determine (a) the total charge enclosed by the volume, (b) the electric field intensity vector for $r \geq a$, and (c) the electric field intensity vector for $r \leq a$. [(a) $\pi k a^4$ C, (b) $\mathbf{E} = ka^4/4\varepsilon_o r^2 \, \mathbf{a}_r$, (c) $\mathbf{E} = kr^2/4\varepsilon_o \, \mathbf{a}_r$]

3.4.2. For the problem of a ring of charge in Problem 3.2.4 and the disk of charge in Problem 3.2.5, can Gauss' law be used to determine the electric field at a distance d from the center and on a line perpendicular to the ring or disk? If not, why not?

3.4.3. A cylindrical volume, $0 \leq z \leq 4$ m and $0 \leq r \leq 2$ m, encloses charge. If the electric field is $\mathbf{E} = zr/\varepsilon_o \, \mathbf{a}_z$, determine the total charge enclosed by the cylinder.

3.4.4. A volume charge density $\rho_v = k/r$ C/m^3 exists in a spherical region $a \leq r \leq b$. Determine the electric field intensity vector. [$E_r = k(b^2 - a^2)/2\varepsilon_o r^2$ for $r \geq b$, $E_r = 0$ for $r \leq a$, $E_r = k/2\varepsilon_o(1 - a^2/r^2)$ for $a \leq r \leq b$]

SECTION 3.5 VOLTAGE

3.5.1. The electric field intensity vector is given in a rectangular coordinate system as $\mathbf{E} = x\mathbf{a}_x + y\mathbf{a}_y + z\mathbf{a}_z$. Show that the electric field is *conservative* by determining the work required to move a charge q around two closed paths consisting of straight-line segments between the points: (a) $(0,0,0) \rightarrow (0,0,1\text{ m}) \rightarrow (0,2\text{ m},1\text{ m}) \rightarrow (0,2\text{ m},0) \rightarrow (0,0,0)$ and (b) $(0,0,0) \rightarrow (0,0,1\text{ m}) \rightarrow (1\text{ m},0,0) \rightarrow (0,0,0)$. [zero for both paths]

3.5.2. A positive 10-μC point charge is located at $(0,3\text{ m},0)$ in a rectangular coordinate system. Determine the voltage of a point $(0,3\text{ m},2\text{ m})$ with respect to the origin.

3.5.3. Two point charges are placed in a rectangular coordinate system as follows. $Q_1 = 10\ \mu$C is placed at $(0,0,3\text{ m})$, and $Q_2 = 5\ \mu$C is placed at $(0,2\text{ m},0)$. Determine the voltage at a point $(0,5\text{ m},5\text{ m})$ with respect to the origin of the coordinate system. $[-28,070\text{ V}]$

3.5.4. Two point charges are placed in a rectangular coordinate system as follows. $Q_1 = 10\ \mu$C is placed at $(0,-2\text{ m},0)$, and $Q_2 = 5\ \mu$C is placed at $(0,3\text{ m},0)$. Determine the voltage at a point $(0,0,5\text{ m})$ with respect to the origin of the coordinate system. $[-35,570\text{ V}]$

3.5.5. Two point charges are placed in a rectangular coordinate system as follows. $Q_1 = 5\ \mu$C is placed at $(3\text{ m},0,0)$, and $Q_2 = -10\ \mu$C is placed at $(0,0,2\text{ m})$. Determine the voltage at a point $(0,5\text{ m},0)$ with respect to the origin of the coordinate system.

3.5.6. Two line charge distributions are placed on the y axis of a rectangular coordinate system and are directed in the x direction. The first charge distribution is $\rho_1 = 5\ \mu$C/m and is placed at $(0,2\text{ m},0)$. The second charge distribution is $\rho_2 = -10\ \mu$ C/m and is placed at $(0,-3\text{ m},0)$. Determine the voltage at $(0,0,5\text{ m})$ with respect to $(0,0,0)$. $[30,477\text{ V}]$

3.5.7. Three point charges $Q = 2\ \mu$C are located on the corners of an equilateral triangle whose sides are of length $l = 3$ m. Determine the absolute potential at the center of the triangle.

3.5.8. A ring of charge of radius a and uniform distribution ρ_l lies in the xy plane of a rectangular coordinate system and is centered on the origin. The electric field intensity vector along the z axis was determined in Problem 3.2.4. Determine an expression for the absolute potential at points along the z axis using (3.44). Determine the electric field from that result, using $\mathbf{E} = -\text{gradient }V$, and show that it agrees with the electric field derived in Problem 3.2.4.

3.5.9. A disk of charge of radius a and uniform distribution ρ_s lies in the xy plane of a rectangular coordinate system and is centered on the origin. The electric field intensity vector along the z axis was determined in Problem 3.2.5. Determine an expression for the absolute potential at points along the z axis using (3.44). Determine the electric field from that result, using $\mathbf{E} = -\text{gradient }V$, and show that it agrees with the electric field derived in Problem 3.2.5. $[V = \rho_s/2\varepsilon_o(\sqrt{z^2 + a^2} - z)]$

3.5.10. Determine the electric field intensity vector for the following potential distributions: (a) $V(x,y,z) = 1/(x^2 + y^2 + z^2)^{1/2}$, (b) $V(r,\phi,z) = re^{-z}\cos\phi$, (c) $V(r,\theta,\phi) = \sin\theta\cos\phi/r^2$.

SECTION 3.6 CAPACITANCE

3.6.1. A parallel-plate capacitor has plate area of 100 cm^2 and plate separation of 1 mm. Mica ($\varepsilon_r = 5.4$) fills the region between the plates. A voltage of 10 V is applied between the plates. Neglecting fringing of the fields, determine the magnitudes of the electric field intensity vector, the electric flux density vector, and the polarization vector. Determine the capacitance. $[E = 10\text{ kV/m}, D = 0.478\ \mu\text{C/m}^2, P = 0.389\ \mu\text{C/m}^2, C = 477.5\text{ pF}]$

3.6.2. Two circular disks of radii 10 cm are separated by air a distance of 1 mm. Assuming the electric field is perpendicular to the plates, determine the capacitance of the structure.

3.6.3. A parallel-plate capacitor shown in Fig. P3.6.3 has two dielectrics filling the region between the plates. Determine the capacitance of this structure and the electric field in the two regions.

Evaluate these for a plate area of 100 cm^2, $V = 10 \text{ V}$, $d_1 = 1 \text{ mm}$, $\varepsilon_{r1} = 2$, $d_2 = 3 \text{ mm}$, $\varepsilon_{r2} = 4$. Hint: Consider this as two parallel-plate capacitors in series.

$$\left[C = \varepsilon_o A \frac{\dfrac{\varepsilon_{r1}\varepsilon_{r2}}{d_1 d_2}}{\dfrac{\varepsilon_{r1}}{d_1} + \dfrac{\varepsilon_{r2}}{d_2}}, \quad C = 70.74 \text{ pF}, \right.$$

$$\left. E_1 = 4000 \text{ V/m}, \ E_2 = 2000 \text{ V/m} \right]$$

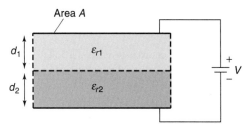

Figure P3.6.3 Problem 3.6.3.

3.6.4. A parallel-plate capacitor shown in Fig. P3.6.4 has two dielectrics filling the region between the plates. Determine the capacitance of this structure and the electric field in the two regions. Evaluate these for plate areas of $A_1 = 100 \text{ cm}^2$ and $A_2 = 400 \text{ cm}^2$, $V = 10 \text{ V}$, $d = 2 \text{ mm}$, $\varepsilon_{r1} = 2$, $\varepsilon_{r2} = 4$. Hint: Consider this as two parallel-plate capacitors in parallel.

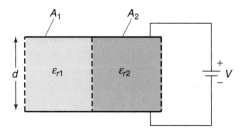

Figure P3.6.4 Problem 3.6.4.

3.6.5. A spherical capacitor (Fig. P3.6.5) consists of an outer conducting sphere of radius b and an inner conducting sphere of radius a. Two dielectrics fill the region between the spheres with ε_{r1} for $a < r < r_1$ and ε_{r2} for $r_1 < r < b$. Determine the capacitance of this structure and

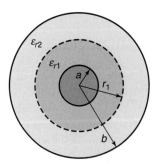

Figure P3.6.5 Problem 3.6.5.

evaluate it for $\varepsilon_{r1} = 2$, $\varepsilon_{r2} = 4$, $a = 1$ mm, $r_1 = 3$ mm, and $b = 5$ mm. Hint: Consider this as two spherical capacitors in series.

$$\left[4\pi\varepsilon_o\dfrac{1}{\dfrac{1}{\varepsilon_{r2}}\left(\dfrac{1}{r_1}-\dfrac{1}{b}\right)+\dfrac{1}{\varepsilon_{r1}}\left(\dfrac{1}{a}-\dfrac{1}{r_1}\right)},\ 0.303\ \text{pF}\right]$$

3.6.6. A coaxial cable is filled with an inhomogeneous dielectric as shown in Fig. P3.6.6. Determine the capacitance per unit length. Evaluate this for $\varepsilon_{r1} = 2$, $\varepsilon_{r2} = 4$, $a = 1$ mm, $r_1 = 3$ mm, and $b = 5$ mm. Hint: Consider this as two coaxial capacitors in series.

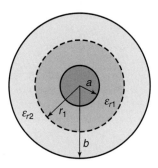

Figure P3.6.6 Problem 3.6.6.

SECTION 3.7 CURRENT AND THE MAGNETIC FLUX DENSITY VECTOR

3.7.1. Determine the resistance of a 1-km length of #20 gauge (radius = 16 mils) copper ($\sigma = 5.8 \times 10^7$ S/m) wire.

3.7.2. A parallel-plate capacitor is filled with a lossy dielectric that has a conductivity σ. The plates have area A and are separated by a distance d. Determine the resistance between the plates and evaluate this for $\sigma = 4$ S/m, $A = 100$ cm^2, and $d = 2$ mm. [$d/\sigma A$ Ω, 50 mΩ]

3.7.3. A spherical capacitor is formed by two concentric spheres of radii a and b with $b > a$. The material between the spheres is a lossy dielectric with conductivity σ. Determine the resistance of this structure. Evaluate this expression for $a = 1$ mm, $b = 3$ mm, and $\sigma = 0.5$ S/m.

3.7.4. A coaxial cable formed by two concentric cylinders of radii a and b with $a < b$ is filled with a lossy dielectric having conductivity σ. Determine the per-unit-length resistance of this structure. Evaluate it for typical coaxial cable dimensions of $a = 16$ mils, $b = 58$ mils, and $\sigma = 2.05$ S/m. [$1/2\pi\sigma\ \ln(b/a)$ Ω/m, 0.1 Ω/m]

3.7.5. A 5-A current flows along a wire of length 10 m. If the wire is situated along the z axis of a rectangular coordinate system and centered at the origin, determine the magnetic flux density at (a) (0,3 m,0) and (b) (0,0,5 m).

3.7.6. An infinitely long current I is situated next to a rectangular loop as shown in Fig. P3.7.6. Determine an expression for the total magnetic flux that penetrates the loop. Evaluate that expression for $I = 10$ A, $l = 10$ cm, $r_1 = 2$ cm, and $r_2 = 4$ cm. Hint: Use the result of Example 3.16. [$\mu_o Il/2\pi\ \ln(r_2/r_1)$ Wb, 0.139 μWb]

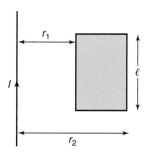

Figure P3.7.6 Problem 3.7.6.

3.7.7. A square loop with side dimensions l lies in the xy plane of a rectangular coordinate system and is centered about the origin. The loop carries a current I that circulates counterclockwise. Determine an expression for the magnetic flux density vector on the z axis. Evaluate that expression at the center of the loop if $l = 2$ m and $I = 10$ A. Hint: Use the result of Example 3.16.

3.7.8. An equilateral triangle of side lengths l is carrying a current I that circulates in the counterclockwise direction. Determine an expression for the magnetic flux density at the center of the triangle. Evaluate that expression for $l = 5$ cm and $I = 4$ A. Hint: Use the result of Example 3.16.

3.7.9. A portion of a circular loop depicted in Fig. P3.7.9 carries a current I. Determine the magnetic flux density at point P. Hint: Use the Biot-Savart law to superimpose the contributions. [$B = \mu_o I \theta / 4\pi \left(1/r_1 - 1/r_2 \right)$ directed into the page]

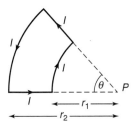

Figure P3.7.9   Problem 3.7.9.

SECTION 3.9 AMPERE'S LAW

3.9.1. Two infinitely long current filaments are parallel and separated a distance d. If the currents of the filaments are I but are oppositely directed, determine the magnetic field intensity vector (a) midway between the currents and (b) at a distance D from the center and in the plane containing the current filaments. Hint: Use the result of Example 3.19.

$$\left[\text{(a)} \ \frac{2I}{\pi d}, \text{(b)} \ H = \frac{I}{2\pi} \frac{d}{(D^2 - d^2/4)} \right]$$

3.9.2. An infinitely long strip of width w carries a linear current density K A/m directed along its length. Determine the magnetic field intensity vector at a distance d from its center and on a line perpendicular to the surface of the current strip. Hint: Treat the strip as filaments of current $K\,dz$ A and superimpose the fields using (3.81) of Example 3.19. You will need the integral $\int 1/(d^2 + z^2)dz = 1/d \tan^{-1}(z/d)$. Let $w \to \infty$ in that result and show that it reduces to the infinite plate result in Example 3.21, Equation (3.86).

3.9.3. A toroid is constructed of a ring of highly permeable material as shown in Fig. P3.9.3. The inner radius of the toroid is a and the outer radius is b and there are N turns wound on it. The windings are tightly wound so that all flux remains in the toroid. Determine an expression for the magnetic field intensity vector for (a) $r < a$, (b) $a < r < b$, and (c) $r > b$. Hint: Use Ampere's law and integrate around a circle of radius r.

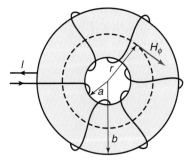

Figure P3.9.3   Problem 3.9.3.

3.9.4. An infinitely long solenoid shown in Fig. P3.9.4 consists of a central core of highly permeable material of radius a and relative permeability μ_r. There are n turns of wire per unit length carrying current I wound on it very tightly so that the magnetic flux is confined to the axial direction in the core. Determine the magnetic flux density in the core. Hint: Construct a rectangular, closed contour that passes along the solenoid a length L and outside the coil a length L and use Ampere's law. $[B = \mu_r\mu_o nI]$

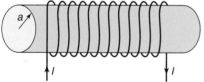

Figure P3.9.4   Problem 3.9.4.

SECTION 3.11 INDUCTANCE

3.11.1.   Determine the per-unit-length self inductance of the solenoid of Problem 3.9.4. Hint: Determine the flux linking one turn and recall that the inductance is the flux linkages per unit of current linking them. Determine the inductance per unit length for an iron ($\mu_r = 1000$) solenoid of radius 1 cm, with 20 turns per cm. $[l = \mu_r\mu_o n^2\pi a^2$ H/m, 1.58 H/m$]$

3.11.2.   A rectangular loop with sides of length l and w with $l \gg w$ carries a current I and is constructed of wires with radii a where $a \ll w$ as shown in Fig. P3.11.2. Determine the approximate inductance of this loop. Hint: Use the result obtained for the per-unit-length inductance of a pair of parallel wires obtained in Example 3.25.

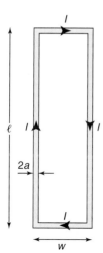

Figure P3.11.2   Problem 3.11.2.

3.11.3.   Determine the per-unit-length self inductance of a planar transmission line consisting of two plates of width w that are infinitely long and carry a linear current density K A/m that is equal and oppositely directed along the infinite dimension. The plate sides are facing, and the plates are separated a distance s. Determine the per-unit-length inductance of this transmission line, assuming that the plate width is much greater that the plate separation, $w \gg s$, so that we can neglect the fringing of the field at the edges and assume the magnetic field between the plates is constant along the width dimension. Hint: Use the result of Example 3.18 for the field of an infinite plate and superimpose the two fields. $[l = \mu_o s/w$ H/m$]$

3.11.4.   Mutual inductance is defined between two loops, one of which is carrying current I. The definition of mutual inductance between the two loops is the ratio of the total flux penetrating the second loop (caused by the current in the first loop) to the current of the first loop, $M = N_2\psi_2/I_1$, where the second loop has N_2 closely spaced loops and ψ_2 is the flux through one of those loops. A very long, parallel-wire transmission line composed of filamentary currents that are equal and oppositely directed and separated a distance s lies adjacent to a rectangular loop of width w and length l as shown in Fig. P3.11.4. The length of the loop, l, is assumed to be very small compared to the length of the transmission line. Both the transmission line and the loop lie in a plane and are separated center-to-center by distance D. Determine the mutual inductance between these two structures. Hint: Superimpose the magnetic flux penetrating the loop due to each of the currents of the line.

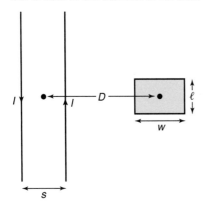

Figure P3.11.4   Problem 3.11.4.

SECTION 3.12 FORCES PRODUCED BY CHARGE AND CURRENT

3.12.1.   A charged particle moving with constant velocity v enters a region where electric field E is set up by vertical deflection plates. Horizontally mounted coils set up a magnetic field B perpendicular to the electric field as shown in Fig. P3.12.1. A metallic screen is located some distance away and a hole is cut into the screen on a line with the original particle velocity. Determine the velocity of charged particles that will pass through the hole. Evaluate this for $E = 2$ kV/m and $B = 1$ mW/m². $[v = E/B, v = 2 \times 10^6$ m/s$]$

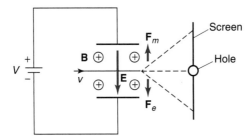

Figure P3.12.1   Problem 3.12.1.

3.12.2.   Two infinitely long wires carry currents I_1 and I_2 that are directed in the same direction and separated a distance s as shown in Fig. P3.12.2. Determine the force per unit length exerted on the wires. $[f = \mu_o I_1 I_2 / 2\pi s$ N/m, causing the wires to attract each other$]$

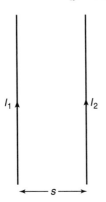

Figure P3.12.2   Problem 3.12.2.

3.12.3. An infinitely long wire carrying current I_1 is adjacent to a rectangular loop that carries a current I_2 as shown in Fig. P3.12.3. Determine the four forces exerted on the sides of the loop.

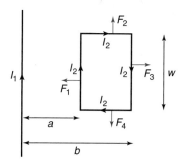

Figure P3.12.3 Problem 3.12.3.

3.12.4. A pair of conducting rails supports a conducting rod that is moving left to right with velocity v in a dc magnetic field B as shown in Fig. P3.12.4. Determine the current I through the resistor with polarity shown.

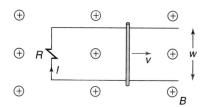

Figure P3.12.4 Problem 3.12.4.

3.12.5. A rectangular loop of width w and length l is rotating at a radian speed of ω radians/s about the z axis as shown in Fig. P3.12.5. A dc magnetic field, B, is directed in the y direction and is uniform across the xz plane. Determine the current through resistor R. $[I = -wl\omega\, B\sin(\omega t)/R]$

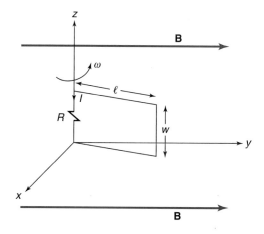

Figure P3.12.5 Problem 3.12.5.

CHAPTER 4

Time-Varying Electromagnetic Fields

In the previous chapter we examined the electromagnetic fields produced by charge distributions that are either fixed in position or are moving at a constant rate (a dc current). The electric and magnetic fields that are produced by these charges and their movement are referred to as static fields. We found that a fixed distribution of charge produced the electric field intensity vector, $\mathbf{E}$, whose units are V/m, and the electric flux density vector, $\mathbf{D}$, whose units are C/m^2. A dc current produces the remaining two field vectors; the magnetic field intensity vector, $\mathbf{H}$, whose units are A/m, and the magnetic flux density vector, $\mathbf{B}$, whose units are Wb/m^2 or equivalently tesla (T). The electric field vectors, $\mathbf{E}$ and $\mathbf{D}$, are independent of the magnetic field vectors, $\mathbf{H}$ and $\mathbf{B}$, except in a lossy medium. In a lossy medium, the current density, $\mathbf{J}$, is related to the electric field intensity vector by the conductivity of the medium as $\mathbf{J} = \sigma \mathbf{E}$. Hence in a lossy medium, $\mathbf{E}$ will produce a current $\mathbf{J}$, which will in turn produce the magnetic field. With this exception, the static electric and magnetic fields are independent of each other.

We now extend our study to the case of charge movement which is not steady, that is, a time-varying current. In this case we will find that the electric and magnetic fields are related; a time-varying magnetic field will produce an electric field and, conversely, a time-varying electric field will produce a magnetic field. This interdependence is the key to the production of electromagnetic *waves*. Radiation of electromagnetic waves is produced when charges are *accelerated*, that is, their velocity is not constant. The remainder of this text will be devoted to the study of those waves. In Chapter 5 we will study the general character of waves, and in Chapter 6 (Transmission Lines) and Chapter 7 (Antennas) we will study the guidance and propagation of those waves.

Chapter Learning Objectives

After completing the contents of this chapter, you should be able to

▶ compute currents and voltages that are induced in electric circuits by a time-varying magnetic field using Faraday's law,

▶ determine the magnetic flux density $\mathbf{B}$ (Wb/m^2 = T) for a given electric field intensity $\mathbf{E}$ (V/m) such that they satisfy Faraday's law,

▶ compute the displacement current caused by a time-varying electric field,

▶ determine the electric flux density $\mathbf{D}$ (C/m^2) for a given magnetic field intensity $\mathbf{H}$ (A/m) such that they satisfy Ampere's law,

▶ understand the meaning of Gauss' laws (unchanged from the static field case),

▷ verify whether the electric flux density **D** (C/m²) and the magnetic flux density **B** (Wb/m² = T) satisfy Gauss' laws,

▷ verify whether the current density **J** (A/m²) satisfies conservation of charge,

▷ determine the power density in an electromagnetic field from a given electric and magnetic field,

▷ determine the electric and magnetic field on one side of a boundary knowing the electric and magnetic fields on the other side of the boundary,

▷ replace an infinite "ground plane" with equivalent charge and current images to simplify a solution,

▷ explain the operation of a current probe.

▷▷ 4.1 FARADAY'S LAW

In the previous chapter we found that a fixed distribution of charge produced an electric field which has the important property that the line integral of the electric field intensity vector around a closed contour c yields a result of zero:

$$\oint_c \mathbf{E} \cdot d\mathbf{l} = 0 \tag{4.1}$$

and the surface integral of the electric flux density vector over a closed surface s yields the net positive charge enclosed by that surface:

$$\oint_s \mathbf{D} \cdot d\mathbf{s} = Q_{\text{enclosed}} \tag{4.2}$$

The first property in (4.1) provides that static electric fields are *conservative*; the net energy expended in moving a charge around a closed path is zero. This allowed the unique definition of voltage between two points as

$$V_{ba} = -\int_a^b \mathbf{E} \cdot d\mathbf{l} \tag{4.3}$$

The path chosen between points a and b is unimportant; we obtain the same result for any path between these two points. The second property of static fields in (4.2) essentially provides that electric field (flux) lines that begin on positive charge end on negative charge.

In this chapter we will find that when the charge distributions are varying with time, (4.1) must be modified as

$$\oint_c \mathbf{E} \cdot d\mathbf{l} = -\frac{d}{dt} \int_s \mathbf{B} \cdot d\mathbf{s} \tag{4.4}$$

This is *Faraday's law*. Observe that when the charge movement is not varying with time (either the charges are fixed in position or are moving at a steady rate), then the **B** field will be constant and the right-hand side is zero, reducing to our previous result in (4.1). Faraday's law is an incredibly powerful result. The generation, transmission, and distribution of electric power, to name one application, would not be possible without this.

Now let us interpret Faraday's law. Observe that the left side has the units of volts and is similar to voltage defined for static fields in (4.3). However, this is a *closed contour or loop, c.* Since the right-hand side is no longer zero, it indicates that *a voltage is induced in the closed loop c.* This has traditionally been called the *electromotive force* or *emf* in reference to its resemblance to a battery:

$$\text{emf} = \oint_c \mathbf{E} \cdot d\mathbf{l} \tag{4.5}$$

The right-hand side of Faraday's law is the *time rate-of-change of the magnetic flux penetrating the surface s whose perimeter is the contour c*:

$$\psi = \int_s \mathbf{B} \cdot d\mathbf{s} \tag{4.6}$$

Hence Faraday's law can be stated as

$$\boxed{\text{emf} = -\frac{d\psi}{dt}} \tag{4.7}$$

Figure 4.1a shows the interpretation of Faraday's law. The *closed* contour *c* and *open* surface *s* are directly related in that *c* bounds surface *s*. It is helpful to think of this as a balloon. The contour *c* is the mouth of the balloon and surface *s* is the surface of the balloon. Since surface *s* is an open surface, we must specify the direction of the flux of **B** through that surface. This is provided by the right-hand-rule. If we place the fingers of our right hand in the direction of contour *c*, the thumb will point in the direction of *s*, which indicates the direction of the flux "through the surface." The minus sign in Faraday's law is referred to as *Lenz's law*. Its significance will be discussed in the remaining paragraphs. But for the moment it is important to observe that *the same result will hold for any number of surfaces s so long as the contour c does not change.* Again this is like the balloon analogy: if we further inflate the balloon but keep the opening constant, we create any number of surface shapes. But Faraday's law gives the same result for all those shapes because the opening (contour *c*) did not change. This is a sensible result because it essentially states that *only those **B** field lines that penetrate the mouth of the surface, bounded by contour c, contribute to flux through surface s.* For example, in Fig. 4.1a we have shown a magnetic flux density line that enters the surface and exits it but does not pass through the mouth. Since it did not enter through the mouth, it contributes nothing to the net flux through the surface.

So Faraday's law essentially provides that *any time-varying magnetic flux that penetrates a surface s bounded by contour c will induce an emf in that contour which is very similar to a voltage source.* The negative sign (Lenz's law) provides that the direction of that induced emf (the induced source) will be to produce a current in contour *c* whose magnetic field will *oppose changes in* the original magnetic field. This induced emf may be inserted in the contour as though it were a voltage source. The key to properly enforcing Faraday's law lies in getting the value *and* polarity of the inserted source correct. To do this, we rewrite (4.4) by moving the negative sign to the left side and defining the voltage of the inserted source as

$$\boxed{\begin{aligned} V_F &= -\oint_c \mathbf{E} \cdot d\mathbf{l} \\ &= \frac{d\psi}{dt} \end{aligned}} \tag{4.8}$$

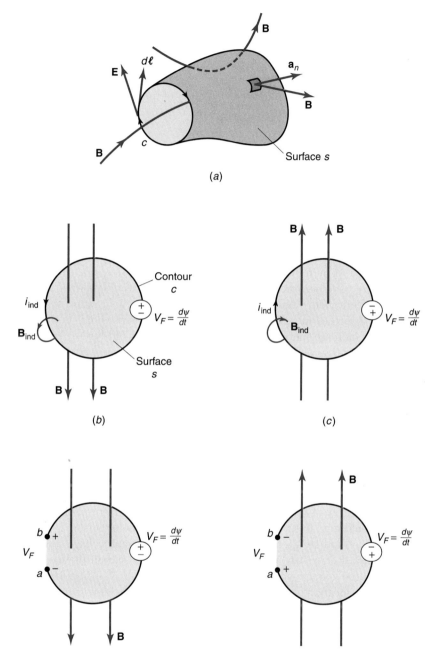

Figure 4.1   Illustration of Faraday's law. (a) General interpretation in terms of the relation of
the electric field induced along contour c due to time-changing magnetic field that penetrates
the surface s bounded by that contour. (b) Induced source in the contour for the magnetic
field directed downward. (c) Induced source in the contour for the magnetic field directed
upward. (d) The open-circuit voltage for the magnetic field directed downward. (e) The open-
circuit voltage for the magnetic field directed upward.

The subscript F is used to indicate "Faraday." Figure 4.1b and c show two examples of this. In Fig. 4.1b we show a circular contour and choose surface s to be the flat surface bounded by that contour. The magnetic field is directed downward. The induced voltage source is inserted in the loop and its value is the time rate-of-change of the flux through s. Observe that the negative sign in Lenz's law is omitted in the value of the source according to (4.8), but we will choose the polarity of the source to take that into account.

The polarity of the induced voltage source is chosen such that it would produce (induce) a current, i_{ind}, which would produce its own magnetic field, $\mathbf{B}_{ind}$, according to the right-hand-rule *which would act in opposition to the original magnetic field.*

Figure 4.1c similarly shows the case where the magnetic field is directed upward; only the polarity of the source is changed. This memory aid for getting the polarity of the source correct and thereby enforcing Lenz's law is a sensible one. For example, suppose the source were inserted with a polarity such that it would produce a current whose resulting magnetic field would aid the original magnetic field rather than opposing it. In this case, a further increase in the total magnetic field would be produced which would further increase the value of the source, which would further increase the total magnetic field, and so on. Hence conservation of energy would not take place.

The above memory aid for determining the correct polarity of the induced source was discussed in terms of generating a current whose magnetic field opposed the original magnetic field. In fact, Faraday's law shows that the source should produce a magnetic field that opposes the *time rate-of-change* of the original magnetic field. However, since the source that is inserted is $d\psi/dt$, then this time rate-of-change will be taken care of. Observe in Fig. 4.1b or c that if $\mathbf{B}$ is increasing with time, then the source *value* will be positive and will generate a current and resulting magnetic field which will be increasing and will oppose the *change* in the original field. On the other hand, suppose the magnetic field in Fig. 4.1b or c were decreasing with time. Then $d\psi/dt$ would be negative, resulting in changing the polarity of the induced source. The induced current would be opposite that shown and would tend to try to prevent the decrease in the original magnetic field.

The induced source, V_F, acts like a Thevenin open-circuit voltage studied in electric circuits courses. Figure 4.1d and e show that if we open-circuit the loops of Fig. 4.1b and c, the source voltage will appear across the terminals. It is important to point out that the precise location of this source in the loop cannot be determined. It simply represents the net accumulation of differential sources (emfs) around the loop.

Faraday's law explains the usual terminal relation for an inductor. Consider Fig. 4.1d and e again. Suppose this represents a loop of wire and hence a one-turn inductor. Also suppose that a current $i(t)$ passes around the loop, which causes the magnetic field $\mathbf{B}$. The terminal relation of the inductor

$$v(t) = L\frac{di(t)}{dt}$$

is essentially the induced Faraday source. To show this, we recall that inductance is the ratio of the magnetic flux penetrating the loop and the current that produces it:

$$L = \frac{\psi}{i}$$

Substituting this into the inductor terminal relation gives

$$v(t) = L\frac{d}{dt}\left(\frac{\psi}{L}\right)$$
$$= \frac{d\psi}{dt}$$
$$= V_F$$

which is the Faraday source.

The following examples illustrate the application of these principles and also demonstrate why voltage cannot be uniquely defined for time-varying applications.

▷ **EXAMPLE 4.1**

Figure 4.2a shows an electric circuit that has a magnetic field of $\mathbf{B} = 10t$ Wb/m^2 directed into the page that penetrates the circuit loop. Determine the voltages V_1 and V_2.

SOLUTION The magnetic flux that penetrates the circuit loop is

$$\psi = \int_s \mathbf{B} \cdot d\mathbf{s}$$
$$= 10t \text{ Wb/m}^2 \times 2 \text{ m}^2$$
$$= 20t \quad \text{Wb}$$

This is simply the product of $\mathbf{B}$ and the area of the loop, since the $\mathbf{B}$ field is assumed to be independent of position across the loop. If the value of the $\mathbf{B}$ field had depended on position at various points over the loop, we would have had to perform the integral to determine the flux. The magnitude of the induced source in the circuit loop is

$$V_F = \frac{d\psi}{dt}$$
$$= 20 \quad \text{V}$$

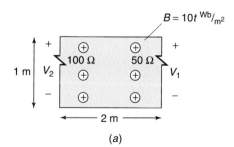

(a)

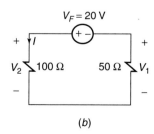

(b)

Figure 4.2 Example 4.1. (a) Physical dimensions of the circuit. (b) Replacing the magnetic field with the induced Faraday voltage source.

The source representing this emf is inserted with a polarity such that it will produce a current counterclockwise around the circuit so as to produce a magnetic field that opposes the change in the original magnetic field. This enforces Lenz's law. Now the problem becomes an ordinary circuit as shown in Fig. 4.2b wherein we can readily determine

$$I = \frac{20 \text{ V}}{100 \ \Omega + 50 \ \Omega}$$
$$= \frac{2}{15} \ \text{A}$$

Hence the voltages are

$$V_2 = 100 \ \Omega \times I$$
$$= 13.33 \ \text{V}$$

and

$$V_1 = -50 \ \Omega \times I$$
$$= -6.67 \ \text{V}$$

Observe that $V_1 \neq V_2$. In lumped-circuit analysis we handle this situation by saying that the **B** field is due to some adjacent circuit and its flux linking this circuit is represented by a mutual inductance between the two circuits. So the concept of mutual inductance is inherently rooted in Faraday's law. ◀

▶ EXAMPLE 4.2

Figure 4.3a shows a circuit wherein a high-impedance voltmeter that draws negligible current is attached across a resistor. In Fig. 4.3b the voltmeter is attached to the same two points but the voltmeter leads are routed differently. Determine the voltage measured by the voltmeter for these two cases. The magnetic field is directed out of the page.

SOLUTION For the case of Fig. 4.3a, the 2 m $\times$ 3 m circuit loop encloses a total magnetic flux of

$$\psi = \int_s \mathbf{B} \cdot d\mathbf{s}$$
$$= 5t^2 \text{ Wb/m}^2 \times 6 \text{ m}^2$$
$$= 30t^2$$

Hence the magnitude of the source induced in that loop is

$$V_F = \frac{d\psi}{dt}$$
$$= 60t \ \text{V}$$

The source representing this emf is inserted with polarity shown in Fig. 4.3c to enforce Lenz's law (the **B** field is out of the page). From that circuit we obtain

$$I = \frac{60t \text{ V}}{100 \ \Omega + 200 \ \Omega}$$
$$= 0.2t \ \text{A}$$

Hence the voltage measured is

$$V = 200I$$
$$= 40t \ \text{V}$$

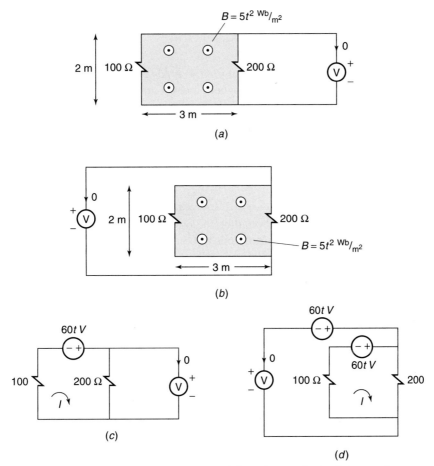

Figure 4.3 Example 4.2. (a) First position of the voltmeter leads. (b) Second position of the voltmeter leads. (c) Replacement of the magnetic field for the first position of the voltmeter leads with the induced Faraday voltage source. (d) Replacement of the magnetic field for the second position of the voltmeter leads with the induced Faraday voltage sources.

The equivalent circuit for Fig. 4.3b is shown in Fig. 4.3d. Observe that now the voltmeter leads also enclose the flux and another voltage source must be inserted in the loop formed by those leads. Hence we obtain

$$I = \frac{60t \text{ V}}{100 \ \Omega + 200 \ \Omega}$$
$$= 0.2t \quad \text{A}$$

but

$$V = 200I - 60t$$
$$= -20t \quad \text{V}$$

This can also be obtained by summing KVL around the inner loop of that circuit:

$$V = -60t + 60t - 100 \ \Omega \times I$$
$$= -100 \ \Omega \times I$$
$$= -20t \quad \text{V}$$

This has shown that for time-varying fields, the position of the voltmeter leads can affect the measured voltage, unlike the case for static fields, where the induced sources are zero because the magnetic fields do not change with time. ◀

Faraday's law as stated in general in (4.4) holds regardless of whether the contour c is stationary with respect to the magnetic field. Faraday's law can be separated into a portion for stationary loops and a portion for loops whose shape and/or position vary with time. For example, Faraday's law can be alternatively written, by taking the time derivative on the right-hand side under the surface integral, as[1]

$$\oint_c \mathbf{E} \cdot d\mathbf{l} = -\int_s \frac{\partial \mathbf{B}}{\partial t} \cdot d\mathbf{s} + \oint_c (\mathbf{v} \times \mathbf{B}) \cdot d\mathbf{l} \qquad (4.9)$$

where c is the closed contour bounding the open surface s. The first part of this is referred to as *transformer emf* for reasons to be discussed:

$$\text{transformer emf} = -\int_s \frac{\partial \mathbf{B}}{\partial t} \cdot d\mathbf{s} \qquad (4.10)$$

Observe that the transformer emf is zero if the magnetic field is constant, that is, not changing with time. The second contribution is called *motional emf*:

$$\text{motional emf} = \oint_c (\mathbf{v} \times \mathbf{B}) \cdot d\mathbf{l} \qquad (4.11)$$

where $\mathbf{v}$ is the velocity vector of the contour movement. The motional emf is zero if the contour that encloses the surface is not moving or changing shape with respect to the applied magnetic field. As in the general case of Faraday's law, these two items will result in two sources induced in the loop. Again, to define these sources we move the negative sign to the left in (4.9) to give

$$\begin{aligned} V_F &= -\oint_c \mathbf{E} \cdot d\mathbf{l} \\ &= V_F^t + V_F^m \end{aligned} \qquad (4.12)$$

where

$$V_F^t = \int_s \frac{\partial \mathbf{B}}{\partial t} \cdot d\mathbf{s} \qquad (4.13)$$

and

$$V_F^m = -\oint_c (\mathbf{v} \times \mathbf{B}) \cdot d\mathbf{l} \qquad (4.14)$$

where the superscripts t and m refer to *transformer* and *motional*, respectively.

---

[1]C.R. Paul and S.A. Nasar, *Introduction to Electromagnetic Fields*, 2nd ed., McGraw Hill, 1987.

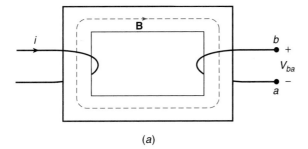

(a)

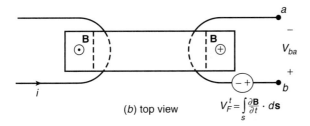

(b) top view $\qquad V_F^t = \int_s \frac{\partial \mathbf{B}}{\partial t} \cdot d\mathbf{s}$

Figure 4.4 Illustration of Faraday's law in the production of a voltage at the secondary of a transformer. (a) Side view showing the magnetic field circulating in the core. (b) Cross-sectional view for computing the magnetic flux linking the secondary.

The transformer induced source is so named because it represents the voltage induction in a transformer. Consider Fig. 4.4a, which shows a transformer having one turn on the "primary side" and one turn on the "secondary side." The core of the transformer is assumed to have a very large value of its permeability (theoretically infinite) so that all the magnetic field is confined to the core and none "leaks out" into the surrounding air. The current i in the primary produces a magnetic field $\mathbf{B}$ that circulates in the core clockwise. This magnetic field penetrates the surface enclosed by the secondary loop, generating a voltage in that loop of

$$V_F^t = \int_s \frac{\partial \mathbf{B}}{\partial t} \cdot d\mathbf{s}$$

where surface s is the cross section of the core that the secondary loop surrounds as shown in Fig. 4.4b. Observe that the polarity of the induced source in the secondary is chosen in that same manner as before: it tends to produce a current in the secondary loop that would produce a magnetic field that opposes the original magnetic field (generated by the current in the primary loop). Hence $V_{ba} = V_F^t$. The directions of the windings relative to each other are important. If the winding on the secondary side were opposite that shown, then the induced source polarity would be opposite that shown. This is taken care of with the "dot convention," where dots on the transformer equivalent circuit in a lumped-circuit model of the transformer denote the polarity of the induced terminal voltage. If the primary consisted of N_1 turns and the secondary consisted of N_2 turns, then the flux in the core would be N_1 times that for one turn and the flux in the secondary would be N_2 times that flux, resulting in an induced voltage $N_1 N_2$ times larger than the induced voltage for one turn on each winding. Observe that when the secondary loop is not moving with respect to the magnetic field, as is the case here,

$$\frac{d}{dt} \int_s \mathbf{B} \cdot d\mathbf{s} = \int_s \frac{\partial \mathbf{B}}{\partial t} \cdot d\mathbf{s}$$

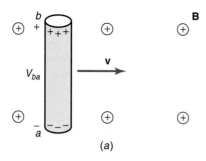

(a)

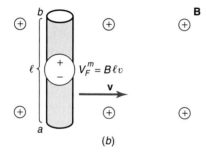

(b)

Figure 4.5 Illustration of motional emf. (a) The wire "cuts" a magnetic field. (b) A voltage source representing this is inserted in the wire with value and polarity shown.

The motional emf provides that if any portion of contour c is moving with respect to the **B** field, a voltage source representing this emf will be induced in the contour. This motional emf was discussed in Section 3.12 of the previous chapter. Here we review that result. Consider Fig. 4.5a, which shows a conductor such as a metallic wire that is *cutting magnetic field lines*. The Lorentz force equation discussed in Section 3.12 of the previous chapter is repeated here:

$$\boxed{\mathbf{F} = q\mathbf{E} + q\mathbf{v} \times \mathbf{B}} \tag{4.15}$$

A charge, q, moving with velocity **v** will experience a force at right angles to the plane containing **v** and **B** according to the right-hand-rule. Positive charge will be forced to the upper end of the wire and negative charge will be forced to the lower end, generating a voltage V_{ba} between the two ends of the wire. (In a metallic conductor, only electrons are free to move but the result is the same.) Hence, it will appear to an observer moving with the conductor that an electric field was responsible for this charge movement:

$$\mathbf{E}_m = \frac{\mathbf{F}}{q}$$

$$= \frac{q\mathbf{v} \times \mathbf{B}}{q}$$

$$= \mathbf{v} \times \mathbf{B}$$

Hence a source would be inserted in the wire length as shown, representing this charge separation as

$$V_F^m = -\int_c \mathbf{E}_m \cdot d\mathbf{l}$$

$$= -\int_c (\mathbf{v} \times \mathbf{B}) \cdot d\mathbf{l}$$

This source is inserted into any portion of the contour c that *cuts across magnetic field lines*.

Determining the value and polarity of the inserted voltage source can be accomplished in the following manner. A frequent case is that of a straight conductor of length l that is moving with velocity v and the conductor; the velocity of propagation and the magnetic field vector are mutually perpendicular as is illustrated in Fig. 4.5b. The value of the inserted voltage source is

$$\boxed{V_F^m = Blv}$$

The polarity of the source is found with the right-hand-rule such that $\mathbf{v} \times \mathbf{B}$ points to the positive terminal.

▶ **EXAMPLE 4.3**

Figure 4.6a shows a metallic bar moving to the right with velocity v along two parallel, conducting rails that are separated by width W. A magnetic field B is perpendicular to the loop formed by the rails and the bar. Determine the induced voltage V_{ba} for the following cases: (a) $B = 2$ Wb/m^2 and $v = 5$ m/s, (b) $B = 2t$ Wb/m^2 and $v = 5$ m/s, and (c) $B = 2t$ Wb/m^2 and $v = 5t$ m/s.

SOLUTION In case (a), both the B field and the velocity of movement of the bar are constant, whereas in case (b) the B field is time varying but the velocity of the bar is constant, and in case (c) the B field is time varying, as is the velocity of the bar. We will compute the voltage induced at the terminals in each of these cases two ways to obtain the same result. The first way is the direct use of the general statement of Faraday's law by inserting the Faraday source given in (4.8), and the second way is in terms of the transformer source in (4.13) and motional source in (4.14).

Case (a): $B = 2$ Wb/m^2 and $v = 5$ m/s. The distance the bar has moved is $L = vt$, where we assume that the bar is at the left end at $t = 0$. The area enclosed by the contour consisting of the rails and the bar is $W \times L = Wvt$. Hence the induced voltage source using (4.8) gives

$$V_F = \frac{d}{dt} \int_s \mathbf{B} \cdot d\mathbf{s}$$

$$= \frac{d}{dt}(BWvt)$$

$$= BWv$$

$$= 10W$$

The result is a source inserted in the loop as shown in Fig. 4.6b, giving

$$V_{ba} = V_F$$
$$= 10W$$

Alternatively, since $\mathbf{B}$ is constant, the transformer source in (4.13) is zero and the motional source is

$$V_F^m = Blv$$
$$= BWv$$
$$= 10W$$

The result is a source inserted in the bar as shown in Fig. 4.6b, once again giving

$$V_{ba} = V_F^m$$
$$= 10W$$

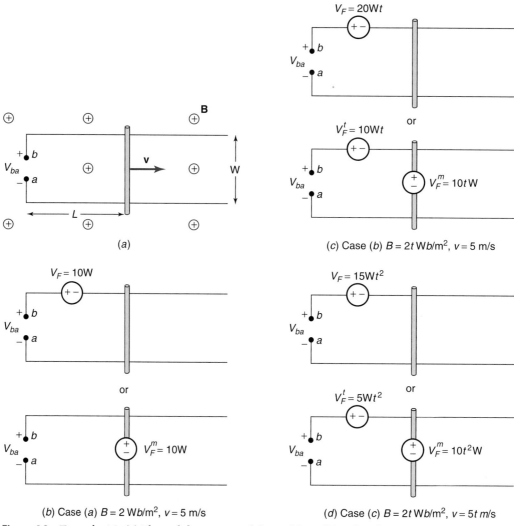

(a)

(b) Case (a) $B = 2$ Wb/m², $v = 5$ m/s

(c) Case (b) $B = 2t$ Wb/m², $v = 5$ m/s

(d) Case (c) $B = 2t$ Wb/m², $v = 5t$ m/s

Figure 4.6 Example 4.3. (a) Physical dimensions of the problem. (b) Induced sources by two methods for constant magnetic field and constant velocity of the moving rail [case (a)]. (c) Induced sources by two methods for a time-varying magnetic field and constant velocity of the moving rail [case (b)]. (d) Induced sources by two methods for a time-varying magnetic field and a time-varying velocity of the moving rail [case (c)].

Case (b): $B = 2t$ Wb/m² and $v = 5$ m/s. Again, the distance the bar has moved is $L = vt$ and the area enclosed by the contour consisting of the rails and the bar is $W \times L = Wvt$. The Faraday source is

$$V_F = \frac{d}{dt} \int_s \mathbf{B} \cdot d\mathbf{s}$$

$$= \frac{d}{dt}(BWvt)$$

$$= \frac{d}{dt}(10Wt^2)$$

$$= 20Wt$$

resulting in an induced source as shown in Fig. 4.6c, giving the voltage at the terminals as

$$V_{ba} = V_F$$
$$= 20Wt$$

Alternatively we may determine

$$V_F^t = \int_s \frac{\partial \mathbf{B}}{\partial t} \cdot d\mathbf{s}$$

$$= \int_s \frac{\partial 2t}{\partial t} ds$$

$$= 2Wvt$$
$$= 10Wt$$

The induced source due to motion of the bar is

$$V_F^m = Blv$$
$$= BWv$$
$$= 10tW$$

These two sources are inserted as shown in the alternative equivalent circuit, giving again

$$V_{ba} = V_F^t + V_F^m$$
$$= 10Wt + 10tW$$
$$= 20Wt$$

Case (c): $B = 2t$ Wb/m^2 and $v = 5t$ m/s. Because the velocity is time varying, we determine the distance moved by the bar in time interval t from

$$L = \int_0^t vdt$$

$$= \int_0^t 5tdt$$

$$= \frac{5t^2}{2}$$

The Faraday source is

$$V_F = \frac{d}{dt} \int_s \mathbf{B} \cdot d\mathbf{s}$$

$$= \frac{d}{dt}(BWL)$$

$$= \frac{d}{dt}(5Wt^3)$$

$$= 15Wt^2$$

resulting in an induced source as shown in Fig. 4.6d, giving the voltage at the terminals as

$$V_{ba} = V_F$$
$$= 15Wt^2$$

Alternatively we may determine

$$V_F^t = \int_s \frac{\partial \mathbf{B}}{\partial t} \cdot d\mathbf{s}$$

$$= \int_s \frac{\partial 2t}{\partial t} ds$$

$$= 2WL$$

$$= 5Wt^2$$

The induced source due to motion of the bar is

$$V_F^m = Blv$$

$$= BWv$$

$$= 10t^2 W$$

These two sources are inserted as shown in the alternative equivalent circuit, giving again

$$V_{ba} = V_F^t + V_F^m$$

$$= 5Wt^2 + 10t^2 W$$

$$= 15Wt^2$$

Whether one uses the complete Faraday source given by (4.8) or separates it into the sum of the transformer source in (4.13) and the motion source in (4.14) is somewhat arbitrary: both methods give the same result. However, in some problems, such as the electric generator in Section 3.13.6 of the previous chapter where the **B** field is constant but there is motion of the contour, it may be simpler to compute only the motional source in (4.14) as was done in that example. To show the equivalence, we will rework that problem using the Faraday source in (4.8).

▶ EXAMPLE 4.4

With reference to Fig. 3.57 of the previous chapter, determine the induced voltage using (4.8).

SOLUTION   The rotating loop is redrawn in Fig. 4.7a. The flux through the loop with direction shown is

$$\psi = \int_s \mathbf{B} \cdot d\mathbf{s}$$

$$= Bwl \cos(\theta)$$

and

$$\theta = \omega t$$

The induced Faraday source using (4.8) is

$$V_F = \frac{d\psi}{dt}$$

$$= -\omega Bwl \sin(\omega t)$$

Hence the voltage induced at the terminals with polarity shown in Fig. 4.7b is

$$V = -V_F$$

$$= \omega BA \sin(\omega t)$$

where $A = wl$ is the area of the loop, as was obtained using the motional emf source in Section 3.13.6 of the previous chapter.

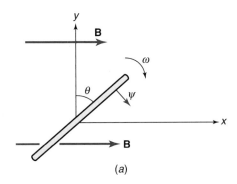

(a)

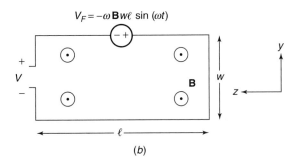

$V_F = -\omega \mathbf{B} w \ell \sin(\omega t)$

(b)

Figure 4.7 Example 4.4; the electric generator of Chapter 3 revisited. (a) Cross-sectional view of the rotating loop. (b) The induced voltage source via Faraday's law.

▶ **QUICK REVIEW EXERCISE 4.1**

In a laboratory experiment, a high-impedance voltmeter is attached across the parallel combination of two resistors as shown in Fig. E4.1. Nearby a 60-Hz power transformer causes a magnetic flux to penetrate the circuit as shown. Determine the voltage read by the voltmeter.

ANSWER $0.88 \sin(120\pi t)$ mV.

◀

▶ **QUICK REVIEW EXERCISE 4.2**

A square loop is moving to the right as shown in Fig. E4.2. A magnetic field, directed into the page, covers the region 2 m in width. Determine the current in the loop versus time, assuming that the right side of the loop enters the magnetic field region at $t = 0$.

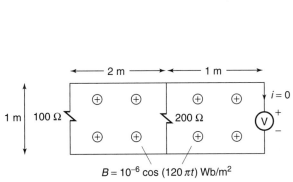

$B = 10^{-6} \cos(120\pi t)$ Wb/m²

Figure E4.1   Quick Review Exercise 4.1.

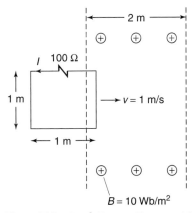

$B = 10$ Wb/m²

Figure E4.2   Quick Review Exercise 4.2.

ANSWER 0.1 A for $0 < t < 1\,\text{s}$, 0 A for $1\,\text{s} < t < 2\,\text{s}$, $-0.1\,\text{A}$ for $2\,\text{s} < t < 3\,\text{s}$, 0 A for $3\,\text{s} < t$. ◀

4.1.1 Faraday's Law in Point Form

Faraday's law has been presented in (4.4) in its *integral form*, which applies to broad regions of space. We will now develop Faraday's law in its *point form* or *differential form*, which applies at discrete points in space. The integral form is most useful in providing insight into the meaning of Faraday's law, whereas the point form is useful in performing computations.

We begin by applying Faraday's law to a rectangular contour in the yz plane as shown in Fig. 4.8. The sides of this rectangular area are of differential lengths Δz and Δy. The intent here is to write Faraday's law in integral form around this contour and then let the area bounded by the contour shrink to zero. This will produce Faraday's law in point form (at the center of the differentially small area). First observe that we have shown the value of the **E** field tangent to the four sides. For example, the value of the **E** field along the left side is E_z but the value along the right side is approximately $E_z + \dfrac{\partial E_z}{\partial y}\bigg|_{y+\Delta y} \Delta y$. The left-hand side of Faraday's law becomes

$$\oint_c \mathbf{E} \cdot d\mathbf{l} = E_y \Delta y + \left(E_z + \frac{\partial E_z}{\partial y}\Delta y \right)\Delta z - \left(E_y + \frac{\partial E_y}{\partial z}\Delta z \right)\Delta y - E_z \Delta z$$

In evaluating this line integral we must observe the direction of each portion of the contour relative to the direction of the **E** field along that portion. After canceling some terms this becomes

$$\oint_c \mathbf{E} \cdot d\mathbf{l} = \frac{\partial E_z}{\partial y}\Delta y\,\Delta z - \frac{\partial E_y}{\partial z}\Delta z\,\Delta y$$

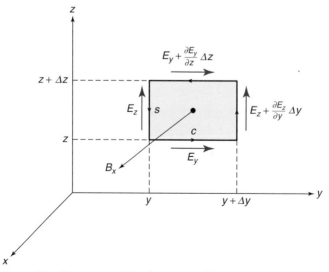

Figure 4.8 Illustration of the derivation of the point form of Faraday's law.

The flux through the enclosed surface (according to the right-hand-rule) is

$$\int_s \mathbf{B} \cdot d\mathbf{s} = B_x \Delta y \Delta z$$

In determining this flux, it is recalled that we will eventually let this surface shrink to zero and hence can simply multiply the B field that is normal to the surface, B_x, by the surface area even though the B field may be varying across that surface. Now we let the surface shrink to zero. In doing so, we first divide both sides of Faraday's law by the surface area, $\Delta s = \Delta y \Delta z$, and take the limit as that surface shrinks to zero, giving

$$\lim_{\Delta s \to 0} \frac{\oint_c \mathbf{E} \cdot d\mathbf{l}}{\Delta y \Delta z} = \frac{\partial E_z}{\partial y} - \frac{\partial E_y}{\partial z} \tag{4.16a}$$

and

$$\lim_{\Delta s \to 0} \frac{\int_s \mathbf{B} \cdot d\mathbf{s}}{\Delta y \Delta z} = B_x \tag{4.16b}$$

Hence, Faraday's law in (4.4) becomes, for this contour and surface in the yz plane,

$$\left(\frac{\partial E_z}{\partial y} - \frac{\partial E_y}{\partial z} \right) = -\frac{\partial B_x}{\partial t} \tag{4.17a}$$

Notice the addition of the negative sign and the time derivative required by the right-hand side of Faraday's law. Similar results are obtained for contours in the xy and xz planes:

$$\left(\frac{\partial E_x}{\partial z} - \frac{\partial E_z}{\partial x} \right) = -\frac{\partial B_y}{\partial t} \tag{4.17b}$$

$$\left(\frac{\partial E_y}{\partial x} - \frac{\partial E_x}{\partial y} \right) = -\frac{\partial B_z}{\partial t} \tag{4.17c}$$

Collecting these three items in a vector form gives

$$\left(\frac{\partial E_z}{\partial y} - \frac{\partial E_y}{\partial z} \right)\mathbf{a}_x + \left(\frac{\partial E_x}{\partial z} - \frac{\partial E_z}{\partial x} \right)\mathbf{a}_y + \left(\frac{\partial E_y}{\partial x} - \frac{\partial E_x}{\partial y} \right)\mathbf{a}_z = -\frac{\partial B_x}{\partial t}\mathbf{a}_x - \frac{\partial B_y}{\partial t}\mathbf{a}_y - \frac{\partial B_z}{\partial t}\mathbf{a}_z \tag{4.18}$$

This gives Faraday's law in *point or differential form* as

$$\boxed{\nabla \times \mathbf{E} = -\frac{\partial \mathbf{B}}{\partial t}} \tag{4.19a}$$

where

$$\boxed{\nabla \times \mathbf{E} = \left(\frac{\partial E_z}{\partial y} - \frac{\partial E_y}{\partial z} \right)\mathbf{a}_x + \left(\frac{\partial E_x}{\partial z} - \frac{\partial E_z}{\partial x} \right)\mathbf{a}_y + \left(\frac{\partial E_y}{\partial x} - \frac{\partial E_x}{\partial y} \right)\mathbf{a}_z} \tag{4.19b}$$

This is written *symbolically* using the "del" operator:

$$\nabla = \frac{\partial}{\partial x}\mathbf{a}_x + \frac{\partial}{\partial y}\mathbf{a}_y + \frac{\partial}{\partial z}\mathbf{a}_z \tag{4.20}$$

This may be formed using the two methods for the cross product discussed in Chapter 2. The first method is to use the determinant mnemonic for evaluating the cross product:

$$\nabla \times \mathbf{E} = \begin{vmatrix} \mathbf{a}_x & \mathbf{a}_y & \mathbf{a}_z \\ \dfrac{\partial}{\partial x} & \dfrac{\partial}{\partial y} & \dfrac{\partial}{\partial z} \\ E_x & E_y & E_z \end{vmatrix} \tag{4.21}$$

The second method (which the author prefers) is to use the cyclic ordering of the axis labels: $x \to y \to z \to x \to \cdots$. Each component is formed as

$$\left(\frac{\partial E_\gamma}{\partial \beta} - \frac{\partial E_\beta}{\partial \gamma} \right)\mathbf{a}_\alpha$$

where the axis labels are cyclic as $\cdots \to \alpha \to \beta \to \gamma \to \cdots$.

The notation "del cross $\mathbf{E}$," $\nabla \times \mathbf{E}$, is said to be the *curl of* $\mathbf{E}$ and is the *circulation of* $\mathbf{E}$ *per unit of surface enclosed*:

$$\nabla \times \mathbf{E} = \lim_{\Delta s \to 0} \frac{\displaystyle\oint_c \mathbf{E} \cdot d\mathbf{l}}{\Delta s} \tag{4.22}$$

The line integral of $\mathbf{E}$ in the numerator amounts to adding the circulation or vorticity of the $\mathbf{E}$ field around the contour. This is like swirls or eddies in a river that indicate circulation of the water about a point. The point form of Faraday's law in (4.19) essentially shows that *a time-varying* $\mathbf{B}$ *field will result in (produce) an* $\mathbf{E}$ *field that circulates about it*. Figure 4.9 illustrates this via a paddle wheel in a fluid. The flow rate of the fluid is represented as the vector $\mathbf{F}$. In Fig. 4.9a, the flow rate is faster at the top,

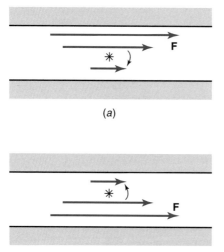

(a)

(b)

Figure 4.9 Illustration of the curl of a vector field with a paddle wheel in a moving fluid.

so that the paddle wheel rotates in the clockwise direction, indicating *circulation* of the fluid in that direction. The larger the disparity in the flow rates from top to bottom, the faster the paddle wheel turns. Similarly, in Fig. 4.9b the flow rate is greater at the bottom, indicating *circulation* in the counterclockwise direction. Hence the paddle wheel turns in the counterclockwise direction. There are three vector components to the curl, which gives the net vector direction of the circulation, which is normal to the plane of circulation, much like the axle of the paddle wheel.

For static fields such as fixed distributions of charges or constant currents, the magnetic field is constant and hence the right-hand side of Faraday's law is zero, giving $\nabla \times \mathbf{E} = 0$. This symbolizes that the static electric field has no circulation or vorticity. This is logical to expect for electric fields that are produced by charges. Observe that for a point charge the electric field lines are radially directed away from the point charge and as such have no circulation. From this we may say that electric field lines that begin on positive charge end on negative charge. But Faraday's law shows that other electric field lines that are caused by a time-changing magnetic field form closed loops.

▷ **EXAMPLE 4.5**

As indicated earlier, the point form of Faraday's law is most useful in computations. To illustrate this, determine the magnetic field if the electric field is given by

$$\mathbf{E} = E_m \cos(\omega t - \beta z)\mathbf{a}_x$$

SOLUTION   The electric field has only an x component. Hence the left side of Faraday's law in (4.19b) becomes

$$\nabla \times \mathbf{E} = \left(\frac{\partial E_z}{\partial y} - \frac{\partial E_y}{\partial z}\right)\mathbf{a}_x + \left(\frac{\partial E_x}{\partial z} - \frac{\partial E_z}{\partial x}\right)\mathbf{a}_y + \left(\frac{\partial E_y}{\partial x} - \frac{\partial E_x}{\partial y}\right)\mathbf{a}_z$$

$$= \frac{\partial E_x}{\partial z}\mathbf{a}_y - \underbrace{\frac{\partial E_x}{\partial y}}_{0}\mathbf{a}_z$$

$$= \frac{\partial E_x}{\partial z}\mathbf{a}_y$$

Observe that this x component of $\mathbf{E}$ depends only on z, so that the z component, $\partial E_x/\partial y$, is zero. Hence

$$\nabla \times \mathbf{E} = \frac{\partial E_x}{\partial z}\mathbf{a}_y$$

$$= \beta E_m \sin(\omega t - \beta z)\mathbf{a}_y$$

Since the left-hand side of Faraday's law has only a y component, the right-hand side, $-\partial \mathbf{B}/\partial t$ can have only a y component. Therefore we obtain

$$\beta E_m \sin(\omega t - \beta z) = -\frac{\partial B_y}{\partial t}$$

Integrating this gives

$$B_y = \frac{\beta E_m}{\omega}\cos(\omega t - \beta z)$$

Therefore the magnetic flux density vector that satisfies Faraday's law is

$$\mathbf{B} = \frac{\beta E_m}{\omega}\cos(\omega t - \beta z)\mathbf{a}_y$$

◀

▶ **QUICK REVIEW EXERCISE 4.3**

Determine whether the following fields satisfy Faraday's law: $\mathbf{E} = E_m \sin x \sin t \, \mathbf{a}_y$, $\mathbf{B} = E_m \cos x \cos t \, \mathbf{a}_z$.

ANSWER Yes.

The above development gives the point form in a rectangular coordinate system. In the cylindrical and spherical coordinate systems, Faraday's law in point form is still written *symbolically* as

$$\nabla \times \mathbf{E} = -\frac{\partial \mathbf{B}}{\partial t} \tag{4.23}$$

but $\nabla \times \mathbf{E}$ is not so simply obtained. It turns out that in cylindrical coordinates the curl is

$$\nabla \times \mathbf{E} = \left(\frac{1}{r}\frac{\partial E_z}{\partial \phi} - \frac{\partial E_\phi}{\partial z}\right)\mathbf{a}_r + \left(\frac{\partial E_r}{\partial z} - \frac{\partial E_z}{\partial r}\right)\mathbf{a}_\phi + \left(\frac{1}{r}\frac{\partial(rE_\phi)}{\partial r} - \frac{1}{r}\frac{\partial E_r}{\partial \phi}\right)\mathbf{a}_z \tag{4.24}$$

and in spherical coordinates the curl is

$$\nabla \times \mathbf{E} = \left(\frac{1}{r\sin\theta}\left[\frac{\partial(\sin\theta E_\phi)}{\partial \theta} - \frac{\partial E_\theta}{\partial \phi}\right]\right)\mathbf{a}_r + \left(\frac{1}{r\sin\theta}\frac{\partial E_r}{\partial \phi} - \frac{1}{r}\frac{\partial(rE_\phi)}{\partial r}\right)\mathbf{a}_\theta$$
$$+ \left(\frac{1}{r}\left[\frac{\partial(rE_\theta)}{\partial r} - \frac{\partial E_r}{\partial \theta}\right]\right)\mathbf{a}_\phi \tag{4.25}$$

▶ 4.2 AMPERE'S LAW

Ampere's law, studied in the previous chapter for dc or static magnetic fields, was stated as

$$\oint_c \mathbf{H} \cdot d\mathbf{l} = I_{enclosed}$$

where

$$I_{enclosed} = \int_s \mathbf{J} \cdot d\mathbf{s}$$

For time-varying fields this must be modified as

$$\boxed{\oint_c \mathbf{H} \cdot d\mathbf{l} = \int_s \mathbf{J} \cdot d\mathbf{s} + \frac{\partial}{\partial t}\int_s \mathbf{D} \cdot d\mathbf{s}} \tag{4.26}$$

The term on the left is referred to as the magnetomotive force or mmf $= \oint_c \mathbf{H} \cdot d\mathbf{l}$ in the same fashion as the corresponding electromotive force or emf $= \oint_c \mathbf{E} \cdot d\mathbf{l}$ on the left-hand side of Faraday's law. The first term on the right-hand side is *conduction current*:

$$I_{conduction} = \int_s \mathbf{J} \cdot d\mathbf{s} \tag{4.27a}$$

and the second term on the right-hand side is called *displacement current*:

$$I_{\text{displacement}} = \frac{\partial}{\partial t} \int_s \mathbf{D} \cdot d\mathbf{s} \tag{4.27b}$$

Hence this modified Ampere's law can be written as

$$\oint_c \mathbf{H} \cdot d\mathbf{l} = I_{\text{conduction}} + I_{\text{displacement}} \tag{4.28}$$

One of James Clerk Maxwell's primary contributions to electromagnetism was the addition of the displacement current term to the existing Ampere's law of that time, which contained only conduction current. Observe that for static (dc) conditions, the displacement current term is zero and hence the revised Ampere's law reduces to the static version.

Observe that the right-hand side of Ampere's law contains the sum of two currents. The conduction current term is simply the flow of free charges, which is a well-recognized form of current. However, the displacement current term indicates that a time-changing electric field or electric flux density can also serve precisely the same role as conduction current. Figure 4.10 shows the interpretation of Ampere's law. As with Faraday's law, the contour c and surface s are related with the right-hand-rule. Placing the fingers of our right hand in the direction of the closed contour c, the thumb will point in the direction of net flux of $\mathbf{J}$ and $\partial \mathbf{D}/\partial t$ through that surface. The remarkable aspect of this revised Ampere's law is that it indicates that a magnetic field, $\mathbf{H}$, can be produced by either true conduction current $\mathbf{J}$ that passes through the enclosed surface or by a time rate-of-change of $\mathbf{D}$ through that surface. So a time-changing electric field can produce a magnetic field in the same fashion as in Faraday's law, where a time-changing magnetic field can produce an electric field. This addition of the displacement current term to Ampere's static law by Maxwell allows, as we will soon see, the combination of Faraday's law and Ampere's law to predict the existence of electromagnetic waves. This revelation occurred when Maxwell published his work, which was verified, experimentally, by Heinrich Hertz in 1887.

As a simple example of the application of this result, consider a capacitor having a sinusoidal voltage source attached to its terminals as shown in Fig. 4.11. The wires attached to the capacitor carry free charges, resulting in a conduction current, I_c. Between

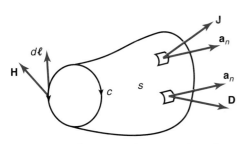

Figure 4.10 Illustration of Ampere's law where magnetic field is induced in a contour c by conduction current and time-varying electric field (displacement current) passing through surface s that is bounded by the contour.

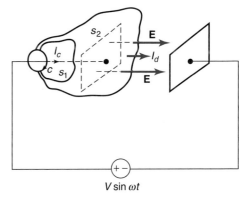

Figure 4.11 Illustration of Ampere's law and conduction current and displacement current for a capacitor.

the plates of the capacitor a time-varying electric field (varying at the rate of the source, ω) is directed between the plates. Let us apply Ampere's law given in (4.26) to this situation. Placing the contour about the wire, we obtain several surfaces bounded by that contour. Again, it is helpful to visualize this as a balloon. If we blow up the balloon so that only the wire penetrates the surface, we obtain

$$\oint_c \mathbf{H} \cdot d\mathbf{l} = \int_{S_1} \mathbf{J} \cdot d\mathbf{s}$$

$$= I_c$$

If we keep the same contour but blow the balloon up so that its surface passes between the capacitor plates, we obtain

$$\oint_c \mathbf{H} \cdot d\mathbf{l} = \frac{\partial}{\partial t} \int_{S_2} \mathbf{D} \cdot d\mathbf{s}$$

$$= \varepsilon \frac{\partial}{\partial t} \int_{S_2} \mathbf{E} \cdot d\mathbf{s}$$

$$= I_d$$

and we have substituted $\mathbf{D} = \varepsilon\mathbf{E}$ where the material between the capacitor plates has permittivity ε. If it were not for the addition of the displacement current term to Ampere's law, we would have an obvious problem here. For the same contour c, two different surfaces would yield different results.

▶ **EXAMPLE 4.6**

In the capacitor situation of Fig. 4.11, a 1-μF capacitor has a sinusoidal voltage source $10\sin(\omega t)$ V applied across its terminals, where the frequency of the source is 1 kHz. Verify that the conduction and displacement currents are equal.

SOLUTION   From our electric circuits course, the conduction current in the attachment wires is

$$I_c = \frac{10\text{ V}}{1/\omega C}$$

$$= 62.8 \quad \text{mA}$$

The capacitance of a parallel-plate capacitor is approximately $C = \varepsilon A/d$, where A is the plate area and d is the plate separation distance. The electric field between the plates is approximately the applied voltage divided by the plate separation distance:

$$E = \frac{10\text{ V}}{d}$$

Hence

$$D = \varepsilon E$$

$$= \varepsilon \frac{10\text{ V}}{d}$$

$$= \frac{C}{A} 10\text{ V}$$

$$= \frac{10^{-5}}{A}$$

Hence the displacement current is

$$I_d = \frac{\partial}{\partial t} \int_s \mathbf{D} \cdot d\mathbf{s}$$

$$= \omega \left(\frac{10^{-5}}{A} A \right)$$

$$= 62.8 \text{ mA}$$

Therefore, the displacement current through a capacitor may be symbolized as a current source between the two plates whose value is the displacement current: $I_d = C(dV/dt)$. This is quite similar to the inductor, wherein the terminal relation $V = L(dI/dt)$ amounts to the Faraday law voltage source which may be inserted in the inductor loop. ◀

▶ EXAMPLE 4.7

Compare the conduction and displacement currents in a copper conductor at a frequency of 1 GHz. The conductivity of copper is 5.8×10^7 S/m. The permittivity of copper, like most metals, is that of free space: $\varepsilon = \varepsilon_o \cong 1/36\pi \times 10^{-9}$.

SOLUTION   The conduction current density is

$$J = \sigma E$$

The displacement current density is

$$\varepsilon_o \frac{\partial E}{\partial t} = \omega \varepsilon_o E$$

Hence the ratio of conduction current to displacement current is

$$\frac{I_c}{I_d} = \frac{\sigma}{\omega \varepsilon_o}$$

$$= 1.04 \times 10^9$$

Therefore in copper and most other metals the conduction current is many orders of magnitude larger than the displacement current. This is why displacement current can be neglected in conductors. ◀

▶ QUICK REVIEW EXERCISE 4.4

Determine the ratio of conduction current to displacement current in seawater at 1 kHz. Seawater has $\varepsilon_r \cong 80$ and $\sigma \cong 4$ at this low frequency.

ANSWER 9×10^5. ◀

4.2.1   Ampere's Law in Point Form

Compare Faraday's law in integral form, given in (4.4), to Ampere's law in integral form in (4.26). There is a considerable amount of *duality* between the two. For example, if we interchange **E** and **H** and interchange **B** and **D**, we obtain very similar forms of equations. Hence we can see what the point form of Ampere's law will become if we carry through a similar development leading to the point form of

Faraday's law:

$$\boxed{\nabla \times \mathbf{H} = \mathbf{J} + \frac{\partial \mathbf{D}}{\partial t}} \tag{4.29}$$

As before, the curl of **H** is the circulation per unit of surface area:

$$\nabla \times \mathbf{H} = \lim_{\Delta s \to 0} \frac{\oint_c \mathbf{H} \cdot d\mathbf{l}}{\Delta s} \tag{4.30}$$

Hence the point form of Ampere's law indicates that current density **J** or the time rate-of-change of **D** will produce a *circulation* of **H** around that point.

▶ **EXAMPLE 4.8**

The magnetic field in free space is

$$\mathbf{H} = H_m \cos(\omega t - \beta z)\mathbf{a}_y$$

In free space, $\sigma = 0$ so that $\mathbf{J} = 0$. Also $\varepsilon = \varepsilon_o$. Determine the corresponding electric field from Ampere's law.

SOLUTION   The curl is expanded as

$$\nabla \times \mathbf{H} = \left(\frac{\partial H_z}{\partial y} - \frac{\partial H_y}{\partial z}\right)\mathbf{a}_x + \left(\frac{\partial H_x}{\partial z} - \frac{\partial H_z}{\partial x}\right)\mathbf{a}_y + \left(\frac{\partial H_y}{\partial x} - \frac{\partial H_x}{\partial y}\right)\mathbf{a}_z$$

$$= -\frac{\partial H_y}{\partial z}\mathbf{a}_x + \underbrace{\frac{\partial H_y}{\partial x}}_{0}\mathbf{a}_z$$

$$= -\frac{\partial H_y}{\partial z}\mathbf{a}_x$$

$$= -\beta H_m \sin(\omega t - \beta z)\mathbf{a}_x$$

Since the curl has only an x component, the right-hand side of Ampere's law can only have an x component, so we obtain

$$-\beta H_m \sin(\omega t - \beta z)\mathbf{a}_x = \varepsilon_o \frac{\partial E_x}{\partial t}\mathbf{a}_x$$

Integrating this gives

$$\mathbf{E} = E_x \mathbf{a}_x$$

$$= \frac{\beta H_m}{\omega \varepsilon_o} \cos(\omega t - \beta z)\mathbf{a}_x$$

◀

▶ **QUICK REVIEW EXERCISE 4.5**

Determine whether the following fields satisfy Ampere's law in free space: $\mathbf{D} = D_m \sin x \sin t\, \mathbf{a}_y$, $\mathbf{H} = D_m \cos x \cos t\, \mathbf{a}_z$.

ANSWER   Yes.

◀

▶ 4.3 THE LAWS OF GAUSS

The laws of Gauss were studied in the previous chapter and remain unchanged for time-varying fields:

$$\oint_s \mathbf{D} \cdot d\mathbf{s} = \underbrace{\int_v \rho_v dv}_{Q_{enclosed}} \tag{4.31}$$

where ρ_v is the volume charge density enclosed by the volume v, which is enclosed by the surface s, and

$$\oint_s \mathbf{B} \cdot d\mathbf{s} = 0 \tag{4.32}$$

Gauss' law for the electric field in (4.31) provides that the net flux of $\mathbf{D}$ out of a closed surface s yields the net positive charge enclosed by that surface. This is a logical result since electric field lines that begin on positive charge must end on negative charge. However, Faraday's law shows that a time-varying $\mathbf{B}$ field can produce an $\mathbf{E}$ field, yet those $\mathbf{E}$ field lines form closed loops which contribute nothing to the net flux of $\mathbf{D}$ out of the closed surface. Gauss' law for the magnetic field in (4.32) provides that there are no isolated sources or sinks for the magnetic field: all magnetic field lines must form closed loops.

4.3.1 Gauss' Laws in Point Form

Gauss' laws in (4.31) and (4.32) are in integral form and apply to broad regions of space. These integral forms are most useful for illustrating the meaning of the laws. For computational purposes we obtain these laws in *point form* or *differential form*, both of which apply to discrete points in space.

   The derivation of the point forms of these laws is very similar to the derivation of the point form of Faraday's law in Section 4.1.1. We consider Gauss' law for the electric field given in (4.31). We form a differential volume of rectangular shape with sides parallel to each axis and of length Δx, Δy, and Δz as shown in Fig. 4.12. The components of $\mathbf{D}$ that are perpendicular to each side are shown. Noting the directions of each component (into or out of the volume enclosed) gives

$$\oint_s \mathbf{D} \cdot d\mathbf{s} = -D_z \Delta x \Delta y + \left(D_z + \frac{\partial D_z}{\partial z} \Delta z \right) \Delta x \Delta y - D_x \Delta y \Delta z + \left(D_x + \frac{\partial D_x}{\partial x} \Delta x \right) \Delta y \Delta z$$

$$-D_y \Delta x \Delta z + \left(D_y + \frac{\partial D_y}{\partial y} \Delta y \right) \Delta x \Delta z$$

Canceling some terms gives

$$\oint_s \mathbf{D} \cdot d\mathbf{s} = \frac{\partial D_z}{\partial z} \Delta z \Delta x \Delta y + \frac{\partial D_x}{\partial x} \Delta x \Delta y \Delta z + \frac{\partial D_y}{\partial y} \Delta y \Delta x \Delta z$$

$$= \left(\frac{\partial D_x}{\partial x} + \frac{\partial D_y}{\partial y} + \frac{\partial D_z}{\partial z} \right) \underbrace{\Delta x \Delta y \Delta z}_{\Delta v}$$

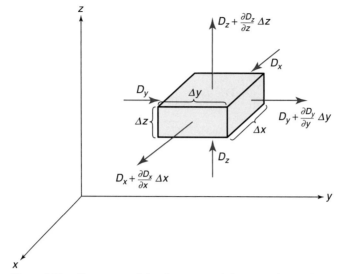

Figure 4.12 Illustration of the derivation of the point form of Gauss' law for the electric field.

Similarly, the right-hand side becomes

$$\int_v \rho_v dv = \rho_v \underbrace{\Delta x \Delta y \Delta z}_{\Delta v}$$

Dividing both sides by the volume enclosed, $\Delta v = \Delta x \Delta y \Delta z$, and allowing that volume to shrink to zero gives the point form as

$$\left(\frac{\partial D_x}{\partial x} + \frac{\partial D_y}{\partial y} + \frac{\partial D_z}{\partial z} \right) = \rho_v \tag{4.33}$$

In terms of the del operator studied earlier, this can be symbolically written as

$$\boxed{\nabla \cdot \mathbf{D} = \rho_v} \tag{4.34}$$

The left-hand side can be computationally determined directly from

$$\nabla \cdot \mathbf{D} = \left(\frac{\partial}{\partial x} \mathbf{a}_x + \frac{\partial}{\partial y} \mathbf{a}_y + \frac{\partial}{\partial z} \mathbf{a}_z \right) \cdot (D_x \mathbf{a}_x + D_y \mathbf{a}_y + D_z \mathbf{a}_z)$$

Hence in rectangular coordinate systems, the left-hand side is

$$\boxed{\nabla \cdot \mathbf{D} = \left(\frac{\partial D_x}{\partial x} + \frac{\partial D_y}{\partial y} + \frac{\partial D_z}{\partial z} \right)} \tag{4.35}$$

The notation $\nabla \cdot \mathbf{D}$ or "del dot D" is referred to as the *divergence of* $\mathbf{D}$. The reason for this name is seen by examining the above derivation and noting that

$$\nabla \cdot \mathbf{D} = \lim_{\Delta v \to 0} \frac{\oint_s \mathbf{D} \cdot d\mathbf{s}}{\Delta v} \tag{4.36}$$

In other words, *the divergence is the net flux of the vector out of the closed surface s per unit of volume as the volume shrinks to zero.* Hence the divergence indicates the strength of a source (or sink) for the vector field lines at that point. If there is a net positive volume charge at the point, we would expect a net outward flux (or divergence) of the fields from that charge.

Gauss' law for the magnetic field given in integral form in (4.32) becomes, in point form,

$$\boxed{\nabla \cdot \mathbf{B} = 0} \tag{4.37}$$

symbolizing that there can be no isolated sources or sinks for the vector field at a point.

▶ **EXAMPLE 4.9**

Show that the fields $\mathbf{E} = E_m \cos(\omega t - \beta z)\mathbf{a}_x$ and $\mathbf{H} = H_m \cos(\omega t - \beta z)\mathbf{a}_y$ satisfy Gauss' law in free space, where $\mu = \mu_o$ and $\varepsilon = \varepsilon_o$ and no free charge is present, $\rho_v = 0$.

SOLUTION   The electric flux density vector is $\mathbf{D} = \varepsilon_o\mathbf{E}$ and the magnetic flux density vector is $\mathbf{B} = \mu_o\mathbf{H}$. Hence we must have $\nabla \cdot \mathbf{D} = 0$ and $\nabla \cdot \mathbf{B} = 0$. Checking the first law, we find

$$\nabla \cdot \mathbf{D} = \left(\frac{\partial D_x}{\partial x} + \frac{\partial D_y}{\partial y} + \frac{\partial D_z}{\partial z} \right)$$

$$= \frac{\partial D_x}{\partial x}$$

$$= \frac{\partial}{\partial x}(\varepsilon_o E_m \cos(\omega t - \beta z))$$

$$= 0$$

Similarly

$$\nabla \cdot \mathbf{B} = \left(\frac{\partial B_x}{\partial x} + \frac{\partial B_y}{\partial y} + \frac{\partial B_z}{\partial z} \right)$$

$$= \frac{\partial B_y}{\partial y}$$

$$= \frac{\partial}{\partial y}(\mu_o H_m \cos(\omega t - \beta z))$$

$$= 0$$

◀

▶ **QUICK REVIEW EXERCISE 4.6**

Determine whether the following fields in free space satisfy Gauss' laws: $\mathbf{D} = D_m \sin x \sin t \, \mathbf{a}_y$, $\mathbf{B} = B_m \cos x \cos t \, \mathbf{a}_z$.

ANSWER   Yes.

◀

The expression for the divergence given in (4.35) is quite simple but is valid only for a rectangular coordinate system. The expressions for divergence in cylindrical and spherical coordinate systems are somewhat more complicated. In a cylindrical coordinate system divergence is

$$\nabla \cdot \mathbf{D} = \frac{1}{r}\frac{\partial(rD_r)}{\partial r} + \frac{1}{r}\frac{\partial(D_\phi)}{\partial \phi} + \frac{\partial(D_z)}{\partial z} \tag{4.38}$$

and in spherical coordinates divergence is

$$\nabla \cdot \mathbf{D} = \frac{1}{r^2}\frac{\partial(r^2 D_r)}{\partial r} + \frac{1}{r \sin\theta}\frac{\partial(\sin\theta\, D_\theta)}{\partial \theta} + \frac{1}{r \sin\theta}\frac{\partial(D_\phi)}{\partial \phi} \tag{4.39}$$

▶ 4.4 CONSERVATION OF CHARGE

Perhaps one of the most fundamental laws of the universe is *conservation of charge*, which provides that charge can be neither created nor destroyed. The mathematical statement of this is

$$\oint_s \mathbf{J} \cdot d\mathbf{s} = -\frac{\partial}{\partial t}\int_v \rho_v dv \tag{4.40}$$

This is an incredibly logical result. The left-hand side is the net flux of current out of the closed surface s. Since current is the (rate of) flow of charge, any outflow of current must equal a *decrease* in the charge enclosed, which is the right-hand side of the law. From our previous results we can immediately obtain the point form of this law as in the preceeding section:

$$\nabla \cdot \mathbf{J} = -\frac{\partial \rho_v}{\partial t} \tag{4.41}$$

This provides that the net outflow of current from a point, that is, the *divergence of* $\mathbf{J}$, must yield the rate of decrease of charge at that point.

For static conditions, $\partial/\partial t = 0$, the right-hand side of the conservation of charge relation is zero, reducing to either $\oint_s \mathbf{J} \cdot d\mathbf{s} = 0$ or $\nabla \cdot \mathbf{J} = 0$. These static laws provide that there is no charge storage within the closed surface or at the point, that is, the current flowing into the closed surface equals the current flowing out of it. This is essentially Kirchhoff's current law (KCL) where the closed surface is a node of a lumped circuit. Similarly, Faraday's law for static conditions reduces to $\oint_c \mathbf{E} \cdot d\mathbf{l} = 0$ or $\nabla \times \mathbf{E} = 0$. This static law provides that the net voltage around a closed loop is zero, which is essentially Kirchhoff's voltage law (KVL) in lumped circuits.

▶ EXAMPLE 4.10

Conducting materials are electrically charge neutral, meaning that the density of free charge is zero or $\rho_v = 0$. The free electrons in the outer valence band are free to move under the influence of an electric field, and the relation between the current density and the electric field is Ohm's law, $\mathbf{J} = \sigma\mathbf{E}$, where σ is the conductivity of the material. If an excess of free charge is introduced in its interior, that charge will eventually move to the conductor surface, rendering the interior once again charge neutral. Determine the length of time required for that excess charge to move to the surface of the conductor.

SOLUTION   We have two laws that relate the desired variables. Conservation of charge with Ohm's law substituted for **J** gives

$$\sigma \nabla \cdot \mathbf{E} = -\frac{\partial \rho_v}{\partial t}$$

and Gauss' law, substituting $\mathbf{D} = \varepsilon \mathbf{E}$:

$$\varepsilon \nabla \cdot \mathbf{E} = \rho_v$$

Combining these two equations by eliminating $\nabla \cdot \mathbf{E}$ yields

$$\frac{\partial \rho_v}{\partial t} + \frac{\sigma}{\varepsilon} \rho_v = 0$$

The solution to this simple first-order differential equation is

$$\rho_v = A e^{-t/T}$$

where the time constant is

$$T = \frac{\varepsilon}{\sigma} \quad \text{s}$$

This is a familiar result obtained in circuits courses for the transient response of RL or RC circuits. The excess charge density will decay to zero but it will take, theoretically, an infinite amount of time to do so. After about five time constants, it will have essentially decayed to zero.   ◄

► QUICK REVIEW EXERCISE 4.7

Determine the time constant for charge decay for copper, $\sigma = 5.8 \times 10^7$ S/m, $\varepsilon_r \cong 1$ and for glass, $\sigma \cong 10^{-14}$ S/m, $\varepsilon_r \cong 6$.

ANSWERS 1.5×10^{-19} s and 5305 s, or about 1.5 hr.   ◄

► 4.5 MAXWELL'S EQUATIONS

The four laws of Faraday, Ampere, and Gauss are collectively know as Maxwell's equations, which are summarized in Table 4.1.

TABLE 4.1   Maxwell's Equations

Law	Integral form	Point form
Faraday	$\oint_c \mathbf{E} \cdot d\mathbf{l} = -\dfrac{d}{dt} \int_s \mathbf{B} \cdot d\mathbf{s}$	$\nabla \times \mathbf{E} = -\dfrac{\partial \mathbf{B}}{\partial t}$
Ampere	$\oint_c \mathbf{H} \cdot d\mathbf{l} = \int_s \mathbf{J} \cdot d\mathbf{s} + \dfrac{d}{dt} \int_s \mathbf{D} \cdot d\mathbf{s}$	$\nabla \times \mathbf{H} = \mathbf{J} + \dfrac{\partial \mathbf{D}}{\partial t}$
Gauss (electric field)	$\oint_s \mathbf{D} \cdot d\mathbf{s} = \int_v \rho_v dv$	$\nabla \cdot \mathbf{D} = \rho_v$
Gauss (magnetic field)	$\oint_s \mathbf{B} \cdot d\mathbf{s} = 0$	$\nabla \cdot \mathbf{B} = 0$

Although not explicitly included in this list, the law of conservation of charge is implicit.

▶ 4.6 POWER DENSITY IN THE ELECTROMAGNETIC FIELD AND THE POYNTING VECTOR

Until now we have not discussed the concept of power or energy in the electromagnetic field. Clearly, there must be stored energy in an electromagnetic field. For example, the electric field about a point charge will provide a force on any other charge that is introduced into its field. Unlike lumped electric circuits where energy and power are localized to the elements, the distinguishing fact about electromagnetic fields is that the stored energy is distributed throughout the field and cannot be localized to a point.

Power also must be carried in the field. For example, waves propagating between two antennas that are used for communication must carry power between those two antennas. How should we quantify this? Observe that the units of the electric field intensity vector, **E**, are V/m while the units of the magnetic field intensity vector, **H**, are A/m. Hence the product of their magnitudes, E and H, has the units of W/m², which is a *power density*. How shall we define the product of these two vector quantities? We have two choices: the dot product or the cross product. Clearly the power flow must have a direction and the dot product has no direction. Therefore we choose the cross product and define the *power density vector* as

$$\boxed{\mathbf{S} = \mathbf{E} \times \mathbf{H} \quad \text{W/m}^2}$$

(4.42)

This is also given the name of the *Poynting vector* after an English physicist, John H. Poynting, who showed that this vector does in fact relate to power density in the EM field. After some vector manipulations it can be shown that

$$-\oint_s \mathbf{S} \cdot d\mathbf{s} = \underbrace{\int_v \sigma |\mathbf{E}|^2 dv}_{P_{\text{dissipated}}} + \underbrace{\int_v \frac{1}{2}\varepsilon \frac{\partial |\mathbf{E}|^2}{\partial t} dv}_{\substack{\text{rate of change of} \\ \text{stored energy in} \\ \text{electric field}}} + \underbrace{\int_v \frac{1}{2}\mu \frac{\partial |\mathbf{H}|^2}{\partial t} dv}_{\substack{\text{rate of change of} \\ \text{stored energy in} \\ \text{magnetic field}}}$$

(4.43)

The term on the left, $-\oint_s \mathbf{S} \cdot d\mathbf{s}$, is the net inward flow of power (because of the negative sign) into the closed surface s. The first term on the right is the power *dissipated* in the volume v that closed surface s encloses. The second and third terms on the right are the time rate-of-change of the energy stored in the electric and magnetic fields, respectively, in that volume. Hence choosing **S** as in (4.42) truly does represent power flow. In an analogy to lumped circuits, $\sigma |\mathbf{E}|^2$ is like V^2/R, which is the power dissipated in a resistor; $\frac{1}{2}\varepsilon |\mathbf{E}|^2$ is like $\frac{1}{2}CV^2$, which is the energy stored in a capacitor; and $\frac{1}{2}\mu |\mathbf{H}|^2$ is like $\frac{1}{2}LI^2$, which is the energy stored in an inductor. However, once again, the difference between EM fields and lumped electric circuits is that the power and stored energy are distributed throughout the EM field and cannot be localized, unlike the lumped circuit. Hence the need for integration throughout the volume v.

► **EXAMPLE 4.11**

We will find in Chapter 7 that the radiated fields at a sufficiently large distance from a dipole antenna are of the form

$$\mathbf{E} = \frac{E_o}{r}\sin\theta\,\sin\omega\!\left(t - \frac{r}{v_o}\right)\mathbf{a}_\theta$$

$$\mathbf{H} = \sqrt{\frac{\varepsilon_o}{\mu_o}}\,\frac{E_o}{r}\sin\theta\,\sin\omega\!\left(t - \frac{r}{v_o}\right)\mathbf{a}_\phi$$

which are specified in a spherical coordinate system and $v_o \cong 3 \times 10^8$ m/s is the velocity of light in free space. Determine the total power radiated by this antenna.

SOLUTION In order to determine this radiated power, we select a closed surface and compute the flux of **S** through that surface. Since the fields are specified in a spherical coordinate system, we choose a sphere of radius R as shown in Fig. 4.13. The Poynting vector is

$$\mathbf{S} = \mathbf{E} \times \mathbf{H}$$

$$= \sqrt{\frac{\varepsilon_o}{\mu_o}}\,\frac{E_o^2}{r^2}\sin^2\theta\,\sin^2\omega\!\left(t - \frac{r}{v_o}\right)\underbrace{\mathbf{a}_\theta \times \mathbf{a}_\phi}_{\mathbf{a_r}}\quad \text{W/m}^2$$

indicating that power flow is in the radial direction away from the antenna. Hence the total power radiated out of that surface (**S** has units of W/m², which when multiplied by the differential surface area gives units of the radiated power of W) is

$$P_{\text{rad}} = \oint_s \mathbf{S} \cdot d\mathbf{s}$$

$$= \int_{\phi=0}^{2\pi}\int_{\theta=0}^{\pi}\sqrt{\frac{\varepsilon_o}{\mu_o}}\,\frac{E_o^2}{R^2}\sin^2\theta\,\sin^2\omega\!\left(t - \frac{r}{v_o}\right)\underbrace{R^2\sin\theta\,d\theta\,d\phi}_{ds}$$

$$= \frac{8\pi}{3}\sqrt{\frac{\varepsilon_o}{\mu_o}}\,E_o^2\sin^2\omega\!\left(t - \frac{r}{v_o}\right)\quad \text{W}$$

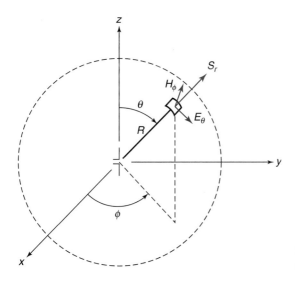

Figure 4.13 Example 4.11; determination of the total power radiated by a dipole antenna.

As is the case in electric circuits, we are interested in the time-average (over one cycle) power (or simply the average power):

$$P_{rad \atop av} = \frac{1}{T} \int_0^T P_{rad}\, dt$$

$$= \frac{4\pi}{3} \sqrt{\frac{\varepsilon_o}{\mu_o}} E_o^2 \quad \text{W}$$

◀

▶ **QUICK REVIEW EXERCISE 4.8**

For the dipole antenna of Example 4.11, the constant in the field expressions is $E_o = 10$. Determine the total average power radiated.

ANSWER 1.11 W.

◀

▶ **4.7 BOUNDARY CONDITIONS**

The above EM field equations in point form are partial differential equations. The solutions to them are in general very difficult to obtain and we will not be so ambitious as to solve them for very many problems. However, it is important to recognize that, like the solution of other differential equations, we need additional information if we did choose to solve them. This is very similar to the solution of ordinary differential equations encountered in lumped-circuits courses. For example, the "solution" to the ordinary differential equation

$$\frac{dy(t)}{dt} + 3y(t) = 0$$

is

$$y(t) = Ae^{-3t}$$

and A is an, as yet, undetermined constant. In order to pin down a specific solution, we need to specify an initial condition. For example, if the value of y at $t = 0$ is $y(0) = 2$, then the specific solution becomes

$$y(t) = 2e^{-3t} \quad t > 0$$

The partial differential equations governing EM fields are no different. In order to pin down the solution for a specific problem, we need to specify the values of the electric and magnetic fields *at the boundaries of the region enclosing those fields*. These are referred to as *boundary conditions* for obvious reasons. First, we obtain the boundary conditions for the electric field intensity vector. Consider Fig. 4.14a, where we show a boundary between two different media. Construct a rectangular contour on both sides of the surface of width Δw and depth Δh. We will integrate Faraday's law around this contour and let the depth go to zero, $\Delta h \to 0$, in order to derive the relation between the electric fields on both sides of the boundary. In the limit as $\Delta h \to 0$, Faraday's law becomes

$$\lim_{\Delta h \to 0} \oint_c \mathbf{E} \cdot d\mathbf{l} = E_{t1}\Delta w - E_{t2}\Delta w$$

$$= \lim_{\Delta h \to 0} \left[-\frac{\partial}{\partial t} \int_s \mathbf{B} \cdot d\mathbf{s} \right]$$

$$= 0$$

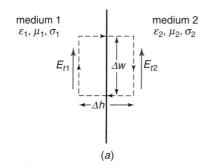

(a)

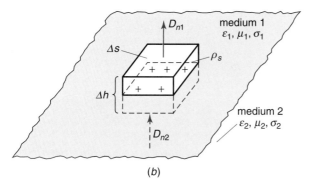

(b)

Figure 4.14 Illustration of the derivation of the boundary conditions. (a) Construction of a contour at the boundary for determining the relationships of the electric field intensity vector on both sides. (b) Construction of a rectangular box on both sides of a boundary for determining the relationships of the electric flux density vector on both sides.

where c is the rectangular contour and s is the flat surface bounded by that contour. The width of the sides, Δw, is infinitesimally small, so that we may assume the electric field is constant over that dimension. Hence the line integral along that dimension is simply the product of the field and the dimension. As $\Delta h \to 0$, the contributions to the line integral from those portions of the contour of length Δh will go to zero and are hence omitted. In the limit as $\Delta h \to 0$, the surface area bounded by this rectangular contour goes to zero and hence the net flux of the magnetic field through the enclosed surface s, the right-hand side of Faraday's law, goes to zero. Dividing out the width, Δw, we obtain

$$\boxed{E_{t1} = E_{t2}} \tag{4.44}$$

Hence the boundary condition on the electric field intensity vector is that *the components of the electric field intensity vector that are tangent to the boundary must be continuous across the boundary*. Similarly we apply Ampere's law around the contour to obtain

$$\lim_{\Delta h \to 0} \oint_c \mathbf{H} \cdot d\mathbf{l} = H_{t1}\Delta w - H_{t2}\Delta w$$

$$= \lim_{\Delta h \to 0} \left[\int_s \mathbf{J} \cdot d\mathbf{s} + \frac{\partial}{\partial t} \int_s \mathbf{D} \cdot d\mathbf{s} \right]$$

$$= 0$$

In the limit as $\Delta h \to 0$, the surface area bounded by this rectangular contour goes to zero and hence the right-hand side of Ampere's law goes to zero. Dividing out the width, Δw, we obtain

$$\boxed{H_{t1} = H_{t2}} \tag{4.45}$$

Hence the boundary condition on the magnetic field intensity vector is that *the components of the magnetic field intensity vector that are tangent to the boundary must be continuous across the boundary.*

Next we obtain the boundary conditions on the electric flux density vector **D** and the magnetic flux density vector **B**. Consider Fig. 4.14b, where we have formed a rectangular box extending on both sides of the boundary. As before, the surface of the box, Δs, as well as the sides, Δh, are infinitesimally small. Applying Gauss' law for the electric field and taking the limit as $\Delta h \to 0$, we obtain

$$\lim_{\Delta h \to 0} \oint_s \mathbf{D} \cdot d\mathbf{s} = D_{n1}\Delta s - D_{n2}\Delta s$$

$$= \lim_{\Delta h \to 0} \int_v \rho_v dv$$

Dividing both sides by the surface area gives

$$D_{n1} - D_{n2} = \lim_{\Delta h \to 0} \frac{\displaystyle\int_v \rho_v dv}{\Delta s}$$

$$= \rho_s \ \mathrm{C/m}^2$$

The right-hand side, $\lim_{\Delta h \to 0} \int_v \rho_v dv / \Delta s = \rho_s \ \mathrm{C/m}^2$, represents any surface charge density residing on the boundary. If no surface charge is intentionally placed on the boundary (as is the usual case), then we obtain

$$\boxed{D_{n1} = D_{n2}} \tag{4.46}$$

Hence the boundary condition on the electric flux density vector is that *the components of the electric flux density vector that are normal to the boundary must be continuous across the boundary.* Similarly, we can show using Gauss' law for the magnetic flux density vector, $\oint_s \mathbf{B} \cdot d\mathbf{s} = 0$, that *the components of the magnetic flux density vector that are normal to the boundary must be continuous across the boundary* or

$$\boxed{B_{n1} = B_{n2}} \tag{4.47}$$

These boundary conditions are summarized in Fig. 4.15. Simply stated, the components of the **E** and **H** field vectors that are *tangent (parallel)* to the boundary must

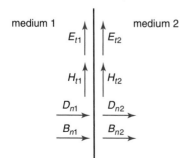

Figure 4.15 Illustration of the boundary conditions; tangential E and H must be continuous across the boundary, and normal D and B must be continuous across the boundary.

be the same (magnitude and direction) on both sides of the boundary, and the components of the **D** and **B** field vectors that are *normal (perpendicular)* to the boundary must be the same (magnitude and direction) on both sides of the boundary.

▶ **EXAMPLE 4.12**

Figure 4.16 shows the boundary between two media. The boundary lies in the *xy* plane. The electric field intensity vector in medium 1 at the boundary is

$$\mathbf{E}_1 = 2\mathbf{a}_x + 3\mathbf{a}_y + 4\mathbf{a}_z \quad \text{V/m}$$

Determine the electric field intensity vector and electric flux density vector in medium 2 at the boundary.

SOLUTION   Recall that we can convert **E** to **D** with $\mathbf{D} = \varepsilon_r \varepsilon_o \mathbf{E}$. The components of **E** that are tangent to the boundary are

$$\mathbf{E}_{t1} = 2\mathbf{a}_x + 3\mathbf{a}_y$$

Hence these are continuous across the boundary, so that

$$\mathbf{E}_{t2} = 2\mathbf{a}_x + 3\mathbf{a}_y$$

The components of **D** that are normal to the boundary are

$$\mathbf{D}_{n1} = \varepsilon_{r1}\varepsilon_o \mathbf{E}_{n1}$$
$$= 9\varepsilon_o(4\mathbf{a}_z)$$
$$= 36\varepsilon_o \mathbf{a}_z$$

Hence these are continuous across the boundary, so that

$$\mathbf{D}_{n2} = 36\varepsilon_o \mathbf{a}_z$$

We have the tangential component of **E** and the normal component of **D** on the other side of the boundary, so we simply use the relation $\mathbf{D} = \varepsilon_r \varepsilon_o \mathbf{E}$ in that medium to convert from one to the other. For example, the normal component of **E** is found from the normal component of **D** as

$$\mathbf{E}_{n2} = \frac{\mathbf{D}_{n2}}{\varepsilon_{r2}\varepsilon_o}$$
$$= \frac{36\varepsilon_o}{4\varepsilon_o}\mathbf{a}_z$$
$$= 9\mathbf{a}_z$$

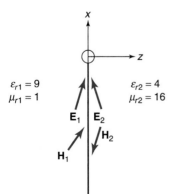

Figure 4.16   Example 4.12.

Similarly, we find the tangential component of **D** from the tangential component of **E** as

$$\mathbf{D}_{t2} = \varepsilon_{r2}\varepsilon_{o}\mathbf{E}_{t2}$$
$$= 4\varepsilon_{o}(2\mathbf{a}_{x} + 3\mathbf{a}_{y})$$
$$= 8\varepsilon_{o}\mathbf{a}_{x} + 12\varepsilon_{o}\mathbf{a}_{y}$$

Now we can form the vectors on the other side of the boundary:

$$\mathbf{E}_{2} = \mathbf{E}_{t2} + \mathbf{E}_{n2}$$
$$= 2\mathbf{a}_{x} + 3\mathbf{a}_{y} + 9\mathbf{a}_{z}$$

and

$$\mathbf{D}_{2} = \mathbf{D}_{t2} + \mathbf{D}_{n2}$$
$$= 8\varepsilon_{o}\mathbf{a}_{x} + 12\varepsilon_{o}\mathbf{a}_{y} + 36\varepsilon_{o}\mathbf{a}_{z}$$

Observe that $\mathbf{D}_{2} = \varepsilon_{r2}\varepsilon_{o}\mathbf{E}_{2}$ where $\varepsilon_{r2} = 4$. ◀

▶ QUICK REVIEW EXERCISE 4.9

In Example 4.12, the magnetic flux density vector in medium 1 at the boundary is

$$\mathbf{B}_{1} = 5\mathbf{a}_{x} + 6\mathbf{a}_{y} + 7\mathbf{a}_{z} \text{ Wb/m}^2$$

Determine the magnetic field intensity vector in medium 2 at the boundary.

ANSWER $\mathbf{H}_{2} = (5/\mu_{o})\mathbf{a}_{x} + (6/\mu_{o})\mathbf{a}_{y} + (7/16\mu_{o})\mathbf{a}_{z}$. ◀

4.7.1 Boundary Conditions at the Surface of a Perfect Conductor

In many boundary condition problems one side of the boundary is a material that is a very good conductor. For example, the conductivity of copper is $\sigma = 5.8 \times 10^7$ S/m. For these problems we often use a *computational simplification* by assuming that the conductivity is infinite, $\sigma \cong \infty$. A material with an infinite conductivity is said to be a *perfect conductor*. Assuming a perfect conductor when the material has a very large conductivity considerably simplifies the mathematics. This is the sole reason for its assumption rather than using the actual (but very large) conductivity of the material.

    All fields in a perfect conductor are zero. To show this we recall that Ohm's law gives the conductivity as the ratio of the conduction current density in the material to the electric field in the material that produces the current: $\sigma = J/E$. If $\sigma = \infty$, then either (1) J in infinite and E is finite but nonzero or (2) J is finite and nonzero and E is zero. Since current, and in particular J, is the *rate of flow of charge*, case (1) would mean that either a finite amount of charge was transported in zero time or that an infinite amount of charge was transported in a finite but nonzero time. Neither of these is possible, so we conclude that *the electric field intensity vector in a perfect conductor must be zero.* Since $\mathbf{D} = \varepsilon\mathbf{E}$, we conclude that **D** must also be zero in a perfect conductor. Considering Faraday's law, $\nabla \times \mathbf{E} = -\partial\mathbf{B}/\partial t$, for $\mathbf{E} = 0$ we conclude that **B** must be a constant (dc). The only known material having an infinite conductivity is a superconductor. Experiments have ruled out the existence of a dc magnetic field in superconductors. Hence we conclude that $\mathbf{B} = 0$. Similarly, if we assume $\mathbf{B} = \mu\mathbf{H}$, then **H** must also be zero.

    Now let us reconsider Fig. 4.14a with medium 2 having an infinite conductivity, $\sigma_2 = \infty$. Conductivity is not involved in Faraday's law. We still obtain (4.44) but the electric field in the perfect conductor is zero, that is, $E_{t2} = 0$, so that (4.44) becomes

$$\boxed{E_{t1} = 0 \qquad \sigma_2 = \infty} \tag{4.48}$$

Hence, *there can be no tangential electric field at the surface of a perfect conductor*. In other words, *the electric field at the surface of a perfect conductor must be perpendicular to that surface*. Hence we say that a perfect conductor "shorts out" the electric field. Ampere's law, $\oint_c \mathbf{H} \cdot d\mathbf{l} = \int_s \mathbf{J} \cdot d\mathbf{s} - \dfrac{\partial}{\partial t} \int_s \mathbf{D} \cdot d\mathbf{s}$, contains conductivity of the second medium in the conduction current term, $\mathbf{J}_2 = \sigma_2 \mathbf{E}_2$. Since the conductivity of the second medium is infinite, the portion of the right-hand side containing $\mathbf{J}$ does not go to zero as we shrink the contour to the surface. Therefore, although the magnetic field in medium 2 is zero and hence $H_{t2} = 0$, we cannot conclude from (4.45) that the tangential magnetic field in the first medium is zero: $H_{t1} \neq 0$. It turns out that a surface current will be induced at the boundary which is numerically equal to the tangential magnetic field. But this is not a useful boundary condition since it does not pin down the value of H; it only allows us to determine this surface current if we know H_{t1}.

Finally we obtain the boundary conditions on the $\mathbf{D}$ and $\mathbf{B}$ fields. In medium 2, which is a perfect conductor, we have $\mathbf{D}_2 = \mathbf{B}_2 = 0$. The derivation of (4.46) assumed that as we shrink the box to the surface, $\Delta h \to 0$, the right-hand side of Gauss' law for the electric field, $\oint_s \mathbf{D} \cdot d\mathbf{s} = \int_v \rho_v dv$, would go to zero. When medium 2 is a perfect conductor, this assumption is no longer valid and a surface charge density will be induced on the boundary so that $D_{n1} = \lim\limits_{\Delta h \to 0} \int_v \rho_v dv = \rho_s \ \text{C/m}^2$.

However, Gauss' law for the magnetic field has the right-hand side equal to zero, $\oint_s \mathbf{B} \cdot d\mathbf{s} = 0$, so that (4.47) applies even when the second medium is a perfect conductor. Setting $B_{n2} = 0$ in (4.47) yields the boundary condition on the magnetic flux density vector:

$$\boxed{B_{n1} = 0 \qquad \sigma_2 = \infty} \tag{4.49}$$

The essential boundary conditions at the surface of a perfect conductor are given by (4.48) and (4.49):

$$\boxed{\left. \begin{array}{l} E_{t1} = 0 \\ B_{n1} = 0 \end{array} \right\} \qquad \sigma_2 = \infty} \tag{4.50}$$

In other words, the electric field intensity vector $\mathbf{E}$ must be *perpendicular* to the surface of a perfect conductor. Assuming in medium 1 that $\mathbf{D}_1 = \varepsilon_1 \mathbf{E}_1$, then the electric flux density vector $\mathbf{D}$ must also be perpendicular to the surface of a perfect conductor. The magnetic flux density vector $\mathbf{B}$ must be *parallel* to the surface of a perfect conductor. Assuming in medium 1 that $\mathbf{B}_1 = \mu_1 \mathbf{H}_1$, then the magnetic field intensity vector $\mathbf{H}$ must also be parallel to the surface of a perfect conductor. These are summarized in Fig. 4.17.

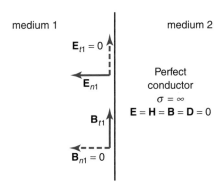

Figure 4.17 The boundary conditions at the surface of a perfect conductor; the electric field intensity vector $\mathbf{E}$ must be normal to the surface, and the magnetic flux density vector $\mathbf{B}$ must be tangent to the surface.

▶ 4.8 THE METHOD OF IMAGES

Quite often we need to solve problems where the charges or currents are situated above a conducting plane that is infinitely large in extent (often called a ground plane). We will assume, as a computational simplification, that the plane is a perfect conductor.

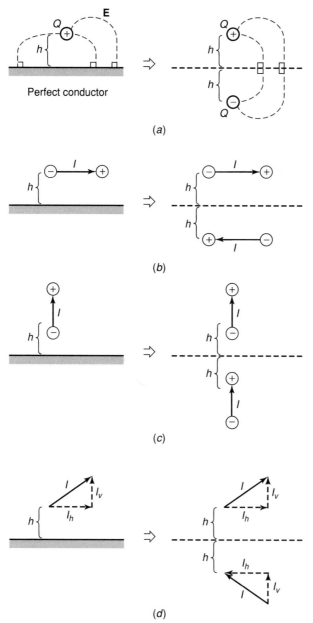

Figure 4.18 Illustration of the method of images. (a) Replacement of a point charge above an infinite perfect conductor with its image resulting in a simpler problem to solve. (b) The image of a current parallel to a perfect conductor. (c) The image of a current perpendicular to a perfect conductor. (d) Resolving a current into its horizontal and vertical components for replacement with its image.

This also applies as a reasonable approximation to a plane that is a very good conductor. The presence of the infinitely large, perfectly conducting plane makes such problems very difficult to solve directly. The *method of images* allows us to replace the infinitely large, perfectly conducting plane with an equivalent charge or current, so that the resulting problem is equivalent to the original problem but is much easier to solve. For example, consider a point charge Q at a height h above an infinite, perfectly conducting plane as shown in Fig. 4.18a. The plane can be replaced with a charge that is equal in magnitude to the original charge but of opposite sign at a depth h *below* the position of the ground plane (which has been removed). This equivalent problem is much easier to solve than is the original problem.

The equivalence of the image problem to the original one can be seen in the fact that the electric field of the image problem is completely perpendicular to the position of the ground plane. Therefore the electric field that is tangent to the ground plane and to the position of the ground plane that was removed is zero. This satisfies the boundary condition in (4.50).

Currents can be imaged in a similar fashion. Consider an element of current I that is parallel to and at a height h above an infinite, perfectly conducting plane as shown in Fig. 4.18b. Current is the flow of charge and hence, for the purposes of determining the images of the current, we can visualize this current as depositing positive charge at the head of the current arrow and negative charge accumulating at its tail. Now if we image those charges, we see that we may replace the plane with another current I that is parallel to the original current but is oppositely directed and is at a depth h below the position of the plane (which has been removed). Similarly, consider a current element I that is directed perpendicular to the plane and at a height h above the plane as shown in Fig. 4.18c. Viewing this as accumulated charges at the ends of the current, we see that we may form an equivalent problem by removing the plane and substituting a current a distance h below the position of the (removed) plane that is also vertically directed and is in the same direction as the original current. In the case of a current that is neither horizontal nor vertical but is inclined at an angle to a infinite, perfectly conducting plane as shown in Fig. 4.18d, we can form an equivalent problem. This can be constructed by resolving the current into vertically directed and horizontally directed components and using the previous results.

The equivalence of the image problem to the original one can again be seen in the fact that the magnetic field of the image problem (recall that the magnetic field about a wire is circumferentially directed according to the right-hand-rule) is completely parallel to the position of the ground plane. Therefore the magnetic field that is perpendicular to the ground plane and to the position of the ground plane that was removed is zero. This again satisfies the boundary condition in (4.50).

The method of images is a powerful computational tool and we shall use it on many occasions.

▷ **EXAMPLE 4.13**

An infinitely long line charge of radius a having a uniform distribution of ρ_l C/m along its length is situated a height h above an infinte, perfectly conducting plane. Determine the per-unit-length capacitance between the line charge and the ground plane.

SOLUTION Replacing the plane with a negative line charge $-\rho_l$ C/m at a depth h below the position of the plane as shown in Fig. 4.19a gives an equivalent problem. The capacitance per

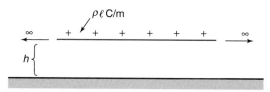

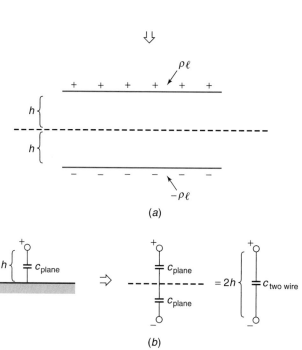

Figure 4.19 Example 4.13; determining the per-unit-length capacitance of a wire above and parallel to a ground plane. (a) Replacing the ground plane with its image. (b) Determining the capacitance in terms of the capacitance between two wires.

unit length of this "two-wire" problem was solved in Chapter 3, Example 3.14, and is

$$c_{\text{two-wire}} = \frac{\pi\varepsilon}{\ln\left(\dfrac{2h}{a}\right)} \quad \text{F/m}$$

where a is the radius of the line charge(s). The original problem of the capacitance between the line charge and the plane, c_{plane}, is related to this problem as shown in Fig. 4.19b. Since capacitances in series add like resistors in parallel, we see from this figure that

$$c_{\text{plane}} = 2c_{\text{two-wire}}$$

$$= \frac{2\pi\varepsilon}{\ln\left(\dfrac{2h}{a}\right)} \quad \text{F/m}$$

▶ **QUICK REVIEW EXERCISE 4.10**

Determine the per-unit-length inductance of a filament of current of radius a that is parallel to and at a distance h above an infinite, perfectly conducting plane. (Hint: See Example 3.25.)

ANSWER $l = \dfrac{\mu}{2\pi}\ln\left(\dfrac{2h}{a}\right)$ H/m.

4.9 SINUSOIDAL VARIATION OF THE FIELDS

The preceeding laws of this chapter are valid for any time variation of the fields. Our primary interest in the future will be in the sinusoidal variation of those fields. This is closely akin to the phasor method developed in electric circuit analysis courses for circuits that are driven by sinusoidal sources and are in the steady state, that is, transients have died down. There are many reasons for this primary interest in the sinusoidal variation of the fields rather than some arbitrary time variation. Perhaps the most important reason is that any waveform can be decomposed into its sinusoidal components using the Fourier series for periodic waveforms or the Fourier transform for nonperiodic waveforms. (See Chapter 1.) Nevertheless we may view a waveform of some arbitrary time variation as being composed of sinusoidal components. If the system, such as an electric circuit, that processes this waveform is *linear*, then we may use superposition to determine the response of the system to the arbitrary waveform by adding the responses of the system to the individual sinusoidal components of that waveform. The medium in which the fields exist will be assumed throughout this text to be exclusively linear so that a similar concept can be used for the fields. Hence if we decompose a field having arbitrary time variation into its sinusoidal components, we can thus determine the response to that waveform by determining the response to the individual sinusoidal components. From a more specific standpoint, this interest in sinusoidal variation of the fields also stems from the fact that radio, TV, and many other similar signals are primarily sinusoidal in nature.

As was done in the phasor method for analyzing the sinusoidal response of electric circuits, we replace the field vectors, $\mathbf{E}(x,y,z,t)$, with their phasor equivalents, $\hat{\mathbf{E}}(x,y,z)$, which is a complex number having a magnitude and a phase. The principle behind this is very simple and is based on Euler's identity:

$$e^{j(\omega t + \theta)} = \cos(\omega t + \theta) + j\sin(\omega t + \theta) \tag{4.51}$$

and $j = \sqrt{-1}$. Phasors and complex numbers will be denoted with a carat ($\wedge$) over the item. The essential idea is that instead of solving for the response to a sine or cosine variation, we replace that with the complex exponential with $e^{j\omega t}$ suppressed:

$$\left.\begin{array}{c}\cos(\omega t + \theta)\\ \sin(\omega t + \theta)\end{array}\right\} \Rightarrow e^{j(\omega t + \theta)} \Rightarrow e^{j\theta}$$

Time derivatives are replaced with $j\omega$:

$$\frac{\partial}{\partial t} \Rightarrow j\omega \tag{4.52}$$

This is because the time derivative of (4.51) is simply $j\omega$ times that quantity:

$$\frac{\partial}{\partial t}e^{j(\omega t + \theta)} = j\omega e^{j(\omega t + \theta)}$$

Hence Maxwell's equations for sinusoidal variation of the fields become

$$
\begin{aligned}
\oint_c \hat{\mathbf{E}} \cdot d\mathbf{l} &= -j\omega \int_s \hat{\mathbf{B}} \cdot d\mathbf{s} \\[2mm]
\oint_c \hat{\mathbf{H}} \cdot d\mathbf{l} &= \int_s \hat{\mathbf{J}} \cdot d\mathbf{s} + j\omega \int_s \hat{\mathbf{D}} \cdot d\mathbf{s} \\[2mm]
\oint_s \hat{\mathbf{D}} \cdot d\mathbf{s} &= \int_v \hat{\rho}_v \, dv \\[2mm]
\oint_s \hat{\mathbf{B}} \cdot d\mathbf{s} &= 0
\end{aligned}
\tag{4.53a}
$$

or in point form:

$$
\begin{aligned}
\nabla \times \hat{\mathbf{E}} &= -j\omega \hat{\mathbf{B}} \\
\nabla \times \hat{\mathbf{H}} &= \hat{\mathbf{J}} + j\omega \hat{\mathbf{D}} \\
\nabla \cdot \hat{\mathbf{D}} &= \hat{\rho}_v \\
\nabla \cdot \hat{\mathbf{B}} &= 0
\end{aligned}
\tag{4.53b}
$$

These phasor field vectors will have a magnitude and a phase angle. For example,

$$
\hat{\mathbf{E}} = E_{mx}\angle\theta_x \mathbf{a}_x + E_{my}\angle\theta_y \mathbf{a}_y + E_{mz}\angle\theta_z \mathbf{a}_z
$$

Once the magnitudes and angles of the components are determined, we return to the time domain by selecting (arbitrarily) a cosine form to give the time-domain result

$$
\mathbf{E} = E_{mx}\cos(\omega t + \theta_x)\mathbf{a}_x + E_{my}\cos(\omega t + \theta_y)\mathbf{a}_y + E_{mz}\cos(\omega t + \theta_z)\mathbf{a}_z
$$

▶ EXAMPLE 4.14

Determine, using the phasor method, whether the following field vectors satisfy Faraday's and Ampere's laws in free space:

$$
\mathbf{E} = A\cos(\omega t - \beta z)\mathbf{a}_x
$$

$$
\mathbf{H} = \frac{A}{\eta}\cos(\omega t - \beta z)\mathbf{a}_y
$$

SOLUTION The field vectors in free space are related in the usual fashion as $\mathbf{D} = \varepsilon_o \mathbf{E}$, $\mathbf{B} = \mu_o \mathbf{H}$, and $\sigma = 0$, so that $\mathbf{J} = 0$. The field vectors become, in phasor form,

$$
\hat{\mathbf{E}} = Ae^{-j\beta z}\mathbf{a}_x
$$

$$
\hat{\mathbf{H}} = \frac{A}{\eta}e^{-j\beta z}\mathbf{a}_y
$$

Faraday's and Ampere's laws become, in phasor form,

$$
\nabla \times \hat{\mathbf{E}} = -j\omega\mu_o \hat{\mathbf{H}}
$$
$$
\nabla \times \hat{\mathbf{H}} = j\omega\varepsilon_o \hat{\mathbf{E}}
$$

First form the left-hand sides of these laws:

$$\nabla \times \hat{\mathbf{E}} = \left(\frac{\partial \hat{E}_z}{\partial y} - \frac{\partial \hat{E}_y}{\partial z}\right)\mathbf{a}_x + \left(\frac{\partial \hat{E}_x}{\partial z} - \frac{\partial \hat{E}_z}{\partial x}\right)\mathbf{a}_y + \left(\frac{\partial \hat{E}_y}{\partial x} - \frac{\partial \hat{E}_x}{\partial y}\right)\mathbf{a}_z$$

$$= \frac{\partial \hat{E}_x}{\partial z}\mathbf{a}_y$$

$$= -j\beta A e^{-j\beta z}\mathbf{a}_y$$

$$\nabla \times \hat{\mathbf{H}} = \left(\frac{\partial \hat{H}_z}{\partial y} - \frac{\partial \hat{H}_y}{\partial z}\right)\mathbf{a}_x + \left(\frac{\partial \hat{H}_x}{\partial z} - \frac{\partial \hat{H}_z}{\partial x}\right)\mathbf{a}_y + \left(\frac{\partial \hat{H}_y}{\partial x} - \frac{\partial \hat{H}_x}{\partial y}\right)\mathbf{a}_z$$

$$= -\frac{\partial \hat{H}_y}{\partial z}\mathbf{a}_x$$

$$= j\beta \frac{A}{\eta} e^{-j\beta z}\mathbf{a}_x$$

Setting these equal to the right-hand sides of the laws gives

$$\underbrace{-j\beta A e^{-j\beta z}\mathbf{a}_y}_{\nabla \times \hat{\mathbf{E}}} = -j\omega\mu_o \underbrace{\frac{A}{\eta} e^{-j\beta z}\mathbf{a}_y}_{\hat{\mathbf{H}}}$$

$$\underbrace{j\beta \frac{A}{\eta} e^{-j\beta z}\mathbf{a}_x}_{\nabla \times \hat{\mathbf{H}}} = j\omega\varepsilon_o \underbrace{A e^{-j\beta z}\mathbf{a}_x}_{\hat{\mathbf{E}}}$$

In order for these to be satisfied we must have

$$\beta = \omega\frac{\mu_o}{\eta}$$

$$\frac{\beta}{\eta} = \omega\varepsilon_o$$

Solving for β and η gives

$$\beta = \omega\sqrt{\mu_o\varepsilon_o}$$

$$\eta = \sqrt{\frac{\mu_o}{\varepsilon_o}}$$

Hence, the field vectors will statisfy Faraday's and Ampere's laws only if the frequency ω and the parameters of the medium, ε_o, μ_o, are such that β and η satisfy the above relations. The fields of this problem are a very important form of *waves* that we will study in the next chapter. ◄

▶ **QUICK REVIEW EXERCISE 4.11**

Determine whether the fields in Example 4.14 satisfy Gauss' laws by using their phasor forms.

ANSWER They do. ◄

▶ **QUICK REVIEW EXERCISE 4.12**

Determine, using their phasor forms, whether the fields $\mathbf{E} = E_m \sin x \sin t\, \mathbf{a}_y$ and $\mathbf{H} = H_m \cos x \cos t\, \mathbf{a}_z$ satisfy all of Maxwell's equations in a nonconducting region where $\sigma = 0$, $\rho_v = 0$. Hint: To convert both to sine or cosine, use $\sin\theta = \cos(\theta - 90°)$.

ANSWER Yes, if $\varepsilon\mu = 1$. ◄

▷ 4.10 THE CURRENT PROBE: COMBINING FARADAY'S AND AMPERE'S LAWS TO MEASURE CURRENT

Consider the measurement of a current in a wire. One way to do this is to insert a resistor R (of very small value so as not to change the current substantially) in series with the wire and measure the voltage developed across that resistor. Then the current is V/R.

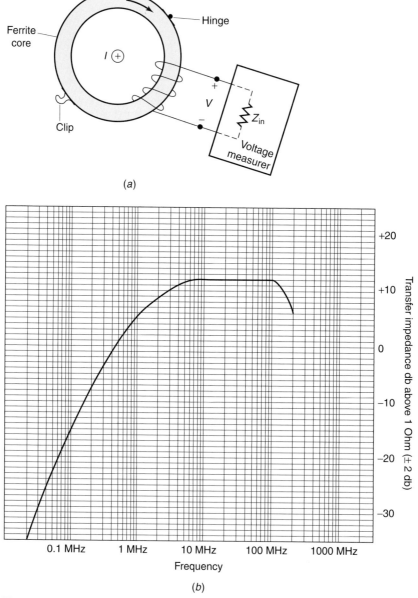

(a)

(b)

Figure 4.20 The current probe. (a) Illustration of its operation in terms of Ampere's and Faraday's laws. (b) A typical current probe transfer impedance as a function of frequency. (c) Photograph of a typical current probe (courtesy of Fischer Custom Communications, Inc).

(c)

Figure 4.20 (*Continued*)

A noninvasive method of measuring the current is with a *current probe*. A current probe is a toroid of ferromagnetic material that can be clamped around a current-carrying wire as shown in Fig. 4.20a. According to Ampere's law, the current will produce a circumferentially directed magnetic field intensity H_ϕ that circulates primarily around the toroid. (The use of a ferromagnetic material to construct the toroid acts to contain the majority of the magnetic field within it.) Several turns of wire are wrapped around the toroid and connected to a voltage measurer. If the current is time varying, then the magnetic field will also be time varying. That magnetic field passes through the turns of wire, generating, according to Faraday's law, an induced voltage V at its terminals which is measured with the attached voltage measurer. The device is calibrated by passing a known value of current through the toroid and measuring the voltage at the frequency of the current. Several frequencies of current are used to provide voltage readings for each of those frequencies. The ratio of the induced (measured) voltage to the current that produced it has the units of ohms and is referred to as the *transfer impedance of the probe*:

$$Z_T = \frac{V}{I} \qquad \Omega$$

where $V = |\hat{V}|$ is the magnitude of the induced voltage, and $I = |\hat{I}|$ is the magnitude of the current being measured. A calibration graph is provided by the manufacturer, showing the transfer impedance versus frequency. The transfer impedance for a typical current probe is shown in Fig. 4.20b in terms of "dB above an ohm" as

$$Z_{T,\mathrm{dB\Omega}} = 20 \log_{10}(Z_T)$$
$$= 20 \log_{10}(V) - 20 \log_{10}(I)$$

Hence the current in dBA (dB above an ampere) is

$$I_{\mathrm{dBA}} = V_{\mathrm{dBV}} - Z_{T,\mathrm{dB\Omega}}$$

and V_{dBV} is dB above a volt. With this calibration graph one can determine the current by using the measured voltage and the transfer impedance. Current probes are available that can measure currents from the low kHz to the GHz range and amplitudes from μA to hundreds of amps. A photograph of a typical current probe is shown in Fig. 4.20c.

► SUMMARY OF IMPORTANT CONCEPTS AND FORMULAE

1. **Faraday's law in integral form:** $\oint_c \mathbf{E} \cdot d\mathbf{l} = -d/dt \int_s \mathbf{B} \cdot d\mathbf{s}$ means that the line integral of the electric field intensity vector around a closed path c will yield the time rate-of-decrease of the magnetic flux that penetrates the surface s bounded by the path. A time-varying magnetic field passing through a loop will induce a voltage source in the loop whose value is the time rate-of-change of the magnetic flux passing through the loop. The polarity of this induced voltage source is such that it would tend to induce a current in the loop whose magnetic field will, by the right-hand-rule, oppose the change in the original magnetic field (Lenz's law).

2. **Faraday's law in point form:** $\nabla \times \mathbf{E} = -\partial \mathbf{B}/\partial t$ where, in rectangular coordinates, $\nabla \times \mathbf{E} = (\partial E_z/\partial y - \partial E_y/\partial z)\mathbf{a}_x + (\partial E_x/\partial z - \partial E_z/\partial x)\mathbf{a}_y + (\partial E_y/\partial x - \partial E_x/\partial y)\mathbf{a}_z$.

3. **Ampere's law in integral form:** $\oint_c \mathbf{H} \cdot d\mathbf{l} = \int_s \mathbf{J} \cdot d\mathbf{s} + \partial/\partial t \int_s \mathbf{D} \cdot d\mathbf{s}$ means that the sum of the conduction current and the displacement current penetrating a surface, s, equals the line integral of the magnetic field around the contour c that encloses the surface.

4. **Ampere's law in point form:** $\nabla \times \mathbf{H} = \mathbf{J} + (\partial \mathbf{D}/\partial t)$ where, in rectangular coordinates, $\nabla \times \mathbf{H} = (\partial H_z/\partial y - \partial H_y/\partial z)\mathbf{a}_x + (\partial H_x/\partial z - \partial H_z/\partial x)\mathbf{a}_y + (\partial H_y/\partial x - \partial H_x/\partial y)\mathbf{a}_z$.

5. **Gauss' law for the electric field:** $\oint_s \mathbf{D} \cdot d\mathbf{s} = \underbrace{\int_v \rho_v dv}_{Q_{\mathrm{enclosed}}}$ where ρ_v is the volume charge density enclosed by the volume v which is enclosed by the surface s. This means that electric field lines that begin on positive charge terminate on negative charge. Electric field lines caused by a time-varying magnetic field, according to Faraday's law, form closed paths. In point form $\nabla \cdot \mathbf{D} = \rho_v$ where, in rectangular coordinates, $\nabla \cdot \mathbf{D} = (\partial D_x/\partial x + \partial D_y/\partial y + \partial D_z/\partial z)$

6. **Gauss' law for the magnetic field:** $\oint_s \mathbf{B} \cdot d\mathbf{s} = 0$, which means that the magnetic field lines must form closed paths; they can have no beginning or end. In point form, $\nabla \cdot \mathbf{B} = 0$ where, in rectangular coordinates, $\nabla \cdot \mathbf{B} = (\partial B_x/\partial x + \partial B_y/\partial y + \partial B_z/\partial z)$.

7. **Conservation of charge:** $\oint_s \mathbf{J} \cdot d\mathbf{s} = -\partial/\partial t \int_v \rho_v dv$. The left-hand side is the net flux of current out of the closed surface s. The right-hand side is the rate of decrease of charge enclosed by volume v that the surface encloses. Since current is the (rate of) flow of charge, any outflow of current must equal a *decrease* in the charge enclosed. In point form, $\nabla \cdot \mathbf{J} = -\partial \rho_v/\partial t$.

8. **Poynting vector and power density:** $\mathbf{S}(\mathrm{W/m^2}) = \mathbf{E}(\mathrm{V/m}) \times \mathbf{H}(\mathrm{A/m})$.

9. **Boundary conditions:** (a) neither side of a boundary is a perfect conductor: $E_{t1} = E_{t2}$, $H_{t1} = H_{t2}$, $D_{n1} = D_{n2}$, $B_{n1} = B_{n2}$, meaning that the tangential components of $\mathbf{E}$ and $\mathbf{H}$ must be

continuous and the normal components of **D** and **B** must be continuous, (b) one side of boundary is a perfect conductor ($\sigma_2 = \infty$): $E_{t1} = 0$, $B_{n1} = 0$, meaning that the tangential components of **E** and the normal components of **B** must be zero at the surface of a perfect conductor.

10. Method of images: can replace an infinite, perfectly conducting plane with equivalent charges and currents below the position of the plane such that the fields above the location of the plane are unchanged and the problem solution is simplified.

▶ PROBLEMS

SECTION 4.1 FARADAY'S LAW

4.1.1. Determine the voltage V in the circuit of Fig. P4.1.1. $[-0.233 \text{ mV}]$.

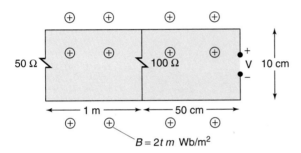

Figure P4.1.1 Problem 4.1.1.

4.1.2. Determine the voltage V in the circuit of Fig. P4.1.2.

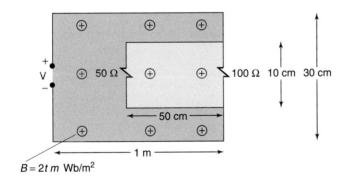

Figure P4.1.2 Problem 4.1.2.

4.1.3. Determine the voltage V in the circuit of Fig. P4.1.3. $[301.6 \cos(2\pi \times 60t) \text{ mV}]$

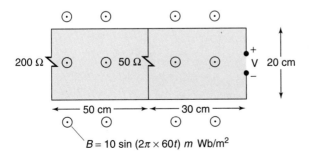

Figure P4.1.3 Problem 4.1.3.

4.1.4. Assuming that current I is positive valued, which of the situations in Fig. P4.1.4 are *not* correct.

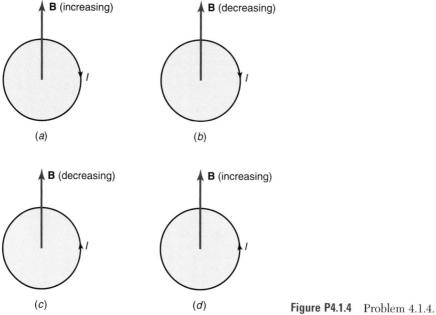

Figure P4.1.4 Problem 4.1.4.

4.1.5. Determine the current I in Fig. P4.1.5 if the magnetic field varies as $B = 2\cos(2\pi \times 60t)$ mWb/m^2 and the velocity of the bar is 10 m/s. $[0.1\cos(2\pi \times 60t) - 37.7t\sin(2\pi \times 60t)$ mA$]$

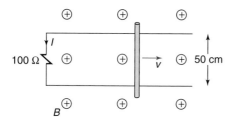

Figure P4.1.5 Problem 4.1.5.

4.1.6. Determine the current I in Fig. P4.1.5 if the magnetic field is constant at $B = 10$ mWb/m^2 and the velocity $v = 100\cos(10t)$ m/s.

4.1.7. A rectangular loop rotates about the z axis at an angular speed of $\omega = 5$ radians/s in a constant magnetic field directed in the y direction of $\mathbf{B} = 10\mathbf{a}_y$ mWb/m^2 as shown in Fig. P4.1.7. Determine the current induced to flow around the loop with direction shown, assuming that the loop is in the x direction at $t = 0$. $[-2.5\cos(5t)$ mA$]$

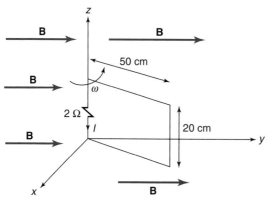

Figure P4.1.7 Problem 4.1.7.

4.1.8. Lightning is particularly detrimental to electronic devices even if those devices do not receive a direct hit. For example, a lightning bolt can be considered as a channel of current, as shown in Fig. P4.1.8, that is very long. Consider modeling the effect of a typical lightning bolt on household wiring. Assume the lightning channel rises linearly to a peak level of 50 kA in 1 μs and then decays linearly to zero at 10 μs as shown in Fig. P4.1.8. Also assume that the house power wiring has the two wires routed around the floor and the ceiling. Determine the induced voltage at the power input to a digital computer that is connected to the power net for a lightning channel 1 km away. [$V = 2585\,\text{V}$ for $0 < t < 1\,\mu\text{s}$ and $V = -287\,\text{V}$ for $1\,\mu\text{s} < t < 10\,\mu\text{s}$].

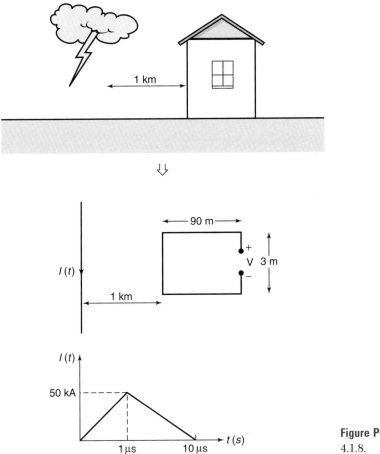

Figure P4.1.8 Problem 4.1.8.

4.1.9. A rectangular loop is moving with velocity v radially away from a wire that carries a dc current I as shown in Fig. P4.1.9. Determine an expression for the current induced in the loop as a function of time.

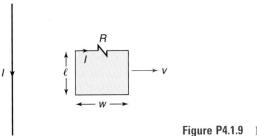

Figure P4.1.9 Problem 4.1.9.

4.1.10. A metal bar of length l rotates about a fixed point with radial speed ω radians/m as shown in Fig. P4.1.10. A dc magnetic field B is oriented perpendicular to the plane of rotation of the bar. Determine the voltage induced in the bar and indicate its polarity with respect to the fixed or rotating end. [$B\omega l^2/2$ positive at the rotating end] This is the basis for the homopolar generator wherein a rotating conducting disk is perpendicular to a magnetic field. Sliding contacts tap off this induced voltage between the rim and the outer edge of the disk.

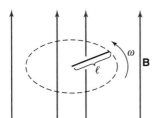

Figure P4.1.10   Problem 4.1.10.

4.1.11. An electric field intensity vector in free space is given in a rectangular coordinate system by $\mathbf{E} = E_m \sin\alpha x \sin(\omega t - \beta z)\mathbf{a}_y$. Determine the magnetic field intensity vector that satisfies Faraday's law.

4.1.12. An electric field intensity vector in free space is given in a rectangular coordinate system by $\mathbf{E} = E_m \sin\beta z \cos(\omega t)\mathbf{a}_x$. Determine the magnetic field intensity vector that satisfies Faraday's law. $\left[\mathbf{H} = -E_m\beta/\omega\mu_o \cos\beta z \sin(\omega t)\mathbf{a}_y\right]$

SECTION 4.2 AMPERE'S LAW

4.2.1. Dry soil has $\varepsilon_r \cong 4$ and a conductivity of $\sigma \cong 10^{-5}$ S/m. Determine the frequency above which the displacement current dominates the conduction current.

4.2.2. A magnetic field intensity vector in free space is given in a rectangular coordinate system as $\mathbf{H} = H_x \sin\alpha x \sin(\omega t - \beta z)\mathbf{a}_x + H_z \cos\alpha x \cos(\omega t - \beta z)\mathbf{a}_z$. Determine the electric field intensity vector that satisfies Ampere's law.

$$\left[\mathbf{E} = \frac{(-\beta H_x + \alpha H_z)}{\omega\varepsilon_o}\sin\alpha x \sin(\omega t - \beta z)\mathbf{a}_y\right]$$

4.2.3. A magnetic field intensity vector in free space is given in a rectangular coordinate system as $\mathbf{H} = H_m \cos\beta z \sin(\omega t)\mathbf{a}_y$. Determine the electric field intensity vector that satisfies Ampere's law.

SECTION 4.3 THE LAWS OF GAUSS

4.3.1. Show that the $\mathbf{E}$ field of Problem 4.1.11 satisfies Gauss' law in free space where, unless some is intentionally introduced, there is no free charge density ($\rho_v = 0$).

4.3.2. Show that the $\mathbf{E}$ field of Problem 4.1.12 satisfies Gauss' law in free space where, unless some is intentionally introduced, there is no free charge density ($\rho_v = 0$).

4.3.3. Show that the $\mathbf{H}$ field of Problem 4.2.2 satisfies Gauss' law in free space.

4.3.4. Show that the $\mathbf{H}$ field of Problem 4.2.3 satisfies Gauss' law in free space.

SECTION 4.6 POWER DENSITY IN THE ELECTROMAGNETIC FIELD AND THE POYNTING VECTOR

4.6.1. A wave traveling in seawater (which is lossy) has its electric and magnetic fields given in a rectangular coordinate system as $\mathbf{E} = 10e^{-4z}\cos(\omega t - 4z)\mathbf{a}_x$ V/m and $\mathbf{H} = 7.15e^{-4z}\cos(\omega t - 4z - \pi/4)\mathbf{a}_y$ A/m. Determine the total power exiting a cube surface having sides of length 1 m and corners at (0,0,0), (1 m,0,0), (0,0,1 m), (1 m,0,1 m), (0,1 m,1 m), (1 m,1 m,1 m), (0,1 m,0), and (1 m,1 m,0). [-25.27 W]

SECTION 4.7 BOUNDARY CONDITIONS

4.7.1. An interface between two media lies in the yz plane at $x = 0$ as shown in Fig. P4.7.1. If the electric field intensity vector in medium 1 at the interface ($x = 0$) is given by $\mathbf{E}_1|_{x=0} = \alpha \mathbf{a}_x + \beta \mathbf{a}_y + \gamma \mathbf{a}_z$ determine $\mathbf{E}_2|_{x=0}$, that is, the electric field intensity vector in the second medium just across the boundary.

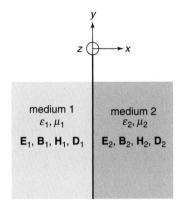

Figure P4.7.1 Problem 4.7.1.

4.7.2. An interface between two media lies in the yz plane at $x = 0$ as shown in Fig. P4.7.1. If the magnetic flux density vector in medium 1 at the interface ($x = 0$) is given by $\mathbf{B}_1|_{x=0} = \alpha \mathbf{a}_x + \beta \mathbf{a}_y + \gamma \mathbf{a}_z$, determine $\mathbf{B}_2|_{x=0}$, that is, the magnetic flux density vector in the second medium just across the boundary. $\left[\mathbf{B}_2|_{x=0} = \alpha \mathbf{a}_x + \beta \mu_2/\mu_1 \mathbf{a}_y + \gamma \mu_2/\mu_1 \mathbf{a}_z \right]$

4.7.3. An interface between two media lies in the yz plane at $x = 0$ as shown in Fig. P4.7.1. If the electric flux density vector in medium 1 at the interface ($x = 0$) is given by $\mathbf{D}_1|_{x=0} = \alpha \mathbf{a}_x + \beta \mathbf{a}_y + \gamma \mathbf{a}_z$, determine $\mathbf{D}_2|_{x=0}$, that is, the electric flux density vector in the second medium just across the boundary. $\left[\mathbf{D}_2|_{x=0} = \alpha \mathbf{a}_x + \beta \varepsilon_2/\varepsilon_1 \mathbf{a}_y + \gamma \varepsilon_2/\varepsilon_1 \mathbf{a}_z \right]$

4.7.4. An interface between two media lies in the yz plane at $x = 0$ as shown in Fig. P4.7.1. If the magnetic field intensity vector in medium 1 at the interface ($x = 0$) is given by $\mathbf{H}_1|_{x=0} = \alpha \mathbf{a}_x + \beta \mathbf{a}_y + \gamma \mathbf{a}_z$, determine $\mathbf{H}_2|_{x=0}$, that is, the magnetic field intensity vector in the second medium just across the boundary.

4.7.5. Figure P4.7.5 shows a boundary between a dielectric medium having parameters ε_1 and μ_1 and a perfect conductor. The electric field intensity vector in medium 1 at the surface of the perfect conductor is given by $\mathbf{E} = \alpha \mathbf{a}_x + \beta \mathbf{a}_y + \gamma \mathbf{a}_z$. Also, the magnetic flux density vector in medium 1 at the surface of the perfect conductor is given by $\mathbf{B} = \sigma \mathbf{a}_x + \delta \mathbf{a}_y + \chi \mathbf{a}_z$. Apply the boundary conditions to determine the values of or constraints between these six parameters.

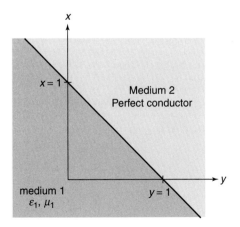

Figure P4.7.5 Problem 4.7.5.

SECTION 4.8 THE METHOD OF IMAGES

4.8.1. A perfect conductor lies in the xy plane of a rectangular coordinate system at $z = 0$ and extends for $z < 0$. A current is above this plane and is described by $\mathbf{I} = 2\mathbf{a}_x - 3\mathbf{a}_y + 4\mathbf{a}_z$ A at $(0,0,2)$. Determine the image current with which the perfect conductor can be replaced so that the fields for $z > 0$ will remain unchanged. $[\mathbf{I}_{image} = -2\mathbf{a}_x + 3\mathbf{a}_y - 4\mathbf{a}_z$ A at $(0,0,-2)]$

SECTION 4.9 SINUSOIDAL VARIATION OF THE FIELDS

4.9.1. Write the phasor or time-domain forms of the following fields as appropriate. The frequency is 10 MHz. (a) $\hat{\mathbf{E}} = -j30\mathbf{a}_x - 10/j\,\mathbf{a}_y$, (b) $\hat{\mathbf{H}} = 10e^{j4\pi/3}\mathbf{a}_z$, (c) $\hat{\mathbf{B}} = 4e^{-2z}e^{-j4\pi z}\mathbf{a}_x$, (d) $\mathbf{E} = 10e^{-3x}\cos(2\pi \times 10^6 t - 3\pi z)\mathbf{a}_y$, (e) $\mathbf{B} = 5\sin(3z)\cos(2\pi \times 10^6 t - 6z)\mathbf{a}_x$. $[$(a) $-30\cos(2\pi \times 10^6 t + \pi/2)\mathbf{a}_x + 10\cos(2\pi \times 10^6 t + \pi/2)\mathbf{a}_y$, (b) $10\cos(2\pi \times 10^6 t + 4\pi/3)\mathbf{a}_z$, (c) $4e^{-2z}\cos(2\pi \times 10^6 t - 4\pi z)\mathbf{a}_x$, (d) $\hat{\mathbf{E}} = 10e^{-3x}e^{-j3\pi z}\mathbf{a}_y$, (e) $\hat{\mathbf{B}} = 5\sin(3z)e^{-j6z}\mathbf{a}_x]$

4.9.2. Determine the magnetic field intensity vector from Faraday's law for the following electric field intensity vector in free space using phasor methods: $\mathbf{E} = E_m\sin\alpha x\cos(\omega t - \beta z)\mathbf{a}_y$.

4.9.3. Show that the time-average Poynting vector can be found from the phasor field vectors as $\mathbf{S}_{av} = 1/2\,\text{Re}(\hat{\mathbf{E}} \times \hat{\mathbf{H}}^\circ)$ where $\hat{\mathbf{A}}^\circ$ is the complex conjugate of the phasor $\hat{\mathbf{A}}$. Hint: Use the relation $\mathbf{E} = \text{Re}(\hat{\mathbf{E}}e^{j\omega t}) = 1/2(\hat{\mathbf{E}}e^{j\omega t} + \hat{\mathbf{E}}^\circ e^{-j\omega t})$.

SECTION 4.10 THE CURRENT PROBE: COMBINING FARADAY'S AND AMPERE'S LAWS TO MEASURE CURRENT

4.10.1. For a current probe illustrated in Fig. 4.20a, assume that the toroid is constructed from a ferromagnetic material having $\mu_r = 200$ and has a mean radius of 2 cm and a core cross-sectional area of 4 cm^2. There are 10 turns wound on the toroid. Determine the transfer impedance for a current of frequency 1 MHz.

CHAPTER 5

Wave Propagation

Faraday's law combined with Ampere's law predicts the existence of electromagnetic waves. Although there are many types of waves, most share many of the properties of *uniform plane waves*, which we will study in this chapter. Waves propagated along two-conductor transmission lines are plane waves. We will study the propagation of waves on transmission lines in the following chapter. Although waves propagated from antennas are truly spherical waves, as when one throws a rock into a body of water, these waves appear, locally, to an observer as plane waves also. We will study the propagation of waves from antennas in the final chapter of this text, Chapter 7. Hence the study of uniform plane waves in this chapter will provide an understanding of waves propagated on or from other structures.

Chapter Learning Objectives

After completing the contents of this chapter, you should be able to

- explain the concept of a wave,
- compute the phase constant, β, the intrinsic impedance, η, the velocity of propagation, v, and the wavelength, λ, of a uniform plane wave (UPW) in a lossless medium,
- write the frequency-domain (phasor) and time-domain forms of the electric and magnetic field intensity vectors for a UPW in a lossless medium,
- repeat the previous two objectives for a UPW in a lossy medium,
- understand that phase shift of a wave in the frequency domain (phasor) is equivalent to time delay of propagation of the wave in the time domain,
- compute the average power density vector of a UPW in a lossless medium and in a lossy medium,
- compute and understand the meaning of skin depth,
- write the forms of the incident, reflected, and transmitted electric and magnetic field intensity vectors for a UPW that is incident normal to a boundary between two lossless media,
- write expressions for the average power density in the incident, reflected, and transmitted waves for a UPW incident normal to the boundary between two lossless media,
- repeat the previous two objectives where the second medium is a good conductor,
- use Snell's laws to compute the angles of incidence, reflection, and transmission for a UPW that is incident at an oblique angle to the boundary between two lossless media,
- cite and explain examples of engineering applications of the principles of this chapter.

► 5.1 THE UNIFORM PLANE WAVE IN LOSSLESS MEDIA

The *uniform plane* wave has two important terms in its title, *uniform* and *plane*. The term *plane* means that the electric and magnetic field intensity vectors both lie in a plane, and all such planes are parallel. Also, the phase of the wave is constant over the plane. The term *uniform* means that these vectors are constant in magnitude and phase over the planes. Figure 5.1a shows such a field configuration where we have chosen, to simplify the notation, the electric field intensity vector to be directed in the x direction. It will turn out that the magnetic field intensity vector will be orthogonal to this and directed in the y direction. These field vectors lie in the xy plane and are *plane* waves. The other part of the name, *uniform*, means that the field vectors are independent of x and y in these planes. Hence the electric field intensity and magnetic field intensity vectors can only be functions of z and time, t, $E_x(z,t)$ and $H_y(z,t)$. For a single-frequency wave, the phasor forms are $\hat{E}_x(z)$ and $\hat{H}_y(z)$. It will also turn out that the wave associated with these field vectors propagates in the z direction as shown in Fig. 5.1b. Plane waves are often referred to as transverse electromagnetic or TEM waves since the electric and magnetic field vectors are perpendicular or transverse to the direction of propagation of the wave (the z direction).

In order to determine the relations between these field vectors, we first write Faraday's law in phasor form as

$$\nabla \times \hat{\mathbf{E}} = -j\omega\mu\hat{\mathbf{H}} \tag{5.1}$$

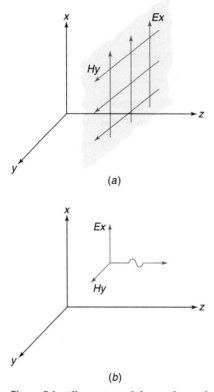

(a)

(b)

Figure 5.1 Illustration of the uniform plane wave. (a) The electric and magnetic field vectors of a uniform plane wave. (b) Propagation direction of a uniform plane wave is perpendicular to the plane containing the electric and magnetic field vectors.

where we have substituted $\mathbf{B} = \mu \mathbf{H}$. Expanding the curl in rectangular coordinates gives

$$\nabla \times \hat{\mathbf{E}} = \underbrace{\left(\frac{\partial \hat{E}_z}{\partial y} - \frac{\partial \hat{E}_y}{\partial z} \right)}_{0} \mathbf{a}_x + \underbrace{\left(\frac{\partial \hat{E}_x}{\partial z} - \frac{\partial \hat{E}_z}{\partial x} \right)}_{0} \mathbf{a}_y + \underbrace{\left(\frac{\partial \hat{E}_y}{\partial x} - \frac{\partial \hat{E}_x}{\partial y} \right)}_{0} \mathbf{a}_z \tag{5.2}$$

$$= \frac{\partial \hat{E}_x}{\partial z} \mathbf{a}_y$$

Since the electric field is assumed to be x-directed, $\hat{E}_z = \hat{E}_y = 0$ so that derivatives with respect to those components are zero. The last derivative in the z component, $\partial \hat{E}_x / \partial y$, is zero since the electric field is uniform over the xy plane, $\hat{E}_x(z)$, and is thus independent of the y coordinate. Hence, the result in (5.2) shows, by matching components on both sides of (5.1), that the magnetic field intensity vector is in the y direction:

$$\frac{\partial \hat{E}_x}{\partial z} \mathbf{a}_y = -j\omega\mu \hat{H}_y \mathbf{a}_y$$

and we obtain the first equation:

$$\boxed{\frac{\partial \hat{E}_x}{\partial z} = -j\omega\mu \hat{H}_y} \tag{5.3}$$

This contains two unknowns, $\hat{E}_x$ and $\hat{H}_y$. To obtain another equation in these two unknowns we apply Ampere's law:

$$\nabla \times \hat{\mathbf{H}} = \hat{\mathbf{J}} + j\omega\varepsilon \hat{\mathbf{E}}$$
$$= \sigma \hat{\mathbf{E}} + j\omega\varepsilon \hat{\mathbf{E}} \tag{5.4}$$

and have substituted $\mathbf{J} = \sigma \mathbf{E}$ and $\mathbf{D} = \varepsilon \mathbf{E}$. Expanding (5.4) gives

$$\nabla \times \hat{\mathbf{H}} = \underbrace{\left(\frac{\partial \hat{H}_z}{\partial y} - \frac{\partial \hat{H}_y}{\partial z} \right)}_{0} \mathbf{a}_x + \underbrace{\left(\frac{\partial \hat{H}_x}{\partial z} - \frac{\partial \hat{H}_z}{\partial x} \right)}_{0} \mathbf{a}_y + \underbrace{\left(\frac{\partial \hat{H}_y}{\partial x} - \frac{\partial \hat{H}_x}{\partial y} \right)}_{0} \mathbf{a}_z$$

$$= -\frac{\partial \hat{H}_y}{\partial z} \mathbf{a}_x$$

Since the magnetic field has only a y component, certain of the terms are zero, $\hat{H}_x = \hat{H}_z = 0$, so that derivatives with respect to those components are zero. Also, since the magnetic field is independent of x and y, $\hat{H}_y(z)$, any derivatives with respect to those coordinates are zero. We will first consider *lossless* media in which $\sigma = 0$. Hence we obtain

$$-\frac{\partial \hat{H}_y}{\partial z} \mathbf{a}_x = j\omega\varepsilon \hat{E}_x \mathbf{a}_x \qquad \sigma = 0$$

Thus we obtain the remaining equation:

$$\boxed{\frac{\partial \hat{H}_y}{\partial z} = -j\omega\varepsilon \hat{E}_x} \tag{5.5}$$

Equations (5.3) and (5.5) are first-order differential equations in $\hat{E}_x$ and $\hat{H}_y$. To obtain separate equations in only $\hat{E}_x$ or $\hat{H}_y$ we differentiate one with respect to z and substitute the other. Also, since $\hat{E}_x$ and $\hat{H}_y$ are only functions of z, the derivatives become ordinary rather than partial derivatives. The result becomes

$$\frac{d^2\hat{E}_x}{d^2z} + \underbrace{\omega^2\mu\varepsilon}_{\beta^2}\, \hat{E}_x = 0 \tag{5.6a}$$

$$\frac{d^2\hat{H}_y}{d^2z} + \underbrace{\omega^2\mu\varepsilon}_{\beta^2}\, \hat{H}_y = 0 \tag{5.6b}$$

The term β is called the *phase constant* and has units of radians/m and is

$$\beta = \omega\sqrt{\mu\varepsilon} \qquad \text{rad/m} \tag{5.7}$$

These are second-order, linear, constant coefficient, ordinary differential equations whose solution we have obtained many times in electric circuit courses and in differential equations courses:

$$\hat{E}_x = \hat{E}_m^+ e^{-j\beta z} + \hat{E}_m^- e^{j\beta z} \tag{5.8a}$$

$$\hat{H}_y = \hat{H}_m^+ e^{-j\beta z} + \hat{H}_m^- e^{j\beta z} \tag{5.8b}$$

The terms $\hat{E}_m^+$, $\hat{E}_m^-$, $\hat{H}_m^+$, and $\hat{H}_m^-$ are, as yet, undetermined constants. The significance of the $\pm$ superscripts on these constants will be explained soon. Although there are four such undetermined constants, two of them are related to the other two. To see this we substitute (5.8) into (5.3) to yield

$$\frac{\partial \hat{E}_x}{\partial z} = -j\beta\hat{E}_m^+ e^{-j\beta z} + j\beta\hat{E}_m^- e^{j\beta z}$$

$$= -j\omega\mu\hat{H}_y$$

$$= -j\omega\mu(\hat{H}_m^+ e^{-j\beta z} + \hat{H}_m^- e^{j\beta z})$$

Matching corresponding exponential terms yields

$$\hat{H}_m^+ = \frac{\hat{E}_m^+}{\left(\dfrac{\omega\mu}{\beta}\right)} = \frac{\hat{E}_m^+}{\left(\sqrt{\dfrac{\mu}{\varepsilon}}\right)} \tag{5.9a}$$

and

$$\hat{H}_m^- = -\frac{\hat{E}_m^-}{\left(\dfrac{\omega\mu}{\beta}\right)} = -\frac{\hat{E}_m^-}{\left(\sqrt{\dfrac{\mu}{\varepsilon}}\right)} \tag{5.9b}$$

Note the similarity of these relations to Ohm's law, $I = V/R$, where we make the analogy of $I \Leftrightarrow H$ and $V \Leftrightarrow E$ by virtue of their similar units. Since $\beta = \omega\sqrt{\mu\varepsilon}$, the denominators simplify as shown. The denominators of the relations in (5.9) between these undetermined constants also have the units of ohms (Ω) and represent the *intrinsic impedance of the medium*:

$$\eta = \sqrt{\frac{\mu}{\varepsilon}} \quad \Omega \tag{5.10}$$

In free space, $\mu = \mu_o = 4\pi \times 10^{-7}$ and $\varepsilon = \varepsilon_o \cong (1/36\pi) \times 10^{-9}$ so that the intrinsic impedance of free space is

$$\eta_o = \sqrt{\frac{\mu_o}{\varepsilon_o}} \cong 120\pi \cong 377 \quad \Omega \tag{5.11}$$

In a medium other than free space we have the relations $\varepsilon = \varepsilon_r \varepsilon_o$ and $\mu = \mu_r \mu_o$. Therefore in a medium other than free space we can relate the intrinsic impedance of that medium to that of free space as

$$\eta = \eta_o \sqrt{\frac{\mu_r}{\varepsilon_r}} \tag{5.12}$$

Tables 5.1 and 5.2 give values of the relative permittivity, ε_r, relative permeability, μ_r, and conductivity, σ, for representative materials. In general, the relative permittivity and relative permeability of these materials vary with frequency.

Observe that most metals are not magnetizeable, that is, $\mu_r \approx 1$. In addition, metals are not dielectric in that $\varepsilon_r \approx 1$.

TABLE 5.1 Relative Permittivities of Various Dielectrics

Material	ε_r
Air	1.0
Polyethylene foam	1.6
Cellular polyethylene	1.8
Teflon	2.1
Polyethylene	2.3
Polystyrene	2.5
Nylon	3.0
Silicon rubber	3.1
Polyvinyl chloride (PVC)	3.5
Epoxy resin	3.6
Quartz (fused)	3.8
Epoxy glass (printed circuit substrate)	4.7
Bakelite	4.9
Glass (pyrex)	5.0
Mylar	5.0
Porcelain	6.0
Neoprene	6.7
Polyurethane	7.0
Silicon	12.0

TABLE 5.2 Relative Permeabilities and Conductivities (Relative to Copper,) of Various Metals

Conductor	σ_r	μ_r
Silver	1.05	1
Copper-annealed	1.00	1
Gold	0.70	1
Aluminum	0.61	1
Brass	0.26	1
Nickel	0.20	1
Bronze	0.18	1
Tin	0.15	1
Steel (SAE 1045)	0.10	2000
Lead	0.08	1
Monel	0.04	1
Stainless steel (430)	0.02	500
Zinc	0.29	1
Iron	0.17	1000
Beryllium	0.10	1
Mumetal (at 1 kHz)	0.03	30,000
Permalloy (at 1 kHz)	0.03	80,000

Hence the final solutions are

$$\hat{E}_x = \hat{E}_m^+ e^{-j\beta z} + \hat{E}_m^- e^{j\beta z} \tag{5.13a}$$

$$\hat{H}_y = \frac{\hat{E}_m^+}{\eta} e^{-j\beta z} - \frac{\hat{E}_m^-}{\eta} e^{j\beta z} \tag{5.13b}$$

We now have only two undetermined constants, $\hat{E}_m^+$ and $\hat{E}_m^-$. These will be determined by the boundary conditions of the particular problem at hand. Observe the important minus sign between the two terms of $\hat{H}_y$.

The above are the phasor or frequency-domain forms of the fields, which have sinusoidal variation at a fixed frequency. Let us now determine the time-domain fields. To do so we simply (1) multiply the phasor result by $e^{j\omega t}$ and (2) take the real part of that. (Recall Euler's identity: $e^{j\theta} = \cos\theta + j\sin\theta$. Hence the real part of $e^{j\theta}$ is $\cos\theta$.) Hence the electric field in (5.13a) becomes, in the time domain

$$
\begin{aligned}
E_x(z,t) &= \text{Re}\big[(E_m^+ e^{-j\beta z} + E_m^- e^{j\beta z})e^{j\omega t}\big] \\
&= \text{Re}\big[E_m^+ e^{j(\omega t - \beta z)} + E_m^- e^{j(\omega t + \beta z)}\big] \\
&= \underbrace{E_m^+ \cos(\omega t - \beta z)}_{\substack{\text{forward-traveling} \\ \text{($+z$ direction) wave}}} + \underbrace{E_m^- \cos(\omega t + \beta z)}_{\substack{\text{backward-traveling} \\ \text{($-z$ direction) wave}}}
\end{aligned} \tag{5.14a}
$$

and, similarly

$$
\begin{aligned}
H_y(z,t) &= \text{Re}\left[\left(\frac{E_m^+}{\eta}e^{-j\beta z} - \frac{E_m^-}{\eta}e^{j\beta z}\right)e^{j\omega t}\right] \\
&= \underbrace{\frac{E_m^+}{\eta}\cos(\omega t - \beta z)}_{\substack{\text{forward-traveling} \\ \text{($+z$ direction) wave}}} - \underbrace{\frac{E_m^-}{\eta}\cos(\omega t + \beta z)}_{\substack{\text{backward-traveling} \\ \text{($-z$ direction) wave}}}
\end{aligned} \tag{5.14b}
$$

and we have assumed the undetermined constants to be real valued, $\hat{E}_m^\pm = E_m^\pm$. The terms involving $\cos(\omega t - \beta z)$ are said to be *forward-traveling waves*, traveling in the $+z$ direction in that as t increases z must increase to keep the argument of the cosine function constant. These are said to be waves like waves on a body of water. In order to track the movement of such waves, we observe the movement of a point on the wave, say, the crest. This amounts to following the movement of a point where $\cos(\omega t - \beta z)$ is constant and hence the argument must remain constant: as t increases, z must also increase. Similarly, the terms involving $\cos(\omega t + \beta z)$ are said to be *backward-traveling waves*, traveling in the $-z$ direction, since in order to keep the argument of the cosine constant, as t increases z must decrease. This is the origin of the $\pm$ superscripts on the undetermined constants: $\hat{E}_m^+$ is associated with a forward-traveling wave in the $+z$ direction, whereas, $\hat{E}_m^-$ is associated with a backward-traveling wave in the $-z$ direction. These can also be written in an alternative form using the relation $\beta = \omega\sqrt{\mu\varepsilon}$ as

$$\cos(\omega t \pm \beta z) = \cos\left(\omega\left(t \pm \frac{z}{v}\right)\right) \tag{5.15}$$

Hence the *velocity of propagation* or *phase velocity* of the wave front where the phase is constant is

$$v = \frac{1}{\sqrt{\mu\varepsilon}} \quad \text{m/s}$$

(5.16)

In free space, $\mu = \mu_o = 4\pi \times 10^{-7}$ and $\varepsilon = \varepsilon_o \cong (1/36\,\pi) \times 10^{-9}$ so that the velocity of propagation in free space is

$$\begin{aligned} v_o &= 2.99792458 \times 10^8 \\ &\cong 3 \times 10^8 \quad \text{m/s} \end{aligned}$$

(5.17)

In a medium other than free space we have the relations $\varepsilon = \varepsilon_r\varepsilon_o$ and $\mu = \mu_r\mu_o$. Therefore in a medium other than free space we can relate the velocity of propagation of the waves in that medium to the velocity of propagation in free space as

$$v = \frac{v_o}{\sqrt{\mu_r\varepsilon_r}}$$

(5.18)

See Tables 5.1 and 5.2 in order to compute the velocity of propagation in other materials. Observe in (5.18) that the velocity of propagation in materials other than air is always less than the velocity of propagation in air, since $\varepsilon_r \geq 1$ and $\mu_r \geq 1$. Observe that the terms in (5.15), $(t - z/v)$ and $(t + z/v)$, amount to a *time delay*. A change at some point in space is felt or observed z/v seconds later. This shows an important point about the relation between the frequency and time domains for waves:

In the frequency domain a wave suffers a phase shift $e^{\pm j\beta z}$ as it travels through a medium in the $\mp z$ direction but this is equivalent in the time domain to a time delay, $\cos[\omega(t \pm z/v)]$.

Consider a forward-traveling wave:

$$\begin{aligned} E_x &= E_m^+ \cos(\omega t - \beta z) \\ H_y &= \frac{E_m^+}{\eta} \cos(\omega t - \beta z) \end{aligned}$$

(5.19)

These fields are plotted versus distance z in Fig. 5.2a for a *fixed time*. Note that the electric and magnetic field vectors have an amplitude distribution in space that is cosinusoidal. As time increases, this envelope moves to the right, the $+z$ direction. The backward-traveling wave:

$$\begin{aligned} E_x &= E_m^- \cos(\omega t + \beta z) \\ H_y &= -\frac{E_m^-}{\eta} \cos(\omega t + \beta z) \end{aligned}$$

(5.20)

has a similar cosinusoidal envelope but moves in the $-z$ direction. Observe the important minus sign in front of H_y in the backward-traveling wave in (5.20). This is necessary because the wave is traveling in the $-z$ direction and hence the power flow given by $\mathbf{E} \times \mathbf{H}$ must also be in the $-z$ direction according to the right-hand-rule.

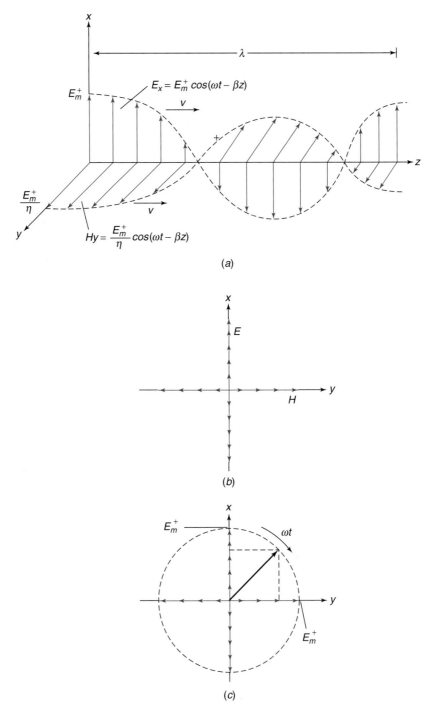

Figure 5.2 Illustration of the propagation of a uniform plane wave. (a) Distribution of the electric and magnetic field vectors in space. (b) Viewing the electric and magnetic field vectors looking in the direction of propagation, indicating linear polarization of the wave. (c) A circularly polarized uniform plane wave.

Another way of looking at this wave is in the xy plane looking in the $+z$ direction as shown in Fig. 5.2b. This represents viewing the wave for a fixed position, z, as time varies. For a fixed position, say, $z = 0$, (5.19) becomes

$$E_x|_{z=0} = E_m^+ \cos(\omega t)$$

$$H_y|_{z=0} = \frac{E_m^+}{\eta} \cos(\omega t)$$

The electric field vector increases and decreases in the x direction, whereas the magnetic field vector increases and decreases in the y direction in concert with the changes in the amplitude of the electric field vector. These waves are said to be *linearly polarized* since the electric and magnetic field vectors vary along lines (the x and y axes). If we combine two linearly polarized uniform plane waves the result is such that the electric and magnetic field vectors can be made to rotate in space as they propagate. For example, suppose we add another wave wherein its electric field has a phase angle θ added:

$$\mathbf{E} = E_{m1}^+ \cos(\omega t - \beta z)\mathbf{a}_x + E_{m2}^+ \cos(\omega t - \beta z + \theta)\mathbf{a}_y$$

Observing this composite wave in the xy plane at $z = 0$ gives

$$\mathbf{E}|_{z=0} = E_{m1}^+ \cos(\omega t)\mathbf{a}_x + E_{m2}^+ \cos(\omega t + \theta)\mathbf{a}_y$$

Observe that the total field is the sum of the x and y components. If we choose the magnitudes to be the same, $E_{m1}^+ = E_{m2}^+ = E_m^+$, and choose the phase angle to be $\theta = -90°$, this composite electric field vector becomes

$$\begin{aligned}\mathbf{E}|_{z=0} &= E_m^+ \cos(\omega t)\mathbf{a}_x + E_m^+ \cos(\omega t - 90°)\mathbf{a}_y \\ &= E_m^+ \cos(\omega t)\mathbf{a}_x + E_m^+ \sin(\omega t)\mathbf{a}_y\end{aligned}$$

Fig. 5.2c shows that the tip of the resultant electric field vector rotates at an angular rate of ωt in the clockwise direction. Hence this is said to represent a *circularly polarized* wave. It is also said to have right-hand circular polarization since when placing the thumb of the right hand in the direction of propagation, the $+z$ direction, the fingers indicate the direction of rotation of the field vector. Choosing the phase angle to be $\theta = 90°$ results in a left-hand circularly polarized wave. Helical antennas are constructed from a spiraled wire and produce circularly polarized waves. We will only investigate linearly polarized waves in this chapter.

There is one final, important property of uniform plane waves that should be discussed. A *wavelength* was discussed in Chapter 1 and is the distance a wave must travel to change phase by 2π radians or 360° as is illustrated in Fig. 5.2a. As a wave propagates it suffers a phase shift of βz radians. Hence a wavelength is

$$\boxed{\lambda = \frac{2\pi}{\beta} \quad \text{m}} \tag{5.21}$$

We may write this alternatively using (5.7) and (5.16) as

$$\boxed{\lambda = \frac{v}{f}} \tag{5.22}$$

The important dimensions of an electromagnetic structure such as a transmission line or an antenna are not physical dimensions such as meters but electrical dimensions in wavelengths. Table 5.3, repeated from Chapter 1, gives the wavelength at various frequencies in air.

TABLE 5.3 Wavelengths at Different Frequencies in Free Space (Air)

Frequency (f)	Wavelength (λ_o)
60 Hz	3107 miles (5000 km)
3 kHz	100 km
30 kHz	10 km
300 kHz	1 km
3 MHz	100 m
30 MHz	10 m
300 MHz	**1 m**
3 GHz	10 cm
30 GHz	1 cm
300 GHz	0.1 cm

This was discussed in Chapter 1. Recall that unless the maximum dimension of an electric circuit is much less than a wavelength, say, $\frac{1}{10}\lambda$, then lumped-circuit models of it are no longer valid. The structure has become a *distributed-parameter* circuit wherein the phase shift between different parts of the circuit are no longer negligible with respect to wavelength. In order to determine the *electrical dimensions* of a structure that is excited at a frequency f, we can simply determine the wavelength at that frequency by scaling the result at 300 MHz where a wavelength is 1 m and realizing that higher frequencies have shorter wavelengths and vice versa. For example, a 2-m length of wire has an electrical dimension of $\lambda_o/2$ at 75 MHz. A 60-Hz power transmission line of length 300 miles can be analyzed using lumped-circuit theory and Kirchhoff's laws since the structure maximum dimension (its length) is $\frac{1}{10}\lambda_o$ at 60 Hz. On the other hand, consider a circuit of a cell phone operating at around 3 GHz. In order to analyze that circuit using lumped-circuit theory and Kirchhoff's laws, the maximum dimension of the circuit must not be larger than about 1 cm!

▶ **EXAMPLE 5.1**

Determine the intrinsic impedance, phase constant, velocity of propagation, and wavelength of a uniform plane wave at 1 GHz in (a) glass-epoxy used to construct printed circuit boards and in (b) silicon used to construct integrated circuits.

SOLUTION For glass-epoxy, $\varepsilon_r = 4.7$, $\mu_r = 1$ so that

$$\eta = \eta_o \sqrt{\frac{\mu_r}{\varepsilon_r}}$$

$$= 120\pi \sqrt{\frac{1}{4.7}}$$

$$= 173.9 \quad \Omega$$

$$\beta = \omega\sqrt{\mu\varepsilon}$$

$$= 2\pi f \frac{\sqrt{\mu_r\varepsilon_r}}{v_o}$$

$$= 2\pi \times 1 \times 10^9 \frac{\sqrt{1 \times 4.7}}{3 \times 10^8}$$

$$= 45.41 \quad \text{radians/m}$$

$$= 2{,}601.5 \quad \text{degrees/m}$$

$$v = \frac{1}{\sqrt{\mu\varepsilon}}$$

$$= \frac{v_o}{\sqrt{\mu_r\varepsilon_r}}$$

$$= \frac{3 \times 10^8}{\sqrt{1 \times 4.7}}$$

$$= 1.38 \times 10^8 \quad \text{m/s}$$

$$\lambda = \frac{2\pi}{\beta}$$

$$= \frac{v}{f}$$

$$= 0.138 \quad \text{m}$$

$$= 13.8 \quad \text{cm}$$

For silicon, $\varepsilon_r = 12$, $\mu_r = 1$ so that

$$\eta = \eta_o \sqrt{\frac{\mu_r}{\varepsilon_r}}$$

$$= 120\pi \sqrt{\frac{1}{12}}$$

$$= 109 \quad \Omega$$

$$\beta = \omega\sqrt{\mu\varepsilon}$$

$$= 2\pi f \frac{\sqrt{\mu_r\varepsilon_r}}{v_o}$$

$$= 2\pi \times 1 \times 10^9 \frac{\sqrt{1 \times 12}}{3 \times 10^8}$$

$$= 72.55 \quad \text{radians/m}$$

$$= 4{,}157 \quad \text{degrees/m}$$

$$v = \frac{1}{\sqrt{\mu\varepsilon}}$$

$$= \frac{v_o}{\sqrt{\mu_r\varepsilon_r}}$$

$$= \frac{3 \times 10^8}{\sqrt{1 \times 12}}$$

$$= 8.66 \times 10^7 \quad \text{m/s}$$

$$\lambda = \frac{2\pi}{\beta}$$

$$= \frac{v}{f}$$

$$= 8.66 \quad \text{cm}$$

▶ **EXAMPLE 5.2**

Write the phasor and time-domain forms of a uniform plane having a frequency of 1 GHz that is traveling in the $+z$ direction in a large block of silicon.

SOLUTION   Using the results of the previous example, the phasor forms are

$$\hat{E}_x = E_m^+ e^{-j72.55z} \quad \text{V/m}$$

$$\hat{H}_y = \frac{E_m^+}{109} e^{-j72.55z} \quad \text{A/m}$$

and the time-domain forms are

$$E_x = E_m^+ \cos(6.28 \times 10^9 t - 72.55z) \quad \text{V/m}$$

$$H_y = \frac{E_m^+}{109} \cos(6.28 \times 10^9 t - 72.55z) \quad \text{A/m}$$

◀

▶ **QUICK REVIEW EXERCISE 5.1**

Determine the intrinsic impedance, phase constant, velocity of propagation, and wavelength of a uniform plane wave at 500 kHz in Teflon. Write the time-domain expressions for the electric and magnetic field intensity vectors. The electric field vector is in the x direction but the wave is propagating in the $-z$ direction.

ANSWERS   260 Ω, 1.52×10^{-2} rad/m, 2.07×10^8 m/s, 414 m, $E_x = E_m^- \cos(3.14 \times 10^6 t + 1.52 \times 10^{-2} z)$ V/m, $H_y = -E_m^-/260 \cos(3.14 \times 10^6 t + 1.52 \times 10^{-2} z)$ A/m. ◀

5.2 UNIFORM PLANE WAVES IN LOSSY MEDIA

We next consider uniform plane waves traveling in a medium where the conductivity is nonzero, $\sigma \neq 0$, that is, a *lossy medium*. Adding the conduction current term to Ampere's law results in modification of (5.5) [see (5.4)] as

$$\frac{\partial \hat{H}_y}{\partial z} = -\sigma \hat{E}_x - j\omega \varepsilon \hat{E}_x$$
$$= -(\sigma + j\omega \varepsilon) \hat{E}_x$$

(5.23)

The result in (5.3) derived from Faraday's law remains unchanged since Faraday's law does not contain conductivity:

$$\frac{\partial \hat{E}_x}{\partial z} = -j\omega \mu \hat{H}_y$$

(5.24)

Equations (5.23) and (5.24) each contain both $\hat{E}_x$ and $\hat{H}_y$. In order to convert these into individual differential equations, each containing only one variable as was done to obtain (5.6) for lossless media, we differentiate one with respect to z and substitute the other to yield

$$\frac{d^2 \hat{E}_x}{d^2 z} - \underbrace{[(j\omega \mu)(\sigma + j\omega \varepsilon)]}_{\hat{\gamma}^2} \hat{E}_x = 0$$

(5.25a)

$$\frac{d^2 \hat{H}_y}{d^2 z} - \underbrace{[(j\omega \mu)(\sigma + j\omega \varepsilon)]}_{\hat{\gamma}^2} \hat{H}_y = 0$$

(5.25b)

Grouping terms as shown, $\hat{\gamma}$ is said to be the *propagation constant* and is

$$\boxed{\begin{aligned} \hat{\gamma} &= \sqrt{(j\omega\mu)(\sigma + j\omega\varepsilon)} \\ &= \alpha + j\beta \end{aligned}}$$

(5.26)

Observe that the propagation constant is a complex number having a real part, α, which is said to be the *attenuation constant* for reasons soon to be apparent, and an imaginary part, β, which is again said to be the *phase constant*. It is very important to observe that for lossy media,

$$\beta \neq \omega\sqrt{\mu\varepsilon}$$

unlike the case for a lossless medium. In fact, we do not know what β will become until we compute the imaginary part of (5.26).

It is appropriate here to discuss the computation of the square root of a complex number as is required in (5.26). The square root of a real number is defined as that number which when multiplied by itself will yield the original number, that is, $a = \sqrt{a}\sqrt{a}$. The same notion applies to the square root of a complex number:

$$\begin{aligned} \sqrt{\hat{A}} &= \sqrt{A\angle\theta_A} \\ &= \sqrt{A}\angle\left(\frac{\theta_A}{2}\right) \end{aligned}$$

(5.27)

In other words, *the square root of a complex number has a magnitude which is the square root of the magnitude of the number and an angle which is one-half of the angle of the number.* For example, consider the complex number $\hat{A} = 2 + j2$. First write the number in *polar form* as $\hat{A} = 2\sqrt{2}\angle 45°$. Hence the square root is $\sqrt{\hat{A}} = \underbrace{\sqrt{2\sqrt{2}}}_{1.68}\angle 22.5°$. That this is the square root can be verified by multiplying it by itself to see if we obtain the original complex number: $\hat{A} = \sqrt{\hat{A}}\sqrt{\hat{A}} = (1.68\angle 22.5°)(1.68\angle 22.5°) = \underbrace{2.83\angle 45°}_{2\sqrt{2}}$. We can then convert the polar form of the square root to rectangular form as is required to obtain α and β in (5.26). For example, $\sqrt{2 + j2} = 1.68\angle 22.5° = 1.55 + j0.644$.

▷ **EXAMPLE 5.3**

Determine the attenuation constant and phase constant in copper ($\sigma = 5.8 \times 10^7$ S/m, $\varepsilon_r = 1, \mu_r = 1$) at 1 MHz. Copper is a common metal used to construct wires and electromagnetic shields in telecommunication systems.

SOLUTION Form the propagation constant:

$$\hat{\gamma} = \sqrt{\left(\underbrace{j2\pi \times 10^6}_{\omega} \times \underbrace{1 \times 4\pi \times 10^{-7}}_{\mu}\right)\left(\underbrace{5.8 \times 10^7}_{\sigma} + \underbrace{j2\pi \times 10^6}_{\omega} \times 1 \times \underbrace{\frac{1}{36\pi} \times 10^{-9}}_{\varepsilon}\right)}$$

$$= \sqrt{-4.39 \times 10^{-4} + j4.58 \times 10^8}$$

$$= \sqrt{4.58 \times 10^8 \angle 90°}$$

$$= 2.14 \times 10^4 \angle 45°$$

$$= \underbrace{1.51 \times 10^4}_{\alpha} + \underbrace{j1.51 \times 10^4}_{\beta}$$

◁

▶ QUICK REVIEW EXERCISE 5.2

Determine the attenuation constant and phase constant in seawater having $\sigma = 4$ S/m, $\varepsilon_r = 81$, $\mu_r = 1$ at 1 MHz.

ANSWERS $\alpha = 3.97$, $\beta = 3.98$ rad/m $= 228$ degrees/m. ◀

The equations in (5.25) governing these fields are again linear, constant coefficient, ordinary differential equations, but the coefficients are complex ($\hat{\gamma}^2$). However this does not change the basic solution, which is, separating the propagation constant as $\hat{\gamma} = \alpha + j\beta$,

$$\hat{E}_x = \hat{E}_m^+ e^{-\alpha z} e^{-j\beta z} + \hat{E}_m^- e^{\alpha z} e^{j\beta z} \tag{5.28a}$$

$$\hat{H}_y = \frac{\hat{E}_m^+}{\hat{\eta}} e^{-\alpha z} e^{-j\beta z} - \frac{\hat{E}_m^-}{\hat{\eta}} e^{\alpha z} e^{j\beta z} \tag{5.28b}$$

The forms of the solutions for lossy media are virtually identical to those for lossless media and consist of forward- and backward-traveling waves. But there are a few important differences. First, observe that the terms $\hat{E}_m^+ e^{-\alpha z}$ and $\hat{E}_m^- e^{+\alpha z}$ are the *amplitudes* of the forward- and backward-traveling waves, respectively. Losses in the medium introduced by a nonzero conductivity, $\sigma \neq 0$, introduce the $e^{\pm \alpha z}$ terms, which *cause the amplitudes of the forward- and backward-traveling waves to decay as they travel through the medium*. This is referred to as *attenuation* of the wave. Second, once again we cannot determine the propagation constant without carrying out the computation in (5.26). In other words, $\beta \neq \omega\sqrt{\mu\varepsilon}$ for the wave in a lossy medium. The third important difference is that the intrinsic impedance is no longer real but is now complex in a lossy medium and is given by

$$\begin{aligned}\hat{\eta} &= \sqrt{\frac{(j\omega\mu)}{(\sigma + j\omega\varepsilon)}} \\ &= \eta \angle \theta_\eta \\ &= \eta\, e^{j\theta_\eta}\end{aligned} \tag{5.29}$$

Hence, the intrinsic impedance has a magnitude and an angle. We have used the important equivalence:

$$M\angle\theta \equiv M e^{j\theta}$$

to write the result in an equivalent form. Therefore we can rewrite (5.28) as

$$\hat{E}_x = \hat{E}_m^+ e^{-\alpha z} e^{-j\beta z} + \hat{E}_m^- e^{\alpha z} e^{j\beta z} \tag{5.30a}$$

$$\hat{H}_y = \frac{\hat{E}_m^+}{\eta} e^{-\alpha z} e^{-j\beta z} e^{-j\theta_\eta} - \frac{\hat{E}_m^-}{\eta} e^{\alpha z} e^{j\beta z} e^{-j\theta_\eta} \tag{5.30b}$$

Observe in (5.30b) that $\hat{\eta}$ is in the denominator and hence when we bring the angle to the numerator, we change the sign:

$$\frac{1}{\eta \angle \theta_\eta} = \frac{1}{\eta} e^{-j\theta_\eta}$$

▷ **EXAMPLE 5.4**

Determine the intrinsic impedance of copper at 1 MHz.

SOLUTION The conductivity of copper is 5.8×10^7 S/m and the relative permittivity and permeability are unity. Hence the intrinsic impedance is

$$\hat{\eta} = \sqrt{\frac{(j\omega\mu)}{(\sigma + j\omega\varepsilon)}}$$

$$= \sqrt{\frac{\left(j\underbrace{2\pi \times 10^6}_{\omega} \times \underbrace{1 \times 4\pi \times 10^{-7}}_{\mu}\right)}{\left(\underbrace{5.8 \times 10^7}_{\sigma} + j\underbrace{2\pi \times 10^6}_{\omega} \times 1 \times \underbrace{\frac{1}{36\pi} \times 10^{-9}}_{\varepsilon}\right)}}$$

$$= \sqrt{1.36 \times 10^{-7} \angle 90°}$$

$$= 3.69 \times 10^{-4} \angle 45°$$

Hence $\eta = 3.69 \times 10^{-4}$ and $\theta_\eta = 45°$. ◀

▷ **QUICK REVIEW EXERCISE 5.3**

Determine the intrinsic impedance of seawater ($\sigma = 4$ S/m, $\varepsilon_r = 81$, $\mu_r = 1$) at 1 MHz.

ANSWER $\hat{\eta} = 1.4 \angle 45°$. ◀

The time-domain fields are obtained from (5.30) in the usual fashion by (1) multiplying by $e^{j\omega t}$ and (2) taking the real part of that result:

$$E_x = \text{Re}[(\hat{E}_m^+ e^{-\alpha z} e^{-j\beta z} + \hat{E}_m^- e^{\alpha z} e^{j\beta z}) e^{j\omega t}]$$
$$= E_m^+ e^{-\alpha z} \cos(\omega t - \beta z) + E_m^- e^{\alpha z} \cos(\omega t + \beta z) \tag{5.31a}$$

$$H_y = \text{Re}\left[\left(\frac{\hat{E}_m^+}{\eta} e^{-\alpha z} e^{-j\beta z} e^{-j\theta_\eta} - \frac{\hat{E}_m^-}{\eta} e^{\alpha z} e^{j\beta z} e^{-j\theta_\eta}\right) e^{j\omega t}\right]$$
$$= \frac{E_m^+}{\eta} e^{-\alpha z} \cos(\omega t - \beta z - \theta_\eta) - \frac{E_m^-}{\eta} e^{\alpha z} \cos(\omega t + \beta z - \theta_\eta) \tag{5.31b}$$

and we have assumed that the undetermined constants are real: $\hat{E}_m^\pm = E_m^\pm$. Again, these fields are very similar to those in a lossless medium in that the general solution may contain forward- and backward-traveling waves. However, the two main distinctions are that (1) the amplitudes decay as the waves travel through the lossly medium due to the presence of the $e^{\pm\alpha z}$ terms, and (2) the magnetic field *lags* the electric field by the phase angle of the intrinsic impedance, θ_η. For example, we found in Example 5.4 that the phase angle of the intrinsic impedance in copper at 1 MHz is 45°. Hence the magnetic field lags the electric field by 45°. Figure 5.3 illustrates this decay in amplitude with distance where we have shown a forward-traveling electric field wave for a fixed time:

$$E_x = E_m^+ e^{-\alpha z} \cos(\omega t - \beta z)$$

$$= E_m^+ e^{-\alpha z} \cos\left(\omega\left(t - \frac{z}{(\omega/\beta)}\right)\right)$$

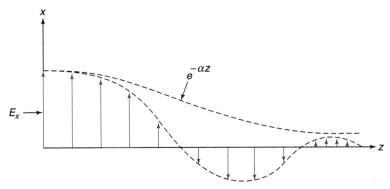

Figure 5.3 Attenuation of the amplitude of a uniform plane wave as it travels through a lossy medium.

Observe that the velocity of propagation in the medium is

$$v = \frac{\omega}{\beta}$$ (5.32)

It is important to observe that for the lossy medium case the velocity of propagation of the wave is not simply

$$v \neq \frac{1}{\sqrt{\mu\varepsilon}}$$

as was the case for lossless media. For a lossy medium we must calculate the propagation constant as the imaginary part of the propagation constant as in (5.26) and then compute the velocity of propagation from (5.32). In a fashion similar to the lossless medium case, we may define a wavelength as the distance the wave must travel to change phase by 2π radians or 360°. Since the phase constant β gives the phase shift per distance, we have

$$\lambda = \frac{2\pi}{\beta} \quad \text{m}$$ (5.33)

Again, like the lossless case, we may compute the wavelength as

$$\lambda = \frac{v}{f}$$ (5.34)

but we must use the velocity of propagation computed from (5.32).

▶ **EXAMPLE 5.5**

Determine the velocity of propagation and wavelength in copper at 1 MHz.

SOLUTION Since $\sigma = 5.8 \times 10^7$ S/m, $\varepsilon_r = 1$, $\mu_r = 1$ in copper, we obtain from Example 5.3, $\beta = 1.51 \times 10^4$ rad/m so that

$$v = \frac{2\pi \times 10^6}{1.51 \times 10^4}$$
$$= 416 \quad \text{m/s}$$

The wavelength is

$$\lambda = \frac{2\pi}{1.51 \times 10^4}$$

$$= 4.16 \times 10^{-4} \quad \text{m}$$

$$= \frac{v}{f}$$

▷ QUICK REVIEW EXERCISE 5.4

Determine the velocity of propagation and wavelength in seawater at 1 MHz.

ANSWERS 1.58×10^6 m/s, 1.58 m.

▷ EXAMPLE 5.6

Write the time-domain expressions for the electric and magnetic field vectors for a wave travel-ing in the $-z$ direction in copper and having a frequency of 1 MHz.

SOLUTION   From (5.31) and the results of the previous example problems, we obtain

$$E_x = E_m^- e^{1.51 \times 10^4 z} \cos(6.28 \times 10^6 t + 1.51 \times 10^4 z)$$

$$H_y = -\frac{E_m^-}{3.69 \times 10^{-4}} e^{1.51 \times 10^4 z} \cos(6.28 \times 10^6 t + 1.51 \times 10^4 z - 45°)$$

Observe that since the fields are stipulated to be traveling in the negative z direction ($-z$), the exponent of the attenuation as well as the sign in front of the phase shift term are both positive. Also observe that the sign of the angle of the intrinsic impedance is always negative since the in-trinsic impedance is in the denominator of the phasor expression for $\hat{H}_y$. Also, the sign of H_y is negative since for this wave traveling in the negative z direction, $\mathbf{E} \times \mathbf{H}$ must be in the negative z direction.

▷ QUICK REVIEW EXERCISE 5.5

Write the time-domain expressions for the electric and magnetic field vectors for a wave travel-ing in the $-z$ direction in seawater and having a frequency of 1 MHz.

ANSWERS $E_x = E_m^- e^{3.97 z} \cos(6.28 \times 10^6 t + 3.98 z)$
$H_y = (-E_m^-/1.4) e^{3.97 z} \cos(6.28 \times 10^6 t + 3.98 z - 45°)$.

▷ 5.3 POWER FLOW IN UNIFORM PLANE WAVES

Consider a forward-traveling uniform plane wave traveling in a *lossless* medium. The phasor fields written in vector form are

$$\hat{\mathbf{E}} = \hat{E}_m e^{-j\beta z} \mathbf{a}_x \tag{5.35a}$$

$$\hat{\mathbf{H}} = \frac{\hat{E}_m}{\eta} e^{-j\beta z} \mathbf{a}_y \tag{5.35b}$$

We will be interested in the flow of power that is carried in this wave. Of particular in-terest will be the *time-average power density* representing the flow of real power. This is similar to the average power delivered to a lumped-circuit element. For example,

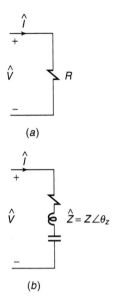

Figure 5.4 Comparison of the calculation of average power density in a uniform plane wave to a similar computation for lumped electric circuits. (a) Analogy to a plane wave in a lossless medium. (b) Analogy to a plane wave in a lossy medium.

consider the lumped-circuit resistor shown in Fig. 5.4a. The phasor voltage and current associated with the terminals are denoted as $\hat{V}$ and $\hat{I}$, respectively. The time-average (or, simply, average) power delivered to the element is

$$P_{AV} = \frac{1}{2} \text{Re}(\hat{V}\hat{I}^{*})$$

$$= \frac{1}{2}|\hat{I}|^2 R \qquad (5.36)$$

$$= \frac{|\hat{V}|^2}{2R} \quad \text{W}$$

where the star superscript indicates the conjugate of the complex number. We can similarly determine the average power density in the forward-traveling wave in (5.35) through the time-average Poynting vector:

$$\mathbf{S}_{AV} = \frac{1}{2} \text{Re}(\hat{\mathbf{E}} \times \hat{\mathbf{H}}^{*}) \qquad \text{W/m}^2 \qquad (5.37)$$

Observe that the units of this are power distributed over a surface area. Substituting (5.35) into this expression yields

$$\mathbf{S}_{AV} = \frac{1}{2} \text{Re}\left(\hat{E}_m e^{-j\beta z}\mathbf{a}_x \times \frac{\hat{E}_m^{*}}{\eta} e^{j\beta z}\mathbf{a}_y \right)$$

$$= \frac{|\hat{E}_m|^2}{2\eta} \mathbf{a}_z \qquad \text{W/m}^2 \qquad (5.38)$$

Observe in this result that the product of a complex number and its conjugate is the magnitude squared: $\hat{E}_m\hat{E}_m^{*} = |\hat{E}_m|^2$. Also note that the exponential in the H field is positive as required by the conjugate of that field vector. Observe that this power density is a vector and is directed in the z direction. This is a sensible result since the wave is traveling in the z direction. This result indicates that average (real) power is carried by the wave and this power is distributed over the plane in which the field vectors are contained, the xy plane. In order to determine the total average power passing through some surface area A of this plane, we write

$$
P_{AV} = \int_s \mathbf{S}_{AV} \cdot d\mathbf{s}
$$
$$
= \frac{|\hat{E}_m|^2}{2\eta} A \quad \text{W}
$$

(5.39)

Observe the similarity of this result to that for a lumped-circuit element in (5.36). The parallels are striking. The primary difference between lumped-circuit elements and electromagnetic fields is that lumped elements are models that assume the fields are confined to within the element and there is no interaction with the rest of the circuit except at the terminals.

Next consider a forward-traveling uniform plane wave traveling in a *lossy* medium. The phasor fields written in vector form are

$$
\hat{\mathbf{E}} = \hat{E}_m e^{-\alpha z} e^{-j\beta z} \, \mathbf{a}_x
$$

(5.40a)

$$
\hat{\mathbf{H}} = \frac{\hat{E}_m}{\eta} e^{-\alpha z} e^{-j\beta z} e^{-j\theta_\eta} \, \mathbf{a}_y
$$

(5.40b)

Observe again that the only differences between fields in a lossy medium and fields in a lossless medium are the presence of the attenuation terms, $e^{-\alpha z}$, and the angle of the intrinsic impedance (which is complex in a lossy medium, $\hat{\eta} = \eta \angle \theta_\eta$), $e^{j\theta_\eta}$. The average-power Poynting vector is similarly obtained as

$$
\mathbf{S}_{AV} = \frac{1}{2} \mathrm{Re}(\hat{\mathbf{E}} \times \hat{\mathbf{H}}^\circ)
$$
$$
= \frac{1}{2} \mathrm{Re}\left(\hat{E}_m e^{-\alpha z} e^{-j\beta z} \, \mathbf{a}_x \times \frac{\hat{E}_m^\circ}{\eta} e^{-\alpha z} e^{j\beta z} e^{j\theta_\eta} \, \mathbf{a}_y \right)
$$
$$
= \frac{|\hat{E}_m|^2}{2\eta} e^{-2\alpha z} \cos(\theta_\eta) \, \mathbf{a}_z \quad \text{W/m}^2
$$

(5.41)

The average power passing through an area A in the xy plane is then

$$
P_{AV} = \frac{|\hat{E}_m|^2}{2\eta} e^{-2\alpha z} \cos(\theta_\eta) A \quad \text{W}
$$

(5.42)

This should be contrasted with the average power delivered to a lumped-circuit element that has a complex impedance, $\hat{Z} = Z \angle \theta_Z$, as shown in Fig. 5.4b:

$$
P_{AV} = \frac{1}{2} \mathrm{Re}(\hat{V}\hat{I}^\circ)
$$
$$
= \frac{1}{2}|\hat{I}|^2 Z \cos(\theta_Z)
$$
$$
= \frac{|\hat{V}|^2}{2Z} \cos(\theta_Z) \quad \text{W}
$$

(5.43)

The significant difference between the two expressions is that the expression for average power in the wave contains the term $e^{-2\alpha z}$, which indicates that power is being dissipated in the medium as the wave passes through it.

▶ **EXAMPLE 5.7**

A 1-MHz, 1-V/m uniform plane wave is traveling through a block of copper. Determine the power dissipated in the copper over a distance of 1 μm with surface area of 2m ^{2}.

SOLUTION The attenuation constant and intrinsic impedance were determined in Examples 5.3 and 5.4 as $\alpha = 1.51 \times 10^4$ and $\hat{\eta} = 3.69 \times 10^{-4} \angle 45°$. Hence the power dissipated in the block is the difference between the power entering the block and the power leaving it using (5.42):

$$P_{\text{dissipated}} = P_{\text{AV}}|_{z=0} - P_{\text{AV}}|_{z=1\,\mu\text{m}}$$

$$= \frac{|\hat{E}_m|^2}{2\eta} \cos(\theta_\eta)A - \frac{|\hat{E}_m|^2}{2\eta} e^{-2\alpha 1\mu\text{m}} \cos(\theta_\eta)A$$

$$= \frac{|\hat{E}_m|^2}{2\eta} \cos(\theta_\eta)A(1 - e^{-2\alpha 1\mu\text{m}})$$

$$= \frac{1}{2 \times 3.69 \times 10^{-4}} \cos(45°)(2 \text{ m}^2)(1 - e^{-2 \times 1.51 \times 10^4 \times 10^{-6}})$$

$$= 57 \quad \text{W}$$

◀

▷ **QUICK REVIEW EXERCISE 5.6**

A 1-MHz, 10-V/m uniform plane wave is traveling through seawater. Determine the power dissipated over a distance of 1 mm with surface area of 2 m².

ANSWER 0.4 W. ◀

▶ **5.4 SKIN DEPTH**

A uniform plane wave traveling through a lossy medium ($\sigma \neq 0$) suffers an attenuation in that its amplitude is diminished as it travels through that medium. This is illustrated in Fig. 5.3 for a forward-traveling wave

$$E_x = E_m^+ e^{-\alpha z} \cos(\omega t - \beta z) \tag{5.44}$$

Skin depth is the distance the wave must travel to have its amplitude reduced by a factor of 1/e or approximately 37%. From the above form of the wave, we see that the skin depth is obtained by setting $\alpha z = 1$, giving

$$\boxed{\delta = \frac{1}{\alpha} \quad \text{m}} \tag{5.45}$$

where we denote the *skin depth* as δ.

We classify media as being either a "good conductor" or a "good dielectric" according to whether the conduction current in Ampere's law is much greater than or much less than the displacement current. Ampere's law in phasor form is

$$\nabla \times \hat{\mathbf{H}} = \underbrace{\sigma \hat{\mathbf{E}}}_{\substack{\text{conduction} \\ \text{current}}} + j \underbrace{\omega \varepsilon \hat{\mathbf{E}}}_{\substack{\text{displacement} \\ \text{current}}} \tag{5.46}$$

Hence we arrive at the characterization

$$\frac{\sigma}{\omega \varepsilon} \gg 1 \quad \text{good conductor} \tag{5.47a}$$

$$\frac{\sigma}{\omega \varepsilon} \ll 1 \quad \text{good dielectric} \tag{5.47b}$$

If the medium is a good conductor, we can determine an approximation for the attenuation constant. The propagation constant for a good conductor is

$$\hat{\gamma} = \alpha + j\beta$$
$$= \sqrt{(j\omega\mu)(\sigma + j\omega\varepsilon)}$$
$$\cong \sqrt{j\omega\mu\sigma} \tag{5.48}$$
$$= \sqrt{\omega\mu\sigma} \angle 45°$$
$$= \underbrace{\sqrt{\frac{\omega\mu\sigma}{2}}}_{\alpha} + j\underbrace{\sqrt{\frac{\omega\mu\sigma}{2}}}_{\beta} \quad \text{good conductor } \sigma \gg \omega\varepsilon$$

Hence we identify the skin depth as

$$\boxed{\begin{aligned} \delta &= \frac{1}{\alpha} \\ &= \frac{1}{\sqrt{\pi f \mu\sigma}} \quad \text{m} \end{aligned}} \tag{5.49}$$

Observe that the skin depth *decreases* with an *increase* in frequency at a rate of the square root of that frequency. The skin depth for several frequencies is given for a copper conductor ($\sigma = 5.8 \times 10^7\,\text{S/m}$) in Table 5.4:

Hence the attenuation and phase constants for a good conductor can be written in terms of the skin depth:

$$\boxed{\begin{aligned} \alpha = \beta &= \frac{1}{\delta} \\ &= \sqrt{\pi f \mu\sigma} \quad \text{good conductor } \sigma \gg \omega\varepsilon \end{aligned}} \tag{5.50}$$

The intrinsic impedance for a good conductor can also be written in terms of the skin depth for a good conductor:

$$\hat{\eta} = \sqrt{\frac{j\omega\mu}{\sigma + j\omega\varepsilon}}$$
$$\cong \sqrt{\frac{j\omega\mu}{\sigma}} \tag{5.51}$$
$$= \sqrt{\frac{\omega\mu}{\sigma}} \angle 45°$$

TABLE 5.4   Skin Depth in Copper

Frequency (f)	Skin depth
60 Hz	8.53 mm
1 MHz	66.1 μm
10 MHz	20.9 μm
100 MHz	6.6 μm
1 GHz	2.09 μm

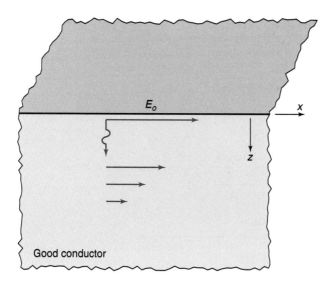

Figure 5.5 Illustration of the attenuation of a wave traveling in a good conductor and skin depth.

Hence

$$
\hat{\eta} = \frac{\sqrt{2}}{\sigma\delta}\angle 45°
$$

$$
= \frac{1}{\sigma\delta}(1+j) \qquad \text{good conductor } \sigma \gg \omega\varepsilon
$$

(5.52)

The practical significance of skin depth is that *currents in conductors tend to be confined predomnantly to a few skin depths of the surface of that conductor.* For example, consider a large block of a good conductor illustrated in Fig. 5.5. A uniform plane wave is propagating downward into the conductor with the electric field tangent to the conductor surface. Hence the electric field is given by

$$
E_x = E_o e^{-\frac{z}{\delta}} \cos\left(\omega t - \frac{z}{\delta}\right) \qquad z \geq 0
$$

(5.53)

where we have substituted $\alpha = \beta = 1/\delta$ into the general form given in (5.44) and E_o is the amplitude at $z = 0$ (the surface). The current associated with this field is

$$
J_x = \sigma E_x
$$

$$
= \sigma E_o e^{-\frac{z}{\delta}} \cos\left(\omega t - \frac{z}{\delta}\right)
$$

(5.54)

Hence the induced current decays with distance into the conductor. The majority of the current exists within a few skin depths of the surface.

This important fact is put to use in numerous practical situations. For example, consider the construction of a 60-Hz high-voltage power line. There are two competing considerations in designing the line conductors. First, the metal chosen must have strength to support the long lengths of line. Second, the chosen metal must have good conductivity to reduce the ohmic losses. From the standpoint of strength, steel is an obvious option but has a conductivity only one-tenth that of copper ($\sigma_{\text{steel}} = 0.1\sigma_{\text{copper}}$) as shown in Table 5.2. The solution to this is to use a core of steel to provide strength and surround that with a sheath of better conductivity such as aluminum as shown in

Fig. 5.6a. The 60-Hz currents will flow primarily in the aluminum sheath so long as its thickness is a few skin depths. The skin depth of aluminum ($\sigma_{\text{aluminum}} = 0.61\sigma_{\text{copper}}$) at 60 Hz is 1.1 cm. Hence for a sheath thickness of a few centimeters, the majority of the current is confined to the aluminum sheath and very little penetrates into the steel core.

Another important application of the concept of skin depth is in the construction of shielded enclosures. "Shields" are used to isolate electronic equipments to prevent the emissions of one from interfering with the other. An example is a shielded room illustrated in Fig. 5.6b. Sensitive electronic equipment is housed within the shielded room in order to protect it from the electromagnetic fields generated, for example, by a radar transmitter that is outside the room. The radar transmitter fields will impinge on the shield wall and induce currents in the wall. However, if the thickness of the conducting wall is on the order of several skin depths, the induced current on the outside will decay significantly before reaching the other side of the wall, hence reducing its level and preventing its reception by the electronics inside the room. If the offending transmitter is a radar transmitter operating at 3 GHz and the room walls are constructed of steel ($\sigma_r = 0.1\sigma_{Cu}$, $\mu_r = 2000$), a thickness of 0.17 μm would constitute 2 skin depths. Of course the wall thickness would no doubt be chosen to be much larger than this for structural reasons, but this illustrates the attenuation of the field due to the concept of skin depth. This idea of "shielding effectiveness" of enclosures will be discussed in more detail in Section 5.7, Engineering Applications.

Another application of the concept of skin depth is illustrated in Fig. 5.7. A wire (circular cross section conductor) of radius r_w is shown. At dc, the current will be uniformly distributed over the wire cross section. Hence the dc resistance per unit length is

$$r_{dc} = \frac{1}{\sigma \pi r_w^2} \quad \Omega/\text{m} \quad r_w \ll \delta$$

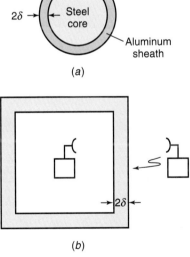

(a)

(b)

Figure 5.6 Illustration of the practical use of the concept of skin depth. (a) Construction of a high-voltage power cable. (b) Protecting sensitive electronics from external electromagnetic fields with a shielded enclosure.

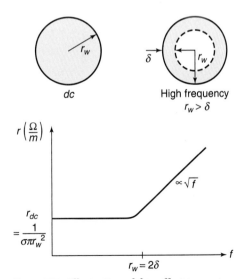

Figure 5.7 Illustration of skin effect in causing the high-frequency resistance of a wire to increase with the square root of frequency.

As the frequency of the current increases, the current will crowd closer to the wire surface as illustrated in Fig. 5.7. At high frequencies where the wire radius is on the order of a skin depth, the cross-sectional area occupied by the current is smaller, giving the high-frequency per-unit-length resistance as

$$r_{hf} = \frac{1}{\sigma 2\pi r_w \delta} \qquad \Omega/\text{m} \qquad r_w \gg \delta$$

Since the skin depth is

$$\delta = \frac{1}{\sqrt{\pi f \mu \sigma}}$$

$$\propto \frac{1}{\sqrt{f}}$$

the high-frequency resistance increases at a rate of $\sqrt{f}$. Hence above a frequency where the wire radius is on the order of two skin depths, the per-unit-length wire resistance begins to dramatically increase, and the high-frequency wire resistance greatly exceeds its dc resistance.

▶ 5.5 NORMAL INCIDENCE OF UNIFORM PLANE WAVES AT PLANE, MATERIAL BOUNDARIES

We now investigate the reflection and transmission of a wave traveling in one medium that is incident normal to the surface of another medium. Consider two *lossless* media shown in Fig. 5.8. The boundary between the two media lies in the xy plane with the z coordinate axis pointing into medium 2 as shown in Fig. 5.8. A uniform plane wave is incident from medium 1, which is characterized by $\varepsilon_1 = \varepsilon_{r1}\varepsilon_o$ and $\mu_1 = \mu_{r1}\mu_o$ normal to the surface of medium 2, which is characterized by $\varepsilon_2 = \varepsilon_{r2}\varepsilon_o$ and $\mu_2 = \mu_{r2}\mu_o$. The phasor form of the fields for this *incident wave* are

$$\hat{E}_x^i = E_m^i e^{-j\beta_1 z} \tag{5.55a}$$

and

$$\hat{H}_y^i = \frac{E_m^i}{\eta_1} e^{-j\beta_1 z} \tag{5.55b}$$

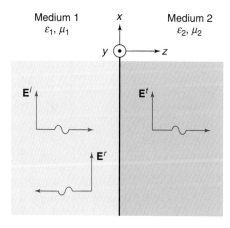

Figure 5.8 Illustration of the reflection and transmission of a uniform plane wave that is incident normal to the surface between two lossless media.

where the phase constant and intrinsic impedance of medium 1 are

$$\beta_1 = \omega\sqrt{\mu_1\varepsilon_1}$$
$$= \frac{\omega}{v_o}\sqrt{\mu_{r1}\varepsilon_{r1}} \tag{5.56a}$$

$$\eta_1 = \sqrt{\frac{\mu_1}{\varepsilon_1}}$$
$$= \eta_o\sqrt{\frac{\mu_{r1}}{\varepsilon_{r1}}} \tag{5.56b}$$

A portion of this incident wave will be reflected at the boundary. We refer to this wave as the *reflected wave* having the phasor form

$$\hat{E}_x^r = E_m^r e^{j\beta_1 z} \tag{5.57a}$$

and

$$\hat{H}_y^r = -\frac{E_m^r}{\eta_1}e^{j\beta_1 z} \tag{5.57b}$$

Observe the sign of the exponential term; it must be positive since this is traveling in the $-z$ direction. Similarly, the magnetic field is directed in the $-y$ direction so that the power flow, $\mathbf{E}^r \times \mathbf{H}^r$ will be in the $-z$ direction. A portion of the incident wave will be transmitted across the boundary into medium 2. We refer to this wave as the *transmitted wave* whose phasor form is

$$\hat{E}_x^t = E_m^t e^{-j\beta_2 z} \tag{5.58a}$$

and

$$\hat{H}_y^t = \frac{E_m^t}{\eta_2}e^{-j\beta_2 z} \tag{5.58b}$$

where the phase constant and intrinsic impedance of medium 2 are

$$\beta_2 = \omega\sqrt{\mu_2\varepsilon_2}$$
$$= \frac{\omega}{v_o}\sqrt{\mu_{r2}\varepsilon_{r2}} \tag{5.59a}$$

$$\eta_2 = \sqrt{\frac{\mu_2}{\varepsilon_2}}$$
$$= \eta_o\sqrt{\frac{\mu_{r2}}{\varepsilon_{r2}}} \tag{5.59b}$$

We reason that there will not be a backward-traveling wave in medium 2 since it extends to infinity, and thus there is no rightmost boundary to produce this reflected wave.

The magnitude of the incident electric field, E_m^i, is assumed known. It is due to some distant source such as a radar and can be computed, as we will see when we consider antennas in Chapter 7. Hence there are only two unknowns, the amplitudes of the reflected and transmitted electric fields, E_m^r and E_m^t. In order to determine these we need two conditions. These conditions are the boundary conditions. Since the tangential electric fields must be continuous across a boundary, we obtain from (5.55a), (5.57a), and (5.58a) evaluated at $z = 0$:

$$E_m^i + E_m^r = E_m^t \tag{5.60}$$

Similarly, the tangential magnetic fields must be continuous across the boundary. Evaluating (5.55b), (5.57b), and (5.58b) at $z = 0$ yields

$$\frac{E_m^i}{\eta_1} - \frac{E_m^r}{\eta_1} = \frac{E_m^t}{\eta_2} \tag{5.61}$$

Solving (5.60) and (5.61) in terms of the magnitude of the incident electric field, E_m^i, yields the *reflection coefficient*

$$\boxed{\begin{aligned} \Gamma &= \frac{\eta_2 - \eta_1}{\eta_2 + \eta_1} \\ &= \frac{E_m^r}{E_m^i} \end{aligned}} \tag{5.62}$$

and the *transmission coefficient*

$$\boxed{\begin{aligned} T &= \frac{2\eta_2}{\eta_2 + \eta_1} \\ &= \frac{E_m^t}{E_m^i} \end{aligned}} \tag{5.63}$$

Observe that the reflection and transmission coefficients are related by

$$\boxed{1 + \Gamma = T} \tag{5.64}$$

Once we compute the reflection and transmission coefficients, we can determine the amplitudes (magnitude and sign) of the reflected and transmitted waves from the amplitude of the incident wave (which is assumed known):

$$E_m^r = \Gamma E_m^i \tag{5.65}$$

and

$$E_m^t = T E_m^i \tag{5.66}$$

The phasor forms of the fields are

$$\boxed{\hat{E}_x^i = E_m^i e^{-j\beta_1 z}} \tag{5.67a}$$

$$\boxed{\hat{H}_y^i = \frac{E_m^i}{\eta_1} e^{-j\beta_1 z}} \tag{5.67b}$$

$$\boxed{\hat{E}_x^r = \Gamma E_m^i e^{j\beta_1 z}} \tag{5.68a}$$

$$\boxed{\hat{H}_y^r = -\Gamma \frac{E_m^i}{\eta_1} e^{j\beta_1 z}} \tag{5.68b}$$

$$\boxed{\hat{E}_x^t = T E_m^i e^{-j\beta_2 z}} \tag{5.69a}$$

$$\boxed{\hat{H}_y^t = T \frac{E_m^i}{\eta_2} e^{-j\beta_2 z}} \tag{5.69b}$$

The complete time-domain forms of the fields are

$$E_x^i = E_m^i \cos(\omega t - \beta_1 z) \qquad \text{V/m} \tag{5.70a}$$

$$H_y^i = \frac{E_m^i}{\eta_1} \cos(\omega t - \beta_1 z) \qquad \text{A/m} \tag{5.70b}$$

$$E_x^r = \Gamma E_m^i \cos(\omega t + \beta_1 z) \qquad \text{V/m} \tag{5.71a}$$

$$H_y^r = -\Gamma \frac{E_m^i}{\eta_1} \cos(\omega t + \beta_1 z) \qquad \text{A/m} \tag{5.71b}$$

$$E_x^t = T E_m^i \cos(\omega t - \beta_2 z) \qquad \text{V/m} \tag{5.72a}$$

$$H_y^t = T \frac{E_m^i}{\eta_2} \cos(\omega t - \beta_2 z) \qquad \text{A/m} \tag{5.72b}$$

Observe the important minus sign in the expression for the reflected magnetic field in (5.68b) and (5.71b). This is essential because the power flow given by $\mathbf{E} \times \mathbf{H}$ must be in the $-z$ direction since the reflected wave is traveling in the $-z$ direction. Also note that the η and β appropriate to the region in which the wave is traveling must be used in these expressions. Observe in (5.62) that if $\eta_2 < \eta_1$, then the reflection coefficient will be a negative number. Hence the reflected electric field in (5.68a) and (5.71a) will be directed in the $-x$ direction. A recommended procedure that will always produce the correct signs is to

1   first compute Γ and T,

2   then write the expressions for the electric fields where the magnitude *and* sign are the products of the reflection and transmission coefficients and the incident electric field as shown in (5.65) and (5.66),

3   and finally, write the expressions for the magnetic fields by dividing the electric fields by the appropriate intrinsic impedance *and* ensure that the sign is such that power flow, $\mathbf{E} \times \mathbf{H}$, is in the appropriate direction for that wave.

▶ **EXAMPLE 5.8**

A 100-V/m, 100-MHz uniform plane wave is traveling in free space and is incident normal to the surface of a material having $\varepsilon_r = 9$ and $\mu_r = 4$. Write complete expressions for the incident, reflected, and transmitted electric and magnetic fields.

SOLUTION   With reference to the previous derivation, the parameters are

$$\eta_1 = \eta_o$$
$$= 377 \ \Omega$$

$$\eta_2 = \eta_o \sqrt{\frac{4}{9}}$$

$$= \frac{2}{3} \eta_o$$

$$= 251.33 \ \Omega$$

$$\beta_1 = 2\pi \times 10^8 \frac{\sqrt{1}}{3 \times 10^8}$$

$$= \frac{2\pi}{3} \quad \text{radians/m}$$

$$\beta_2 = 2\pi \times 10^8 \frac{\sqrt{4 \times 9}}{3 \times 10^8}$$

$$= 4\pi \quad \text{radians/m}$$

The reflection and transmission coefficients are then computed as

$$\Gamma = \frac{\frac{2}{3}\eta_o - \eta_o}{\frac{2}{3}\eta_o + \eta_o}$$

$$= -\frac{1}{5}$$

$$T = \frac{2\frac{2}{3}\eta_o}{\frac{2}{3}\eta_o + \eta_o}$$

$$= \frac{4}{5}$$

Observe that $1 + \Gamma = T$. Hence the incident fields are

$$E_x^i = 100 \cos\left(\omega t - \frac{2\pi}{3}z\right)$$

$$= 100 \cos(6.28 \times 10^8 t - 2.09z) \quad \text{V/m}$$

$$H_y^i = \frac{100}{\eta_o} \cos\left(\omega t - \frac{2\pi}{3}z\right)$$

$$= 0.27 \cos(6.28 \times 10^8 t - 2.09z) \quad \text{A/m}$$

The reflected and transmitted fields are

$$E_x^r = \Gamma E_x^i$$

$$= -\frac{1}{5} \times 100 \cos\left(\omega t + \frac{2\pi}{3}z\right)$$

$$= -20 \cos(6.28 \times 10^8 t + 2.09z) \quad \text{V/m}$$

$$H_y^r = -\frac{E_m^r}{\eta_o} \cos\left(\omega t + \frac{2\pi}{3}z\right)$$

$$= -\frac{-20}{377} \cos\left(\omega t + \frac{2\pi}{3}z\right)$$

$$= 0.053 \cos(6.28 \times 10^8 t + 2.09z) \quad \text{A/m}$$

$$E_x^t = T E_x^i$$

$$= \frac{4}{5} \times 100 \cos(\omega t - 4\pi z)$$

$$= 80 \cos(6.28 \times 10^8 t - 12.57z) \quad \text{V/m}$$

$$H_y^t = \frac{E_m^t}{\eta_2} \cos(\omega t - 4\pi z)$$

$$= \frac{80}{\frac{2}{3}\eta_o} \cos(\omega t - 4\pi z)$$

$$= 0.32 \cos(6.28 \times 10^8 t - 12.57z) \quad \text{A/m}$$

▷ **QUICK REVIEW EXERCISE 5.7**

A 10-V/m, 5-MHz uniform plane wave traveling in free space is incident normal to the surface of a lossless material having $\varepsilon_r = 1$ and $\mu_r = 16$. Write complete expressions for the incident, reflected, and transmitted waves.

ANSWERS

$E_x^i = 10 \cos(3.14 \times 10^7 t - 0.105z)$ V/m, $H_y^i = 0.0265 \cos(3.14 \times 10^7 t - 0.105z)$ A/m,

$E_x^r = 6 \cos(3.14 \times 10^7 t + 0.105z)$ V/m, $H_y^r = -0.0159 \cos(3.14 \times 10^7 t + 0.105z)$ A/m,

$E_x^t = 16 \cos(3.14 \times 10^7 t - 0.419z)$ V/m, $H_y^t = 0.0106 \cos(3.14 \times 10^7 t - 0.419z)$ A/m. ◁

According to conservation of energy, time-average power flow across the boundary must be conserved. This can be demonstrated rather easily. First, write the phasor expression for the *total* fields in the first medium, $z < 0$, in vector form as

$$
\begin{aligned}
\hat{\mathbf{E}}_1 &= [\hat{E}_x^i + \hat{E}_x^r]\mathbf{a}_x \\
&= E_m^i[e^{-j\beta_1 z} + \Gamma e^{j\beta_1 z}]\mathbf{a}_x
\end{aligned}
\tag{5.73a}
$$

and

$$
\begin{aligned}
\hat{\mathbf{H}}_1 &= \hat{H}_y^i + \hat{H}_y^r \\
&= \frac{E_m^i}{\eta_1}[e^{-j\beta_1 z} - \Gamma e^{j\beta_1 z}]\mathbf{a}_y
\end{aligned}
\tag{5.73b}
$$

Similarly, write the phasor expressions for the *total* fields in the second medium, $z > 0$, in vector form as

$$
\begin{aligned}
\hat{\mathbf{E}}_2 &= \hat{E}_x^t \mathbf{a}_x \\
&= T E_m^i e^{-j\beta_2 z}\mathbf{a}_x
\end{aligned}
\tag{5.74a}
$$

and

$$
\begin{aligned}
\hat{\mathbf{H}}_2 &= \frac{\hat{E}_x^t}{\eta_2}\mathbf{a}_y \\
&= T\frac{E_m^i}{\eta_2}e^{-j\beta_2 z}\mathbf{a}_y
\end{aligned}
\tag{5.74b}
$$

The average-power Poynting vector in the first medium is

$$
\begin{aligned}
\mathbf{S}_{AV,1} &= \frac{1}{2}\mathrm{Re}[\hat{\mathbf{E}}_1 \times \hat{\mathbf{H}}_1^*] \\
&= \frac{1}{2}\mathrm{Re}\left[E_m^i[e^{-j\beta_1 z} + \Gamma e^{j\beta_1 z}]\mathbf{a}_x \times \frac{E_m^i}{\eta_1}[e^{j\beta_1 z} - \Gamma e^{-j\beta_1 z}]\mathbf{a}_y\right] \\
&= \frac{(E_m^i)^2}{2\eta_1}\mathrm{Re}\left[1 + \Gamma\underbrace{(e^{j2\beta_1 z} - e^{-j2\beta_1 z})}_{2j\sin 2\beta_1 z} - \Gamma^2\right]\mathbf{a}_z \\
&= \frac{(E_m^i)^2}{2\eta_1}[1 - \Gamma^2]\mathbf{a}_z \qquad \mathrm{W/m^2}
\end{aligned}
\tag{5.75}
$$

Observe that the middle term, $j2\Gamma \sin(2\beta_1 z)$, is imaginary so that when we take the real part, this term is discarded. The average-power Poynting vector in the second medium is

$$
\begin{aligned}
\mathbf{S}_{AV,2} &= \frac{1}{2}\mathrm{Re}\left[\hat{\mathbf{E}}_2 \times \hat{\mathbf{H}}_2^*\right] \\
&= \frac{1}{2}\mathrm{Re}\left[TE_m^i e^{-j\beta_2 z}\mathbf{a}_x \times T\frac{E_m^i}{\eta_2}e^{j\beta_2 z}\mathbf{a}_y\right] \\
&= \frac{T^2(E_m^i)^2}{2\eta_2}\mathbf{a}_z \qquad \mathrm{W/m^2}
\end{aligned}
\tag{5.76}
$$

It is a simple matter to show that

$$
\frac{T^2}{\eta_2} = \frac{1}{\eta_1}[1 - \Gamma^2]
\tag{5.77}
$$

and hence the average power flow across the boundary is conserved as it should be.

▶ **EXAMPLE 5.9**

Show that average power is conserved for the problem of Example 5.8.

SOLUTION In Example 5.8 we determined that $\Gamma = -1/5$, $T = 4/5$, $\eta_1 = 377\ \Omega$, $\eta_2 = 251.33\ \Omega$. The incident electric field has magnitude of 100 V/m. Hence the average power density in the incident wave is

$$
\begin{aligned}
S_{AV}^i &= \frac{(E_m^i)^2}{2\eta_1} \\
&= 13.26 \qquad \mathrm{W/m^2}
\end{aligned}
$$

The average power density in the reflected wave is

$$
\begin{aligned}
S_{AV}^r &= \Gamma^2\frac{(E_m^i)^2}{2\eta_1} \\
&= 0.53 \qquad \mathrm{W/m^2}
\end{aligned}
$$

Hence the total average power density in this region is

$$
\begin{aligned}
S_{AV,1} &= S_{AV}^i - S_{AV}^r \\
&= 13.26 - 0.53 \\
&= 12.73 \qquad \mathrm{W/m^2} \\
&= \frac{(E_m^i)^2}{2\eta_1}[1 - \Gamma^2]
\end{aligned}
$$

Observe that the total power density is the sum of the incident and reflected wave power densities *but* the reflected power density must be subtracted since this wave is traveling in the $-z$ direction. The total average power density in the second region is just that contained in the transmitted wave:

$$
\begin{aligned}
S_{AV,2} &= S_{AV}^t \\
&= \frac{T^2(E_m^i)^2}{2\eta_2} \\
&= 12.73 \qquad \mathrm{W/m^2}
\end{aligned}
$$

Verify that conservation of average power is achieved for Quick Review Exercise Problem 5.7. ◀

5.5.1 Normal Incidence on Good Conductors

In the previous section we considered a boundary between two *lossless* materials. In this section we will modify those results for the second medium being a good conductor having parameters ε_{r2}, μ_{r2} and conductivity σ_2 such that $\sigma_2 \gg \omega\varepsilon_2$ as illustrated in Fig. 5.9. The only modifications from the previous results are that (1) the propagation constant in medium 2 has a real part representing attenuation of the transmitted wave and (2) the intrinsic impedance of medium 2 is now complex. These were obtained for good conductors in Equations (5.50) and (5.52):

$$\alpha_2 = \beta_2 = \frac{1}{\delta_2} = \sqrt{\pi f \mu_2 \sigma_2} \tag{5.78a}$$

$$\hat{\eta}_2 = \frac{\sqrt{2}}{\sigma_2 \delta_2} \angle 45° \tag{5.78b}$$

Hence

$$\hat{\eta}_2 = \eta_2 \angle \theta_{\eta_2} \tag{5.79}$$

where $\eta_2 = \sqrt{2}/(\sigma_2 \delta_2)$ and $\theta_{\eta_2} = 45°$. The reflection and transmission coefficients are now complex:

$$\begin{aligned}
\hat{\Gamma} &= \frac{\hat{\eta}_2 - \eta_1}{\hat{\eta}_2 + \eta_1} \\
&= \Gamma \angle \theta_\Gamma \\
&= \Gamma e^{j\theta_\Gamma}
\end{aligned} \tag{5.80}$$

$$\begin{aligned}
\hat{T} &= \frac{2\hat{\eta}_2}{\hat{\eta}_2 + \eta_1} \\
&= T \angle \theta_T \\
&= T e^{j\theta_T}
\end{aligned} \tag{5.81}$$

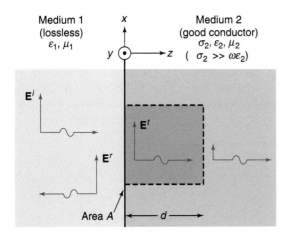

Figure 5.9   Illustration of the reflection and transmission of a uniform plane wave that is incident normal to the surface between a lossless medium and a good conductor.

The phasor forms of the transmitted wave becomes

$$
\boxed{
\begin{aligned}
\hat{E}_x^t &= \hat{T} E_m^i e^{-\alpha_2 z} e^{-j\beta_2 z} \\
&= T E_m^i e^{-\frac{z}{\delta_2}} e^{-j\frac{z}{\delta_2}} e^{j\theta_T}
\end{aligned}
}
\tag{5.82a}
$$

$$
\boxed{
\begin{aligned}
\hat{H}_y^t &= \hat{T} \frac{E_m^i}{\hat{\eta}_2} e^{-\alpha_2 z} e^{-j\beta_2 z} \\
&= T \frac{E_m^i}{\eta_2} e^{-\frac{z}{\delta_2}} e^{-j\frac{z}{\delta_2}} e^{j\theta_T} e^{-j\theta_{\eta_2}}
\end{aligned}
}
\tag{5.82b}
$$

Observe some important differences between these expressions and those for a lossless second medium given in (5.69). Since we are considering the second medium to be a good conductor which is lossy, these expressions contain the attenuation of the amplitude as $e^{-\alpha_2 z} = e^{-z/\delta_2}$, which is written in terms of the skin depth for that medium. Second, because the transmission coefficient is now complex (because the intrinsic impedance of the second medium is complex), it has an angle which is represented in these expressions by the complex exponential $e^{j\theta_T}$. Third, because the intrinsic impedance is complex, it too has an angle which is represented by the complex exponential $e^{-j\theta_{\eta_2}}$ in the expression for the magnetic field.

The expressions for the incident and reflected waves in the first medium (lossless) are essentially unchanged from before except that the reflection coefficient is complex with an angle. This change is represented by the exponential $e^{j\theta_\Gamma}$ in the reflected wave expressions. Hence,

$$
\boxed{\hat{E}_x^i = E_m^i e^{-j\beta_1 z}}
\tag{5.83a}
$$

$$
\boxed{\hat{H}_y^i = \frac{E_m^i}{\eta_1} e^{-j\beta_1 z}}
\tag{5.83b}
$$

$$
\boxed{\hat{E}_x^r = \Gamma E_m^i e^{j\beta_1 z} e^{j\theta_\Gamma}}
\tag{5.84a}
$$

$$
\boxed{\hat{H}_y^r = -\Gamma \frac{E_m^i}{\eta_1} e^{j\beta_1 z} e^{j\theta_\Gamma}}
\tag{5.84b}
$$

Observe that the transmitted fields are attenuated as they propagate through the second medium as indicated in Fig. 5.9. Hence power is being dissipated in the surface of that medium. We can determine the power lost in a block of that surface of area A and depth d by computing the average power entering the block and subtracting the average power leaving the block. The average power density vector in the good conductor is

$$
\boxed{
\begin{aligned}
\mathbf{S}_{AV}^t &= \frac{1}{2} \mathrm{Re}\left[\hat{E}_x^t \mathbf{a}_x \times \hat{H}_y^{t,*} \mathbf{a}_y \right] \\
&= \frac{1}{2} \mathrm{Re}\left[\left(T E_m^i e^{-\frac{z}{\delta_2}} e^{-j\frac{z}{\delta_2}} e^{j\theta_T} \right)\left(T \frac{E_m^i}{\eta_2} e^{-\frac{z}{\delta_2}} e^{j\frac{z}{\delta_2}} e^{-j\theta_T} e^{j\theta_{\eta_2}} \right) \right] \mathbf{a}_z \\
&= T^2 \frac{(E_m^i)^2}{2\eta_2} e^{-2\frac{z}{\delta_2}} \cos(\theta_{\eta_2}) \qquad \text{W/m}^2
\end{aligned}
}
\tag{5.85}
$$

Hence, the average power dissipated in this block of material of surface area A and depth d is

$$
\begin{aligned}
P_{\text{dissipated}} &= P_{\text{AV}}\big|_{z=0} - P_{\text{AV}}\big|_{z=d} \\
&= T^2 \frac{(E_m^i)^2}{2\eta_2} \cos(\theta_{\eta_2}) A - T^2 \frac{(E_m^i)^2}{2\eta_2} e^{-2\frac{d}{\delta_2}} \cos(\theta_{\eta_2}) A \\
&= T^2 \frac{(E_m^i)^2}{2\eta_2} \cos(\theta_{\eta_2}) A \left(1 - e^{-2\frac{d}{\delta_2}}\right) \qquad \text{W}
\end{aligned}
\tag{5.86}
$$

▷ **EXAMPLE 5.10**

A 10-V/m, 1-MHz uniform plane wave is traveling in free space and strikes a large block of copper normal to its surface. Determine the power dissipated in a block of that copper having a surface area of 2 m² and a depth of 1 skin depth.

SOLUTION   In Examples 5.3 and 5.4 we computed the attenuation and phase constants and the intrinsic impedance of copper at 1 MHz as

$$
\alpha_2 = \beta_2 = 1.51 \times 10^4
$$
$$
\hat{\eta}_2 = 3.69 \times 10^{-4} \angle 45°
$$

and $\eta_1 = \eta_o = 377\ \Omega$. Hence, the transmission coefficient is

$$
\begin{aligned}
\hat{T} &= \frac{2\hat{\eta}_2}{\hat{\eta}_2 + \eta_1} \\
&= 1.96 \times 10^{-6} \angle 45°
\end{aligned}
$$

Hence, the average power expression is

$$
\begin{aligned}
P_{\text{AV}} &= T^2 \frac{(E_m^i)^2}{2\eta_2} e^{-2\frac{z}{\delta_2}} \cos(\theta_{\eta_2}) A \\[2mm]
&= 7.34 \times 10^{-7} e^{-2\frac{z}{\delta_2}} \qquad \text{W}
\end{aligned}
$$

Hence, the power dissipated is

$$
\begin{aligned}
P_{\text{dissipated}} &= 7.34 \times 10^{-7} (1 - e^{-2}) \\
&= 0.637 \qquad \mu\text{W}
\end{aligned}
$$

◁

▷ **QUICK REVIEW EXERCISE 5.9**

A 100-V/m, 1-MHz uniform plane wave is traveling in free space and strikes the ocean surface normal to its surface. Determine the power dissipated in a block of that seawater having a surface area of 10 m² and a depth of 1 skin depth.

ANSWER   1.2 W.

◁

If the second medium is a good conductor, the intrinsic impedance is very small. Hence the reflection coefficient will be approximately

$$
\begin{aligned}
\hat{\Gamma} &= \frac{\hat{\eta}_2 - \eta_1}{\hat{\eta}_2 + \eta_1} \\
&\cong -1 \qquad \hat{\eta}_2 \ll \eta_1
\end{aligned}
\tag{5.87}
$$

Hence almost all the incident electric field is reflected and only a small part is transmitted

$$\hat{E}^r_x \cong -\hat{E}^i_x \qquad \hat{\eta}_2 \ll \eta_1 \tag{5.88}$$

The total electric field in the first medium is the sum of the incident and reflected waves:

$$\begin{aligned}
\hat{\mathbf{E}}_1 &= (\hat{E}^i_x + \hat{E}^r_x)\mathbf{a}_x \\
&= E^i_m(e^{-j\beta_1 z} + \hat{\Gamma}e^{j\beta_1 z})\mathbf{a}_x \\
&\cong E^i_m\underbrace{(e^{-j\beta_1 z} - e^{j\beta_1 z})}_{-2j\sin\beta_1 z}\mathbf{a}_x \\
&= -2jE^i_m \sin\beta_1 z\,\mathbf{a}_x
\end{aligned} \tag{5.89}$$

The time-domain field is

$$\begin{aligned}
\mathbf{E}_1 &= \operatorname{Re}\left[\hat{\mathbf{E}}_1 e^{j\omega t}\right] \\
&= \operatorname{Re}\left[-2jE^i_m \sin\beta_1 z\, e^{j\omega t}\right]\mathbf{a}_x \\
&= 2E^i_m \sin\beta_1 z\, \operatorname{Re}(-j(\cos\omega t + j\sin\omega t))\mathbf{a}_x \\
&= 2E^i_m \sin\beta_1 z\, \operatorname{Re}(\sin\omega t - j\cos\omega t)\mathbf{a}_x \\
&= 2E^i_m \sin\beta_1 z\, \sin\omega t\, \mathbf{a}_x
\end{aligned} \tag{5.90}$$

Observe in this time-domain result that the electric field is distributed with distance as an envelope $2E^i_m \sin\beta_1 z$ and the envelope oscillates as $\sin\omega t$. This is said to be a *standing wave* in the sense that there is no movement; the envelope simply pulsates. This is plotted in Fig. 5.10a. The envelope of the wave can be written, substituting $\beta_1 = 2\pi/\lambda_1$, as $\sin[2\pi(z\lambda_1)]$. This envelope has nulls at multiples of a half wavelength away from the boundary: $z = 0, -\lambda_1/2, -\lambda_1, -3\lambda_1/2, \cdots$ The maxima of $2E^i_m$ occur at odd multiples of

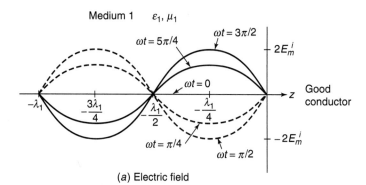

(a) Electric field

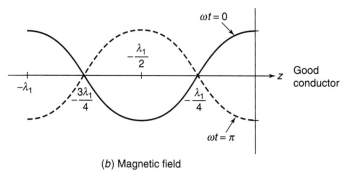

(b) Magnetic field

Figure 5.10 Illustration of standing waves created by the incidence of a uniform plane wave normal to the surface of a perfect conductor. (a) The total electric field is zero at the surface and goes to zero again at multiples of a half wavelength from the boundary. (b) The total magnetic field is zero at odd multiples of a quarter wavelength from the boundary.

a quarter wavelength away from the boundary: $z = -\lambda_1/4, -3\lambda_1/4, \cdots$. The phasor magnetic field is

$$
\begin{aligned}
\hat{\mathbf{H}}_1 &= (\hat{H}_x^i + \hat{H}_x^r)\mathbf{a}_y \\
&= \frac{E_m^i}{\eta_1}(e^{-j\beta_1 z} - \hat{\Gamma}e^{j\beta_1 z})\mathbf{a}_y \\
&\cong \frac{E_m^i}{\eta_1}\underbrace{(e^{-j\beta_1 z} + e^{j\beta_1 z})}_{2\cos\beta_1 z}\mathbf{a}_y \\
&= 2\frac{E_m^i}{\eta_1}\cos\beta_1 z\,\mathbf{a}_y
\end{aligned}
\tag{5.91}
$$

and the time-domain field is

$$
\mathbf{H}_1 = 2\frac{E_m^i}{\eta_1}\cos\beta_1 z\,\cos\omega t\,\mathbf{a}_y
\tag{5.92}
$$

The magnetic field is also a standing wave, and its envelope varies with distance away from the boundary as $2(E_m^i/\eta_1)\cos\beta_1 z = 2(E_m^i/\eta_1)\cos(2\pi z/\lambda_1)$. Hence the envelope has nulls at odd multiples of a quarter wavelength away from the boundary, $z = -\lambda_1/4, -3\lambda_1/4, \cdots$, and maxima at multiples of a half wavelength from the boundary, $z = 0, -\lambda_1/2, -\lambda_1, -3\lambda_1/2, \cdots$. The behavior of the magnetic field envelope is the opposite of the electric field envelope; where a null (maximum) occurs for the electric field envelope, a maximum (null) occurs for the magnetic field envelope.

▷ **QUICK REVIEW EXERCISE 5.10**

A uniform plane wave traveling in free space is incident normal to the surface of a large sheet of copper. The total electric field is zero at a distance of 3 cm from the surface of the copper sheet. Determine the possible frequencies of the wave.

ANSWER   5 GHz, 10 GHz, etc.   ◀

▷ **5.6 SNELL'S LAWS**

The previous discussion assumed that the wave was incident *normal* to the interface between the two media. In this section we will investigate uniform plane waves that are incident on the interface between two *lossless* media at some arbitrary angle of incidence. Consider Fig. 5.11, which shows this. The incident wave has an angle of incidence, θ_i, that is measured from a line perpendicular to the boundary. Similarly, the angles of the reflected wave, θ_r, and the transmitted wave, θ_t, are also measured from the perpendicular to the boundary. Snell's laws relate these angles. Recall that uniform plane waves have planes of constant phase (the wavefront) that are perpendicular to the direction of propagation of the wave. These wavefronts are shown as dashed lines in the figure. When each point on these wavefronts strikes the interface, a reflected and a transmitted wavefront are produced. A point on the incident wave strikes the interface at point O. Later a point on the wavefront strikes the interface at point O'. Reflected and transmitted waves are produced from these two points. The wavefronts of the incident, reflected, and transmitted waves are denoted as OD^i, $O'D^r$, and $O'D^t$, respectively. The time that it takes the incident wave to propagate from D^i to O' is the same as the time it takes the reflected wave to propagate from O to D^r, which is also the time it takes the transmitted wave to propagate from O to D^t. The incident and reflected

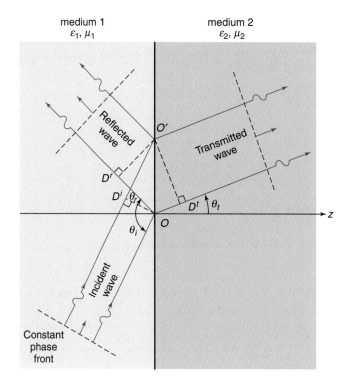

Figure 5.11 Illustration of Snell's laws for a uniform plane wave that is incident at an oblique angle to the boundary between two lossless media.

waves travel in medium 1 with velocity $v_1 = v_o/\sqrt{\mu_{r1}\varepsilon_{r1}}$ and the transmitted wave travels in medium 2 with velocity $v_2 = v_o/\sqrt{\mu_{r2}\varepsilon_{r2}}$. Hence

$$\frac{D^iO'}{v_1} = \frac{OD^r}{v_1} = \frac{OD^t}{v_2} \tag{5.93}$$

where D^iO', OD^r, and OD^t are the distances between the respective points. These distances can be related to a common distance, OO', as

$$D^iO' = OO' \cos(90° - \theta_i) = OO' \sin\theta_i \tag{5.94a}$$
$$OD^r = OO' \cos(90° - \theta_r) = OO' \sin\theta_r \tag{5.94b}$$
$$OD^t = OO' \cos(90° - \theta_t) = OO' \sin\theta_t \tag{5.94c}$$

Hence we obtain Snell's laws:

$$\boxed{\theta_i = \theta_r \qquad \text{(Snell's law of reflection)}} \tag{5.95a}$$

and

$$\boxed{\frac{\sin\theta_t}{\sin\theta_i} = \frac{v_2}{v_1} = \sqrt{\frac{\mu_{r1}\varepsilon_{r1}}{\mu_{r2}\varepsilon_{r2}}} \qquad \text{(Snell's law of refraction)}} \tag{5.95b}$$

Snell's laws are commonly used in optics (since light is an electromagnetic wave). In optics it is common to characterize each material with its *index of refraction*. The index of refraction of a lossless medium is the ratio of the speed of light to the velocity of propagation in that medium:

$$\boxed{n = \frac{v_o}{v} = \sqrt{\mu_r\varepsilon_r}} \tag{5.96}$$

Most materials of interest are nonmagnetic, $\mu_r = 1$, and for these materials, Snell's law of refraction can be written in terms of the index of refraction of the material as

$$\frac{\sin\theta_t}{\sin\theta_i} = \frac{\sqrt{\varepsilon_{r1}}}{\sqrt{\varepsilon_{r2}}} = \frac{n_1}{n_2} \qquad \text{(Snell's law of refraction, } \mu_1 = \mu_2\text{)} \qquad (5.97)$$

Hence Snell's laws state that the angles of incidence and reflection are equal, and the angle of incidence and the angle of transmission are related by the indices of refraction of the two materials. A material is said to be more *dense* than another if its index of refraction is larger than that of the other. Snell's law of refraction in (5.97) shows that *an incident wave is bent toward the normal,* $\theta_t < \theta_i$, *if it passes into a more dense medium,* $n_2 > n_1$. This is the case for transmission of light into water, glass, etc. This is illustrated in Fig. 5.12. An important special case occurs when the angle of transmission is 90°, that is, the transmitted wave becomes a *surface wave* that travels along the interface and none of it penetrates into the second medium as illustrated in Fig. 5.13. The is called the *critical angle* and is denoted as θ_c. From (5.97) we see that this critical angle is given by

$$\sin(\theta_i = \theta_c) = \frac{n_2}{n_1} = \sqrt{\frac{\varepsilon_{r2}}{\varepsilon_{r1}}} \qquad (5.98)$$

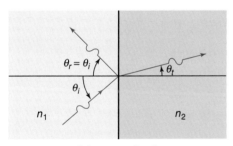

(a) $n_2 > n_1$, $\theta_t < \theta_i$

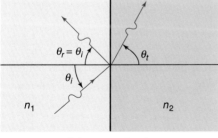

(a) $n_2 < n_1$, $\theta_t > \theta_i$

Figure 5.12 Illustration of the bending of the transmitted wave. (a) The wave transmitted into a more dense medium is bent toward the normal to the surface. (b) The wave transmitted into a less dense medium is bent away from the normal to the surface.

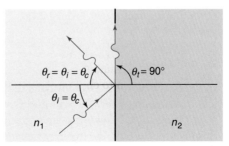

Figure 5.13 Illustration of the critical angle of incidence.

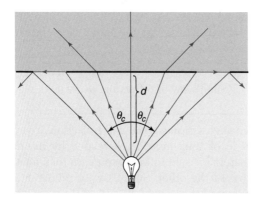

Figure 5.14   Example 5.11; transmission of an underwater light source.

▶ **EXAMPLE 5.11**

Consider the problem of using a submerged light source to illuminate a body of water as illustrated in Fig. 5.14. If the light is placed at a depth of 2 m below the surface, determine the surface area of light seen on the surface. The relative permittivity of water at optical frequencies is 1.77.

SOLUTION   Only light within a cone angle equal to the critical angle will be transmitted. Light incident on the surface at an angle greater than the critical angle will be totally reflected back into the water. Thus (5.98) gives the critical angle as

$$\sin(\theta_c) = \frac{1}{\sqrt{1.77}}$$

so that

$$\theta_c = 48.7°$$

Hence the illuminated surface area is

$$\pi(d \tan\theta_c)^2 = 16.3 \quad \text{m}^2$$

◀

The detailed results for the electric and magnetic fields of UPWs that are incident on the boundary at oblique angles are given in Appendix A.

▶ **5.7 ENGINEERING APPLICATIONS**

In this section we will show numerous practical applications of the concepts concerning uniform plane wave propagation.

5.7.1   Transmission Lines

In the next chapter we will investigate the propagation of waves on two-conductor, lossless transmission lines. We will find that there are a great number of parallels between waves transmitted down transmission lines and the uniform plane wave we studied in this chapter. In fact, the analogy between the two will provide several memory aids for the properties of transmission lines.

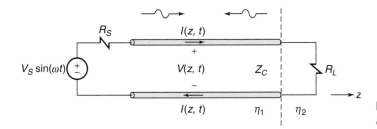

Figure 5.15 A transmission line carrying plane waves.

Consider a pair of parallel conductors (such as wires) that connect a source to a load as shown in Fig. 5.15. The electric and magnetic field intensity vectors will lie in a plane perpendicular or transverse to the line (z) axis. Hence these are TEM waves also. We will find that the voltage and current on this line consist of a forward- and a backward-traveling wave. The phasor voltage between the two conductors can be written as

$$\hat{V}(z) = \hat{V}_m^+ e^{-j\beta z} + \hat{V}_m^- e^{j\beta z} \tag{5.99a}$$

and the current (going down one conductor and returning on the other) can be written as

$$\hat{I}(z) = \frac{\hat{V}_m^+}{Z_C} e^{-j\beta z} - \frac{\hat{V}_m^-}{Z_C} e^{j\beta z} \tag{5.99b}$$

where Z_C is said to be the *characteristic impedance* of the transmission line. Compare (5.99) to the general solution for uniform plane waves in a lossless medium given in (5.13). The parallels are striking:

$$\hat{E}_x \Leftrightarrow \hat{V}$$
$$\hat{H}_y \Leftrightarrow \hat{I}$$
$$\eta \Leftrightarrow Z_C$$

At the load a portion of the incoming wave will be reflected. This is determined by the load reflection coefficient:

$$\Gamma_L = \frac{R_L - Z_C}{R_L + Z_C}$$

Contrast this with the reflection coefficient at the interface between two lossless media given in (5.62):

$$\Gamma = \frac{\eta_2 - \eta_1}{\eta_2 + \eta_1}$$

In Fig. 5.15 we have drawn a vertical line separating the transmission line from the load. If we make the analogies

$$\eta_1 \Leftrightarrow Z_C$$
$$\eta_2 \Leftrightarrow R_L$$

the correspondence between the two reflection coefficients becomes apparent. The important result here is that now that we have invested the time in understanding uniform plane waves, we may essentially transfer that understanding, along with the associated formulae, to provide an immediate understanding of transmission lines.

5.7.2 Antennas

We will study the propagation of waves from antennas in Chapter 7. At a sufficient distance from the antenna we will find that waves propagated from antennas resemble uniform plane waves. The waves propagated from antennas are actually spherical waves, as when we throw a rock in a body of water. However, locally they appear to an observer to be uniform plane waves as illustrated in Fig. 5.16. The outward-traveling waves are, in spherical coordinates, of the form

$$\hat{E}_\theta = \hat{E}_m \frac{e^{-j\beta r}}{r} \tag{5.100a}$$

$$\hat{H}_\phi = \frac{\hat{E}_m}{\eta} \frac{e^{-j\beta r}}{r} \tag{5.100b}$$

The electric field is in the θ direction and the magnetic field is in the ϕ direction, which is orthogonal to the electric field. Also observe that power flow is in the $\hat{\mathbf{E}} \times \hat{\mathbf{H}} \Leftrightarrow \mathbf{a}_r$ or radial direction, showing that power is being radiated away from the antenna. Also the electric field and the magnetic field are related by the intrinsic impedance of the medium in which the wave is propagating, which is normally free space. Once again we will find that many of the properties of radiation from antennas will have already been learned now that we have invested the time in understanding the properties of uniform plane waves.

5.7.3 Communication with Submarines

Submarines must stay submerged for long periods of time to prevent their detection. This poses a problem for communication with them. Radio transmissions to submarines take place at very low frequencies in the kHz range and below because the ocean attenuates higher frequencies more strongly. For example, consider the problem shown in Fig. 5.17, where a wave associated with a radio transmission is transmitted into the ocean. We wish to determine the attenuation of the signal (transmitted wave) received by the

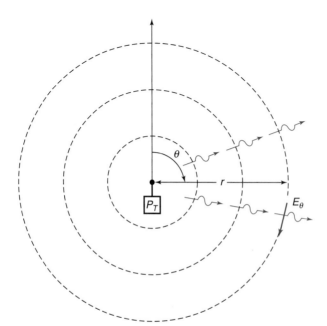

Figure 5.16 Waves radiated from an antenna are spherical waves but appear to a local observer as uniform plane waves.

submarine, which is at a depth d below the surface of the ocean. In other words, we want to determine the ratio of the power in the wave at the location of the submarine and the power in the wave just below the ocean surface. The important parameters that must be computed are the attenuation constant and the intrinsic impedance of the ocean. At kHz frequencies, seawater has the parameters $\varepsilon_r = 81$, $\mu_r = 1$, and $\sigma = 4$ S/m. At 1 kHz, the ratio of conduction current to displacement current in the seawater is

$$\left.\frac{\sigma}{\omega\varepsilon}\right|_{@1\,\text{kHz}} = 8.89 \times 10^5$$

and seawater may be considered a good conductor. Similarly, at 1 MHz the ratio is

$$\left.\frac{\sigma}{\omega\varepsilon}\right|_{@1\,\text{MHz}} = 8.89 \times 10^2$$

so that seawater may still be considered a good conductor. Hence we may use the approximations for a good conductor given in (5.50) and (5.52) in terms of the skin depth. At 1 kHz the skin depth is

$$\delta = \left.\frac{1}{\sqrt{\pi f \mu \sigma}}\right|_{@1\,\text{kHz}} = 7.96 \quad \text{m}$$

At a depth d, the power density of the transmitted wave is

$$S_{\text{AV}}^t = \frac{(E_m^t)^2}{2\eta} e^{-2\frac{d}{\delta}} \cos(\theta_\eta)$$

Hence the ratio of the power density just below the ocean surface $(d = 0)$ and the power density at the location of the submarine (d) is

$$\frac{S_{\text{AV}}|_{d=0}}{S_{\text{AV}}|_d} = e^{2\frac{d}{\delta}}$$

In decibels, this attenuation is

$$\text{Attenuation (dB)} = 10 \log_{10}\left(e^{2\frac{d}{\delta}}\right)$$

$$= 20\frac{d}{\delta} \log_{10}(e)$$

$$= 8.6859\frac{d}{\delta} \quad \text{dB}$$

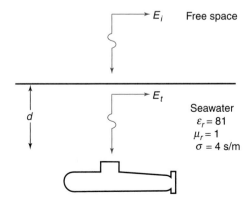

Free space

Seawater
$\varepsilon_r = 81$
$\mu_r = 1$
$\sigma = 4$ s/m

Figure 5.17 Illustration of communication with submarines. Low frequencies are used to avoid the large attenuation of higher frequencies by the seawater.

For a wave of frequency 1 kHz and a depth of $d = 30$ m, the attenuation is

$$\text{Attenuation (1 kHz, } d = 30 \text{ m)} = 32.75 \quad \text{dB}$$

Hence, the power density at a depth of 30 m is reduced by a factor of 1882 from that just below the ocean surface. For a wave of frequency 100 kHz the skin depth is 0.796 m, and at a depth of $d = 30$ m the attenuation is

$$\text{Attenuation (100 kHz, } d = 30 \text{ m)} = 327.5 \quad \text{dB}$$

Hence, the power density at a depth of 30 m is reduced by a factor of 5.56×10^{32} from that just below the ocean surface! This rather dramatically demonstrates why communication with submerged submarines requires the use of very low frequencies.

In addition to attenuation of the wave as it passes through the seawater, some of the power in the incident wave is reflected at the air–ocean interface. If we assume that the wave incident on the ocean surface is a uniform plane wave, we can calculate the reflection and transmission coefficients as

$$\hat{\Gamma}\Big|_{@1 \text{ kHz}} = \frac{\hat{\eta}_2 - \eta_o}{\hat{\eta}_2 + \eta_o}$$
$$= -0.9998$$

and

$$\hat{T}\Big|_{@1 \text{ kHz}} = \frac{2\hat{\eta}_2}{\hat{\eta}_2 + \eta_o}$$
$$= 2.357 \times 10^{-4}\angle 45°$$

where the intrinsic impedance of seawater at 1 kHz is

$$\hat{\eta}_2 = \sqrt{\frac{j\omega\mu_2}{\sigma_2}}$$
$$= \sqrt{\frac{j2\pi \times 10^3 \times 4\pi \times 10^{-7}}{4}}$$
$$= 4.443 \times 10^{-2}\angle 45°$$

Hence, $|\hat{\Gamma}|^2 = 0.9997$ or 99.97% of the incident power is reflected at the interface. This further reduces the power received by the submarine.

5.7.4 Design of Radomes

Modern jet aircraft rely on weather radar to safely maneuver around bad weather. The weather radar is usually housed in the nose of the aircraft and operates at frequencies in the low GHz range. A plastic radome covers the antenna and provides streamlining of the airplane. It is apparent that we need to minimize the effect of the plastic housing on the transmitted and received signal of the radar. The thickness of the radome can be chosen such that it is transparent to the radar signals. We can model the radome as a material boundary of thickness d as shown in Fig. 5.18. The incident wave from the radar antenna is written in phasor form in the usual fashion as

$$\hat{E}_x^i = E_m^i e^{-j\beta_o z}$$
$$\hat{H}_y^i = \frac{E_m^i}{\eta_o} e^{-j\beta_o z}$$

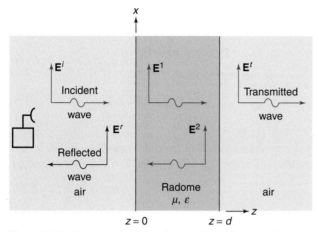

Figure 5.18 Illustration of the design of a radome to house an airborne radar. The radome thickness is chosen such that it is transparent at the radar frequency.

and the reflected wave and transmitted wave are written in the usual fashion as

$$\hat{E}_x^r = E_m^r e^{j\beta_o z}$$

$$\hat{H}_y^r = -\frac{E_m^r}{\eta_o} e^{j\beta_o z}$$

and

$$\hat{E}_x^t = E_m^t e^{-j\beta_o z}$$

$$\hat{H}_y^t = \frac{E_m^t}{\eta_o} e^{-j\beta_o z}$$

where the medium on both sides of the boundary is assumed to be air, so that the phase constants are

$$\beta_o = \omega\sqrt{\mu_o \varepsilon_o}$$

and the intrinsic impedances are

$$\eta_o = \sqrt{\frac{\mu_o}{\varepsilon_o}}$$

So far this is quite similar to the problem of normal incidence on a boundary. But here we have two boundaries, the front and back sides of the plastic radome. Internal to the radome there will also be forward- and backward-traveling waves, which we write in the usual fashion as

$$\hat{E}_x^1 = E_m^1 e^{-j\beta z}$$

$$\hat{H}_y^1 = \frac{E_m^1}{\eta} e^{-j\beta z}$$

$$\hat{E}_x^2 = E_m^2 e^{j\beta z}$$

$$\hat{H}_y^2 = -\frac{E_m^2}{\eta} e^{j\beta z}$$

where the medium in this middle region is the plastic of the radome, so that the phase constant is

$$\beta = \omega \sqrt{\mu \varepsilon}$$

and the intrinsic impedance of this region is

$$\eta = \sqrt{\frac{\mu}{\varepsilon}}$$

The boundary conditions are once again that the tangential electric field and tangential magnetic field be continuous across each boundary. For the first boundary at $z = 0$ we have

$$E_m^i + E_m^r = E_m^1 + E_m^2 \tag{a}$$

and

$$\frac{E_m^i}{\eta_o} - \frac{E_m^r}{\eta_o} = \frac{E_m^1}{\eta} - \frac{E_m^2}{\eta} \tag{b}$$

At the second boundary, $z = d$, we again match the tangential electric and magnetic fields to give

$$E_m^1 e^{-j\beta d} + E_m^2 e^{j\beta d} = E_m^t e^{-j\beta_o d} \tag{c}$$

and

$$\frac{E_m^1}{\eta} e^{-j\beta d} - \frac{E_m^2}{\eta} e^{j\beta d} = \frac{E_m^t}{\eta_o} e^{-j\beta_o d} \tag{d}$$

Eliminating E_m^1 and E_m^2 from (a), (b), (c), and (d) gives

$$E_m^i + \left(\frac{\eta_o - \eta}{\eta_o + \eta} \right) E_m^r = E_m^t e^{-j\beta_o d} e^{j\beta d} \tag{5.101a}$$

$$E_m^i + \left(\frac{\eta_o + \eta}{\eta_o - \eta} \right) E_m^r = E_m^t e^{-j\beta_o d} e^{-j\beta d} \tag{5.101b}$$

Combining these, we obtain the reflection coefficient at the left boundary

$$\frac{E_m^r}{E_m^i} = \frac{2j \sin(\beta d)}{\left(\dfrac{\eta_o - \eta}{\eta_o + \eta} \right) e^{-j\beta d} - \left(\dfrac{\eta_o + \eta}{\eta_o - \eta} \right) e^{j\beta d}} \tag{5.102}$$

and the transmission coefficient

$$\frac{E_m^t}{E_m^i} = \frac{4\eta_o \eta}{(\eta_o + \eta)^2} \frac{e^{j\beta_o d} e^{-j\beta d}}{1 - \left(\dfrac{\eta_o - \eta}{\eta_o + \eta} \right)^2 e^{-j2\beta d}} \tag{5.103}$$

If

$$\beta d = n\pi \qquad n = 1,2,3,\cdots \tag{5.104}$$

then $\sin(\beta d) = \sin(n\pi) = 0$ and $e^{-j2\beta d} = e^{-j2n\pi} = 1$ and (5.102) and (5.103) become

$$\frac{E_m^r}{E_m^i} = 0 \qquad \beta d = n\pi \tag{5.105a}$$

and

$$\frac{E_m^t}{E_m^i} = \pm e^{j\beta_o d} \qquad \beta d = n\pi \tag{5.105b}$$

The reflected field at the left boundary is zero, and the transmitted field is the incident radar field shifted in phase. Substituting the relation between the phase constant and wavelength, $\beta = 2\pi/\lambda$, yields

$$d = n\frac{\lambda}{2} \qquad n = 1,2,3,\cdots \tag{5.106}$$

Hence if we make the thickness of the radome a multiple of one-half of a wavelength where the wavelength is computed in the radome material,

$$d = n\frac{v_o}{2f\sqrt{\mu_r\varepsilon_r}} \qquad n = 1,2,3,\cdots \tag{5.107}$$

then the radome will be transparent to the radar signal. For example, for a radar signal of 10 GHz and a radome material having $\mu_r = 1$, $\varepsilon_r = 6$, we obtain transparent thicknesses of $d = 6.12$ mm, 12.25 mm, $\cdots$. For structural reasons, the thickness will usually be chosen to be many multiples of a half wavelength.

5.7.5 Shielding of Electronic Equipment

Shielded enclosures are used to (1) prevent a signal outside the enclosure from interfering with electronic equipment inside the enclosure or to (2) prevent a signal inside the enclosure from interfering with electronic equipment outside the enclosure as shown in Fig. 5.19.

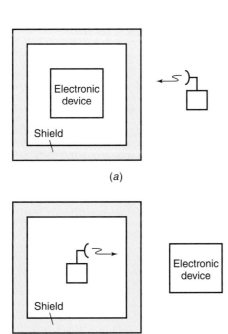

(a)

(b)

Figure 5.19 Illustration of using a shielded enclosure to protect sensitive electronics from electromagnetic fields that would interfere with their operation where (a) the interference source is outside the shield and (b) the interference source is inside the shield. (c) Photograph of a shielded room used for electromagnetic compatibility (EMC) testing (courtesy of ETS-Lindgren, Inc.).

(c)

Figure 5.19 *(Continued)*

A photograph of a shielded room is shown in Fig. 5.19c. Consider a wall of the enclosure that is a good conductor and has thickness t as shown in Fig. 5.20. This is similar to the radome design in the previous section except that here we want to have a considerable reflection of the incident field and to minimize the transmitted field. Once again we postulate an incident field that is in the form of a uniform plane wave from some distant transmitter:

$$\hat{E}_x^i = E_m^i e^{-j\beta_o z}$$

$$\hat{H}_y^i = \frac{E_m^i}{\eta_o} e^{-j\beta_o z}$$

There will be a reflected field at the surface of the conducting barrier

$$\hat{E}_x^r = \hat{E}_m^r e^{j\beta_o z}$$

$$\hat{H}_y^r = -\frac{\hat{E}_m^r}{\eta_o} e^{j\beta_o z}$$

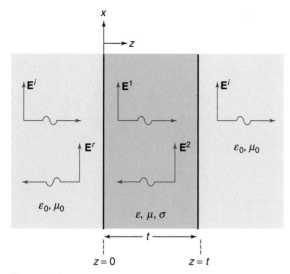

Figure 5.20 Illustration of the use of a conducting barrier in constructing a shield to prevent transmission of a potentially interfering, high-frequency electromagnetic wave.

and a transmitted field exiting the conducting barrier.

$$\hat{E}_x^t = \hat{E}_m^t e^{-j\beta_o z}$$

$$\hat{H}_y^t = \frac{\hat{E}_m^t}{\eta_o} e^{-j\beta_o z}$$

Since the barrier is conductive, the undetermined constants, $\hat{E}_m^r$ and $\hat{E}_m^t$, will, in general, be complex (have an angle). In the barrier there will be forward- and backward-traveling waves like the case of a radome. However, there is one important difference between the radome problem and the shield. The radome was assumed lossless, but the shield is lossy with conductivity σ. Hence we write the forward- and backward-traveling waves in the lossy conducting barrier in the usual form as

$$\hat{E}_x^1 = \hat{E}_m^1 e^{-\hat{\gamma} z}$$

$$\hat{H}_y^1 = \frac{\hat{E}_m^1}{\hat{\eta}} e^{-\hat{\gamma} z}$$

$$\hat{E}_x^2 = \hat{E}_m^2 e^{\hat{\gamma} z}$$

$$\hat{H}_y^2 = -\frac{\hat{E}_m^2}{\hat{\eta}} e^{\hat{\gamma} z}$$

where the propagation constant in the conducting barrier is now complex as

$$\hat{\gamma} = \sqrt{j\omega\mu(\sigma + j\omega\varepsilon)}$$
$$= \alpha + j\beta$$

Assuming that the conducting barrier is a "good conductor," the attenuation and phase constants can be written in terms of the skin depth as before as

$$\alpha = \beta = \frac{1}{\delta} = \sqrt{\pi f \mu \sigma}$$

and the intrinsic impedance of this region is

$$\hat{\eta} = \sqrt{\frac{j\omega\mu}{\sigma + j\omega\varepsilon}}$$

$$= \frac{\sqrt{2}}{\sigma\delta}\angle 45°$$

We wish to obtain the *shielding effectiveness* of the enclosure, which is the ratio of the magnitudes of the transmitted field to the incident field:

$$SE = \left|\frac{\hat{E}_m^t}{E_m^i}\right| \tag{5.108}$$

It is common to express this in decibels (dB) as

$$SE_{dB} = 20\log_{10}\left|\frac{E_m^i}{\hat{E}_m^t}\right| \tag{5.109}$$

Note that in dB, the shielding effectiveness is commonly defined to be the ratio of the incident field to the transmitted field as opposed to (5.108). This is done to give shielding effectiveness numbers as positive dBs, since the incident field will be larger than the transmitted field. A shielding effectiveness of 40 dB means that the incident field has been reduced by a factor of 100 as the wave passes through (exits) the barrier.

We can obtain this result from the results of the radome design in the previous section except that we replace the following quantities since the barrier is lossy:

$$\eta \Rightarrow \hat{\eta}$$

$$j\beta \Rightarrow \hat{\gamma} = \frac{1}{\delta} + j\frac{1}{\delta}$$

$$d \Rightarrow t$$

Hence Equations (5.101) in the radome design problem become

$$E_m^i + \left(\frac{\eta_o - \hat{\eta}}{\eta_o + \hat{\eta}}\right)\hat{E}_m^r = \hat{E}_m^t e^{-j\beta_o d}e^{\hat{\gamma}t} \tag{5.110a}$$

$$E_m^i + \left(\frac{\eta_o + \hat{\eta}}{\eta_o - \hat{\eta}}\right)\hat{E}_m^r = \hat{E}_m^t e^{-j\beta_o d}e^{-\hat{\gamma}t} \tag{5.110b}$$

The ratio of the incident and transmitted fields in (5.103) becomes

$$\frac{E_m^i}{\hat{E}_m^t} = \frac{(\eta_o + \hat{\eta})^2}{4\hat{\eta}\eta_o}\left[1 - \left(\frac{\eta_o - \hat{\eta}}{\eta_o + \hat{\eta}}\right)^2 e^{-2\frac{t}{\delta}}e^{-j2\frac{t}{\delta}}\right]e^{-j\beta_o t}e^{\frac{t}{\delta}}e^{j\frac{t}{\delta}} \tag{5.111}$$

Taking the magnitude of this yields

$$\left|\frac{E_m^i}{\hat{E}_m^t}\right| = \left|\frac{(\eta_o + \hat{\eta})^2}{4\hat{\eta}\eta_o}\right|\left|1 - \left(\frac{\eta_o - \hat{\eta}}{\eta_o + \hat{\eta}}\right)^2 e^{-2\frac{t}{\delta}}e^{-j2\frac{t}{\delta}}\right|e^{\frac{t}{\delta}} \tag{5.112}$$

To obtain a simpler result, we observe that the intrinsic impedance of the (good conductor) barrier is much less than the intrinsic impedance of free space, $\hat{\eta} \ll \eta_o$, and hence the result in (5.112) can be simplified to

$$\left|\frac{E_m^i}{\hat{E}_m^t}\right| \cong \left|\frac{\eta_o}{4\hat{\eta}}\right|\left|1 - e^{-2\frac{t}{\delta}}e^{-j2\frac{t}{\delta}}\right|e^{\frac{t}{\delta}} \qquad \hat{\eta} \ll \eta_o, \text{ good conductor} \tag{5.113}$$

A further simplification can be made if, as is the usual practice, the barrier is several skin depths thick:

$$t \gg \delta \tag{5.114}$$

Hence (5.113) simplifies to

$$\left|\frac{E_m^i}{\hat{E}_m^t}\right| \cong \left|\frac{\eta_o}{4\hat{\eta}}\right| e^{\frac{t}{\delta}} \qquad \hat{\eta} \ll \eta_o, \text{ good conductor, } t \gg \delta \tag{5.115}$$

In decibels, the shielding effectiveness becomes

$$\text{SE}_{\text{dB}} = \underbrace{20 \log_{10}\left|\frac{\eta_o}{4\hat{\eta}}\right|}_{\text{R}_{\text{dB}}} + \underbrace{20 \log_{10} e^{\frac{t}{\delta}}}_{\text{A}_{\text{dB}}} \tag{5.116}$$

The result shows that the total shielding effectiveness (in dB) is the sum of a *reflection* term:

$$\text{R}_{\text{dB}} = 20 \log_{10}\left|\frac{\eta_o}{4\hat{\eta}}\right| \tag{5.117}$$

and an *absorption* term:

$$\text{A}_{\text{dB}} = 20 \log_{10} e^{\frac{t}{\delta}} \tag{5.118}$$

The reflection term accounts for reflections at the left and right boundaries, while the absorption term accounts for attentuation of the waves, $\hat{E}^1$ and $\hat{E}^2$, as they travel through the barrier. Observe that only the absorption term is affected by the thickness of the shield. Figure 5.21a shows the shielding effectiveness of a 20-mil (1 mil = $\frac{1}{1000}$ in.) copper barrier. Observe that the reflection term dominates below 2 MHz and the absorption term dominates above that. Figure 5.21b shows the shielding effectiveness of a 20-mil (1 mil = $\frac{1}{1000}$ in.) sheet steel barrier. Observe that the reflection term dominates below 20 kHz and the absorption term dominates above that. Hence low-frequency shielding relies on reflection at the two boundaries, whereas high-frequency shielding relies on attenuation of the waves as they pass through the barrier and hence requires barrier thicknesses much larger than a skin depth to provide this attenuation.

This shielding effectiveness is a somewhat ideal quantity. Any penetrations, such as wires, holes, and seams, can drastically reduce the shielding effectiveness of a shielded enclosure. This is why penetrations of a shield such as wires are treated with filters to prevent unintended exiting of the shielded enclosure by electronic signals within the enclosure.

5.7.6 Microwave Health Hazards

Electromagnetic fields can have adverse effects on the human body. In the GHz range of frequencies there are numerous sources of waves that may have deleterious effects on the human body. Microwave ovens operate at frequencies around 2 GHz, whereas high-power radars transmit kilowatts of power in the GHz frequency range. The effects of these signals on the human body is predominantly caused by the heating of the skin. Regulatory bodies in the United States generally set "safe" levels of the power densities of these fields in the GHz range that range from 1 mW/cm^2 to 10 mW/cm^2. A power density of 10 mW/cm^2 is equivalent to a power density of 100 W/m^2. Assuming the incident wave is a uniform plane wave carrying this power density, the magnitude of the electric field vector would be 274.6 V/m.

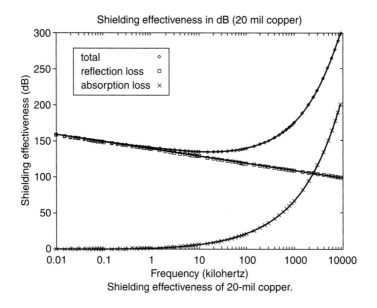

Shielding effectiveness of 20-mil copper.

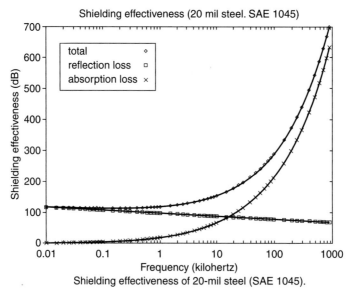

Shielding effectiveness of 20-mil steel (SAE 1045).

Figure 5.21 Plots of the shielding effectiveness separated into the reflection loss and the absorption loss for (a) a 20-mil (0.51-mm) copper shield and (b) a 20-mil shield made of steel.

As an example, consider a 2-GHz uniform plane wave having this amplitude of 274.6 V/m that is incident on the body surface. We will treat this problem as the interface between free space and a semi-infinite plane representing the body and determine the power dissipated in the body. At this frequency, the body parameters are on the order of $\sigma \cong 1.5$ S/m, $\varepsilon_r \cong 50$, and $\mu_r = 1$. The problem is essentially that described earlier in Fig. 5.9, where medium 2 is the human body. However, the human body at this frequency is not a good conductor, as evidenced by

$$\left.\frac{\sigma}{\omega\varepsilon}\right|_{@2\,\text{GHz}} = 0.27$$

Hence we must directly compute the intrinsic impedance of the body (medium 2 in Fig. 5.9) as

$$\hat{\eta}_2 = \sqrt{\frac{j\omega\mu_o}{(\sigma + j\omega\varepsilon_o\varepsilon_r)}}$$
$$= 52.38\angle7.55° \quad \Omega$$

The incident wave is in free space, so we compute the transmission coefficient as

$$\hat{T} = \frac{2\hat{\eta}_2}{\hat{\eta}_2 + \eta_o}$$
$$= 0.244\angle6.6°$$

The propagation constant in the body (medium 2 in Fig. 5.9) is

$$\hat{\gamma}_2 = \sqrt{j\omega\mu_o(\sigma + j\omega\varepsilon_o\varepsilon_r)}$$
$$= 39.63 + j298.83$$

Hence we identify the attenuation constant as

$$\alpha_2 = 39.63$$

A skin depth in the body becomes

$$\delta = \frac{1}{\alpha_2}$$
$$= 2.5 \text{ cm}$$

From the results of Section 5.5.1 we obtain the power density dissipated in a block of body tissue of depth d as

$$S_{AV,\text{dissipated}} = S_{AV}|_{z=0} - S_{AV}|_{z=d}$$
$$= T^2 \frac{(E_m^i)^2}{2\eta_2} \cos(\theta_{\eta_2})\left[1 - e^{-2\frac{d}{\delta}}\right]$$

To determine the power dissipated in the semi-infinite body, we let $d \to \infty$. For an incident wave of amplitude $E_m^i = 274.6$ V/m we obtain

$$S_{AV,\text{dissipated}} = 42.48 \quad \text{W/m}^2$$
$$= 4.248 \quad \text{mW/cm}^2$$

Hence, not all the power in the incident wave is absorbed by the body; some is lost to reflection at the interface.

5.7.7 Fiber-Optic Cables

Convential transmission lines such as parallel wires or coaxial cables are limited in their use at microwave frequencies because of (1) their bandwidths (information carrying capacity) and (2) excessive losses. Fiber-optic cables overcome most of these deficiencies. A fiber-optic cable consists of a glass fiber core having relative permittivity of ε_{rf} surrounded by a cladding having relative permittivity of ε_{rc} as shown in Fig. 5.22a. Infrared signals ($\lambda = 1$ μm–2 μm) are presented to one end of the fiber by the use of a light-emitting diode or a laser and extracted from the receiving end by a photocell or a

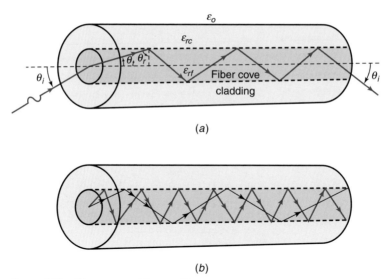

(a)

(b)

Figure 5.22 Illustration of how a fiber-optic cable guides light waves. (a) Illustration of the reflection at the core-cladding interface. (b) Illustration of dispersion caused by different paths.

phototransistor. The incident wave is incident at an angle θ_i to the axis of the core. According to Snell's laws, the angle of the transmission into the core, θ_t, is related to the angle of incidence as

$$\frac{\sin\theta_t}{\sin\theta_i} = \frac{n_o}{n_f}$$

where the indices of refraction are

$$n_o = 1$$

and

$$n_f = \sqrt{\varepsilon_{rf}}$$

This transmitted wave strikes the interface between the fiber and the cladding with an angle of incidence relative to a normal to that interface, θ_r. If this angle is greater than or equal to the critical angle, all the incident signal will be reflected at the interface, causing the beam to produce a "zigzag" pattern continuously being reflected at the fiber-cladding interface as it travels down the core. Hence we must choose the fiber core and cladding properties such that

$$\sin\theta_r \geq \sin\theta_{\text{critical}} = \frac{n_c}{n_f}$$

where the index of refraction of the cladding is

$$n_c = \sqrt{\varepsilon_{rc}}$$

We can also relate the angles θ_t and θ_r using trigonometry as

$$\cos\theta_t = \sin\theta_r$$

$$= \sqrt{1 - \sin^2\theta_t}$$

Hence we require that

$$\cos\theta_t = \sqrt{1 - \left(\frac{n_o}{n_f}\right)^2 \sin^2\theta_i}$$

$$\geq \frac{n_c}{n_f}$$

Solving this gives the angle of the *cone of acceptance*

$$\boxed{\sin\theta_i \leq \sqrt{n_f^2 - n_c^2}}$$ (5.119)

where we have substituted $n_o = 1$. Any signal that is incident on the end of the fiber within this angle of incidence will be transmitted into the fiber core where it will propagate down the fiber by being reflected at the fiber core-cladding interface. For example, consider a fiber core having $\varepsilon_{rf} = 2.3$, giving an index of refraction of $n_f = 1.52$ and a cladding having $\varepsilon_{rc} = 2.1$, giving an index of refraction of $n_c = 1.45$. The cone of acceptance becomes $\theta_i = 26.57°$.

The light wave can travel down the fiber using a number of paths, each reflecting off the fiber-cladding interface as shown in Fig. 5.22b. This potential for numerous paths causes the various paths to have different transit times to complete the trip down the cable. This is called *dispersion* and results in a smearing out of digital signals that are being propagated in the fiber as they reach the end of the cable. This dispersion imposes a limit on the maximum information transmission rate.

▶ SUMMARY OF IMPORTANT CONCEPTS AND FORMULAE

1. **Uniform plane waves (UPW) in lossless media:**

phasor form:

$$\hat{E}_x = \hat{E}_m^+ e^{-j\beta z} + \hat{E}_m^- e^{j\beta z}$$

$$\hat{H}_y = \frac{\hat{E}_m^+}{\eta} e^{-j\beta z} - \frac{\hat{E}_m^-}{\eta} e^{j\beta z}$$

time-domain form:

$$E_x(z,t) = \underbrace{E_m^+ \cos(\omega t - \beta z)}_{\substack{\text{forward-traveling} \\ \text{(+z direction) wave}}} + \underbrace{E_m^- \cos(\omega t + \beta z)}_{\substack{\text{backward-traveling} \\ \text{(−z direction) wave}}}$$

$$H_y(z,t) = \underbrace{\frac{E_m^+}{\eta} \cos(\omega t - \beta z)}_{\substack{\text{forward-traveling} \\ \text{(+z direction) wave}}} - \underbrace{\frac{E_m^-}{\eta} \cos(\omega t + \beta z)}_{\substack{\text{backward-traveling} \\ \text{(−z direction) wave}}}$$

2. **Parameters of UPWs in lossless media:** phase constant: $\beta = \omega\sqrt{\mu\varepsilon} = \omega\sqrt{\mu_r\varepsilon_r}/v_o$, intrinsic impedance: $\eta = \sqrt{\mu/\varepsilon} = \eta_o\sqrt{\mu_r/\varepsilon_r}$, intrinsic impedance of free space: $\eta_o \cong 120\pi \cong 377\,\Omega$, velocity of propagation: $v = 1/\sqrt{\mu\varepsilon} = v_o/\sqrt{\mu_r\varepsilon_r}$, velocity of propagation in free space: $v_o \cong 3 \times 10^8$ m/s, wavelength: $\lambda = 2\pi/\beta = v/f$.

3. **UPW in a lossy medium:**

phasor form:

$$\hat{E}_x = \hat{E}_m^+ e^{-\alpha z} e^{-j\beta z} + \hat{E}_m^- e^{\alpha z} e^{j\beta z}$$

$$\hat{H}_y = \frac{\hat{E}_m^+}{\eta} e^{-\alpha z} e^{-j\beta z} e^{-j\theta_\eta} - \frac{\hat{E}_m^-}{\eta} e^{\alpha z} e^{j\beta z} e^{-j\theta_\eta}$$

time-domain form:

$$E_x = E_m^+ e^{-\alpha z} \cos(\omega t - \beta z) + E_m^- e^{\alpha z} \cos(\omega t + \beta z)$$

$$H_y = \frac{E_m^+}{\eta} e^{-\alpha z} \cos(\omega t - \beta z - \theta_\eta) - \frac{E_m^-}{\eta} e^{\alpha z} \cos(\omega t + \beta z - \theta_\eta)$$

4. **Parameters of UPWs in lossy media:** propagation constant:
$\hat{\gamma} = \sqrt{(j\omega\mu)(\sigma + j\omega\varepsilon)} = \alpha + j\beta$, intrinsic impedance:
$\hat{\eta} = \sqrt{\dfrac{(j\omega\mu)}{(\sigma + j\omega\varepsilon)}} = \eta\angle\theta_\eta = \eta e^{j\theta_\eta}$, velocity of propagation: $v = \dfrac{\omega}{\beta} \neq \dfrac{1}{\sqrt{\mu\varepsilon}}$,
wavelength: $\lambda = \dfrac{2\pi}{\beta} = \dfrac{v}{f}$.

5. **Average power density of UPW in lossless media:**
$$\mathbf{S}_{AV} = \frac{|\hat{E}_m|^2}{2\eta}\ \mathbf{a}_z \qquad \text{W/m}^2.$$

6. **Average power density of UPW in lossy medium:**
$$\mathbf{S}_{AV} = \frac{|\hat{E}_m|^2}{2\eta}e^{-2\alpha z}\cos(\theta_\eta)\ \mathbf{a}_z \qquad \text{W/m}^2.$$

7. **Skin depth:**
$$\delta = \frac{1}{\alpha} = \frac{1}{\sqrt{\pi f\mu\sigma}}\ \text{m (good conductor)}.$$

8. **UPW parameters for good conductor($\sigma/\omega\varepsilon \gg 1$):**
$$\alpha = \beta = \frac{1}{\delta} = \sqrt{\pi f\mu\sigma},$$
$$\hat{\eta} = \sqrt{\frac{\omega\mu}{\sigma}}\angle 45° = \frac{\sqrt{2}}{\sigma\delta}\angle 45°\quad \Omega$$

9. **Reflection and transmission coefficients for UPW incident normal to boundary between two lossless media:**
$$\Gamma = \frac{\eta_2 - \eta_1}{\eta_2 + \eta_1} = \frac{E_m^r}{E_m^i}, \quad T = \frac{2\eta_2}{\eta_2 + \eta_1} = \frac{E_m^t}{E_m^i}$$

10. **Electric and magnetic fields for UPW incident normal to boundary between two lossless media:**
$$\hat{E}_x^i = E_m^i e^{-j\beta_1 z}$$
$$\hat{H}_y^i = \frac{E_m^i}{\eta_1}e^{-j\beta_1 z}$$
$$\hat{E}_x^r = \Gamma E_m^i e^{j\beta_1 z}$$
$$\hat{H}_y^r = -\Gamma\frac{E_m^i}{\eta_1}e^{j\beta_1 z}$$
$$\hat{E}_x^t = T E_m^i e^{-j\beta_2 z}$$
$$\hat{H}_y^t = T\frac{E_m^i}{\eta_2}e^{-j\beta_2 z}$$

11. **Reflection and transmission coefficients for UPW incident normal to the surface of a lossy medium:**
$$\hat{\Gamma} = \frac{\hat{\eta}_2 - \eta_1}{\hat{\eta}_2 + \eta_1} = \Gamma\angle\theta_\Gamma = \Gamma e^{j\theta_\Gamma}, \quad \hat{T} = \frac{2\hat{\eta}_2}{\hat{\eta}_2 + \eta_1} = T\angle\theta_T = T e^{j\theta_T}$$

12. **Electric and magnetic fields for UPW incident normal to the surface of a good conductor:**
$$\hat{E}_x^i = E_m^i e^{-j\beta_1 z}$$
$$\hat{H}_y^i = \frac{E_m^i}{\eta_1}e^{-j\beta_1 z}$$

$$\hat{E}_x^r = \hat{\Gamma}E_m^i e^{j\beta_1 z} = \Gamma E_m^i e^{j\beta_1 z} e^{j\theta_\Gamma}$$

$$\hat{H}_y^r = \frac{-\hat{\Gamma}E_m^i}{\eta_1} e^{j\beta_1 z} = -\Gamma \frac{E_m^i}{\eta_1} e^{j\beta_1 z} e^{j\theta_\Gamma}$$

$$\hat{E}_x^t = \hat{T}E_m^i e^{-\alpha_2 z} e^{-j\beta_2 z} = TE_m^i e^{-\frac{z}{\delta_2}} e^{-j\frac{z}{\delta_2}} e^{j\theta_T}$$

$$\hat{H}_y^t = \hat{T}\frac{E_m^i}{\hat{\eta}_2} e^{-\alpha_2 z} e^{-j\beta_2 z} = T\frac{E_m^i}{\eta_2} e^{-\frac{z}{\delta_2}} e^{-j\frac{z}{\delta_2}} e^{j\theta_T} e^{-j\theta_{\eta_2}}$$

13. **Snell's laws for UPW incident on boundary at an oblique angle:**

$$\theta_i = \theta_r \qquad \text{(Snell's law of reflection)}$$

$$\frac{\sin\theta_t}{\sin\theta_i} = \frac{v_2}{v_1} = \sqrt{\frac{\varepsilon_{r1}}{\varepsilon_{r2}}} \qquad \text{(Snell's law of refraction)}$$

$$n = \frac{v_o}{v} = \sqrt{\varepsilon_r} \qquad \text{(index of refraction)}$$

$$\sin(\theta_i = \theta_c) = \frac{n_2}{n_1} = \sqrt{\frac{\varepsilon_{r2}}{\varepsilon_{r1}}} \qquad \text{(critical angle of incidence)}$$

▷ PROBLEMS

SECTION 5.1 THE UNIFORM PLANE WAVE IN LOSSLESS MEDIA

5.1.1. Show that the equations governing the *phasor* fields of a uniform plane wave and given in (5.3) and (5.5) for a lossless medium convert in the time domain to

$$\frac{\partial E_x(z,t)}{\partial z} = -\mu\frac{\partial H_y(z,t)}{\partial t} \quad \text{and} \quad \frac{\partial H_y(z,t)}{\partial z} = -\varepsilon\frac{\partial E_x(z,t)}{\partial t}$$

5.1.2. Show that the equations for a uniform plane wave in a lossless medium given in the time domain in (5.14) satisfy the equations in Problem 5.1.1.

5.1.3. A 10-MHz uniform plane wave is traveling in the following lossless dielectric ($\mu_r = 1$) media (a) polyvinyl chloride ($\varepsilon_r = 3.5$), (b) Teflon ($\varepsilon_r = 2.1$), (c) Mylar ($\varepsilon_r = 5$), and (d) polyurethane ($\varepsilon_r = 7$). Determine the phase constant β, the intrinsic impedance η, the phase velocity of propagation v, and the wavelength λ. [(a) $\beta = 0.392$ rad/m, $\eta = 202\,\Omega$, $v = 1.6 \times 10^8$ m/s, $\lambda = 16$ m, (b) $\beta = 0.304$ rad/m, $\eta = 260\,\Omega$, $v = 2.07 \times 10^8$ m/s, $\lambda = 20.7$ m, (c) $\beta = 0.468$ rad/m, $\eta = 169\,\Omega$, $v = 1.34 \times 10^8$ m/s, $\lambda = 13.4$ m, (d) $\beta = 0.554$ rad/m, $\eta = 142\,\Omega$, $v = 1.13 \times 10^8$ m/s, $\lambda = 11.3$ m.]

5.1.4. Write the phasor and time-domain expressions for a 5-MHz uniform plane wave traveling in free space. The 10-V/m electric field intensity vector is directed in the $+z$ direction and the wave is propagating in the $-y$ direction. Hint: Draw a sketch and get the direction of **H** such that $\mathbf{E} \times \mathbf{H}$ is in the direction of propagation.

5.1.5. Suppose that a uniform plane wave is traveling in the x direction in a lossless dielectric ($\mu_r = 1$) with the 100-V/m electric field in the z direction. If the wavelength is 25 cm and the velocity of propagation is 2×10^8 m/s, determine the frequency of the wave and the relative permittivity of the medium. Write complete time-domain expressions for the electric and magnetic field vectors. Hint: Draw a sketch and get the direction of **H** such that $\mathbf{E} \times \mathbf{H}$ is in the direction of propagation. [800 MHz, $\varepsilon_r = 2.25$, $\mathbf{E} = 100\cos(16\pi \times 10^8 t - 25.1x)\mathbf{a}_z$, $\mathbf{H} = -0.398\cos(16\pi \times 10^8 t - 25.1x)\mathbf{a}_y$]

5.1.6. Write the time-domain expression for the electric field of a uniform plane wave traveling in silicon ($\varepsilon_r = 12$) if the magnetic field is given by $\mathbf{H} = 0.1\cos(8\pi \times 10^7 t - 2.9z)\mathbf{a}_x$. Hint: Draw a sketch and get the direction of **E** such that $\mathbf{E} \times \mathbf{H}$ is in the direction of propagation.

5.1.7. A 2-GHz uniform plane wave is traveling in a medium that is characterized by $\varepsilon_r = 4$ and $\mu_r = 9$ in the $-z$ direction. The magnetic field intensity vector is directed in the y direction and the magnitude is 0.02 A/m. Write time-domain expressions for the electric and magnetic field vectors. Hint: Draw a sketch and get the direction of **H** such that **E** × **H** is in the direction of propagation. [$\mathbf{E} = -11.3\cos(4\pi \times 10^9 t + 251z)\mathbf{a}_x$, $\mathbf{H} = 0.02\cos(4\pi \times 10^9 t + 251z)\mathbf{a}_y$]

✳ **5.1.8.** A uniform plane wave has a wavelength of 2 cm in free space and 1 cm in a dielectric ($\mu_r = 1$). Determine the relative permittivity of the dielectric.

SECTION 5.2 UNIFORM PLANE WAVES IN LOSSY MEDIA

5.2.1. A uniform plane wave is propagating in a lossy medium having $\varepsilon_r = 36$, $\mu_r = 4$, and $\sigma = 1$ S/m. The electric field is given by $\mathbf{E} = 100e^{-\alpha x}\cos(10\pi \times 10^8 t - \beta x)\mathbf{a}_z$. Determine α and β and $\hat{\eta}$, and write a time-domain expression for the associated magnetic field vector.

5.2.2. Determine the phase velocity of propagation, attenuation constant, phase constant, and intrinsic impedance of a uniform plane wave traveling in wet, marshy soil ($\sigma \cong 10^{-2}$ S/m, $\varepsilon_r \cong 15$, $\mu_r = 1$) at 60 Hz (power frequency), 1 MHz (AM radio broadcast frequency), 100 MHz (FM radio broadcast frequency), and 10 GHz (microwave radio relay frequency). [(a) $\alpha = 1.54 \times 10^{-3}$, $\beta = 1.54 \times 10^{-3}$ rad/m, $v = \frac{\omega}{\beta} = 2.45 \times 10^5$ m/s, $\hat{\eta} = 0.22\angle 45°$, (b) $\alpha = 0.19$, $\beta = 0.21$ rad/m, $v = \frac{\omega}{\beta} = 3.03 \times 10^7$ m/s, $\hat{\eta} = 28.05\angle 42.62°$, (c) $\alpha = 0.49$, $\beta = 8.13$ rad/m, $v = \frac{\omega}{\beta} = 7.73 \times 10^7$ m/s, $\hat{\eta} = 96.99\angle 3.42°$, (d) $\alpha = 0.49$, $\beta = 811.2$ rad/m, $v = \frac{\omega}{\beta} = 7.75 \times 10^7$ m/s, $\hat{\eta} = 97.34\angle 0.03°$]

5.2.3. If a material has $\sigma = 2$ S/m, $\varepsilon_r = 9$, and $\mu_r = 16$ at a frequency of 1 GHz, calculate the attenuation constant, phase constant, velocity of propagation, and intrinsic impedance.

5.2.4. Write a time-domain expression for the electric field of a uniform plane wave in a lossy dielectric ($\mu_r = 1$) if the magnetic field is given by $\mathbf{H} = 0.1e^{-200y}\cos(2\pi \times 10^{10}t - 300y)\mathbf{a}_x$. [$\mathbf{E} = 21.9e^{-200y}\cos(2\pi \times 10^{10}t - 300y + 33.69°)\mathbf{a}_z$]

SECTION 5.3 POWER FLOW IN THE UNIFORM PLANE WAVES

5.3.1. A 10-MHz uniform plane wave is traveling in the $+z$ direction in a lossless dielectric having $\varepsilon_r = 5$. The 10-V/m electric field is directed in the $+y$ direction. Determine the average power crossing a surface in the xy plane bounded by (3,2,2), (3,−1,2), (−1,2,2), (−1,−1,2) where the coordinates are in meters.

5.3.2. A 200-MHz uniform plane wave is traveling in the $+z$ direction in a lossless dielectric having $\varepsilon_r = 9$. The 0.2-A/m magnetic field is directed in the $+y$ direction. Determine the average power crossing a surface in the xy plane bounded by (0,0,0), (0,5,0), (3,5,0), (3,0,0) where the coordinates are in meters. [37.7 W]

5.3.3. A uniform plane wave is propagating in a lossy medium having $\varepsilon_r = 36$, $\mu_r = 4$, and $\sigma = 1$ S/m. The electric field is given by $\mathbf{E} = 100e^{-\alpha x}\cos(10\pi \times 10^8 t - \beta x)\mathbf{a}_z$. Determine the average power lost in propagating through a rectangular volume bounded by (0,0,0), (0,2 m,0), (20 mm,2 m,0), (20 mm,0,0), (20 mm,0,3 m), (20 mm,2 m,3 m), (0,2 m,3 m), (0,0,3 m).

5.3.4. A 1-GHz, 1-V/m uniform plane wave is propagating in a lossy material having parameters of $\sigma = 2$ S/m, $\varepsilon_r = 9$, and $\mu_r = 16$ at a frequency of 1 GHz. Determine the average power lost in propagating across a 100-cm² surface perpendicular to it and along a depth in the material of 5 mm. [15.2 μW]

SECTION 5.4 SKIN DEPTH

5.4.1. Compare the distances required for a uniform plane wave to travel in seawater ($\sigma = 4$ S/m, $\mu_1 = 1$, $\varepsilon_r = 81$) in order that the amplitude of the wave is reduced by 80 dB (a factor of 10,000) at the following frequencies: (a) 1 kHz, (b) 10 kHz, (c) 100 kHz, (d) 1 MHz, (e) 10 MHz, (f) 100 MHz. Show that the attenuation of the amplitude in dB is $-8.69\alpha d$. State whether seawater is a good conductor or a good dielectric at each frequency. This illustrates why communication with submarines uses very low frequencies in the kHz range. [Good conductor for all frequencies, (a) 73.3 m, (b) 23.2 m, (c) 7.33 m, (d) 2.32 m, (e) 0.733 m, (f) 23.2 cm]

✱ **5.4.2.** A uniform plane wave is propagating in wet, marshy soil ($\sigma \cong 10^{-2}$ S/m, $\varepsilon_r \cong 15$, $\mu_r = 1$) at (a) 60 Hz (power frequency), (b) 1 MHz (AM radio broadcast frequency), (c) 100 MHz (FM radio broadcast frequency), and (d) 10 GHz (microwave relay frequency). Determine the distance at each frequency for the wave to travel such that its amplitude is attenuated by 20 dB (a reduction of $\frac{1}{10}$).

5.4.3. A 1-V/m, uniform plane wave is traveling in seawater ($\sigma = 4$ S/m, $\mu_1 = 1$, $\varepsilon_r = 81$). Determine the power dissipated in a block of seawater having a surface area of 5 m^2 as the wave travels a distance of one skin depth for frequencies of (a) 1 kHz, (b) 10 kHz, and (c) 100 kHz. [(a) 34.5 W, (b) 10.9 W, (c) 3.45 W]

5.4.4. Lossy materials have previously been characterized through a conductivity σ. However, a more common method of characterizing them is in terms of a loss tangent defined as $\tan\phi = \sigma/\omega\varepsilon$. Observe that for good dielectrics where $\sigma \ll \omega\varepsilon$; the loss tangent is less than one, whereas for good conductors where $\sigma \gg \omega\varepsilon$ the loss tangent is greater than unity. The origin of this name stems from the observation that the right-hand side of Ampere's law involves conduction and displacement currents via a term $\sigma + j\omega\varepsilon$. The conduction and displacement current terms are 90° out of phase and when drawn in the complex plane, the tangent of the angle between the hypotenuse and the real part, σ, is the loss tangent. Numerous handbooks tabulate the loss tangent for materials at various frequencies rather than giving σ at those frequencies. Show that the attenuation and phase constants can be written in terms of the loss tangent as

$$\alpha = \frac{\omega\sqrt{\mu\varepsilon}}{\sqrt{2}}\sqrt{(\sqrt{1 + \tan^2\phi} - 1)}$$

and

$$\beta = \frac{\omega\sqrt{\mu\varepsilon}}{\sqrt{2}}\sqrt{(\sqrt{1 + \tan^2\phi} + 1)}$$

SECTION 5.5 NORMAL INCIDENCE OF UNIFORM PLANE WAVES AT PLANE, MATERIAL BOUNDARIES

5.5.1. With reference to Fig. 5.8, medium 1 has $\varepsilon_{r1} = 4$, $\mu_{r1} = 1$ and medium 2 has $\varepsilon_{r2} = 9$, $\mu_{r2} = 4$. Write time-domain expressions for the fields if the incident electric field is $\mathbf{E}^i = 100\cos(\omega t - 6\pi z)\mathbf{a}_x$. Determine the average power transmitted through a 2-m^2 area of the surface.

5.5.2. With reference to Fig. 5.8, medium 1 has $\varepsilon_{r1} = 4$, $\mu_{r1} = 16$ and medium 2 has $\varepsilon_{r2} = 9$, $\mu_{r2} = 1$. Write time-domain expressions for the fields if the incident electric field is $\mathbf{E}^i = 10\cos[\omega t - (8\pi/3z)\mathbf{a}_x]$. Determine the average power transmitted through a 5-m^2 area of the surface.

$$\left[\mathbf{E}^i = 10\cos\left(10\pi \times 10^7 t - \frac{8\pi}{3}z\right)\mathbf{a}_x, \right.$$

$$\mathbf{E}^r = -\frac{50}{7}\cos\left(10\pi \times 10^7 t + \frac{8\pi}{3}z\right)\mathbf{a}_x, \mathbf{E}^t = \frac{20}{7}\cos(10\pi \times 10^7 t - \pi z)\mathbf{a}_x,$$

$$\mathbf{H}^i = \frac{10}{754}\cos\left(10\pi \times 10^7 t - \frac{8\pi}{3}z\right)\mathbf{a}_y, \mathbf{H}^r = \frac{50}{7 \times 754}\cos\left(10\pi \times 10^7 t + \frac{8\pi}{3}z\right)\mathbf{a}_y,$$

$$\left. \mathbf{H}^t = \frac{20}{7 \times 126}\cos(10\pi \times 10^7 t - \pi z)\mathbf{a}_y, 162 \text{ mW} \right].$$

5.5.3. With reference to Fig. 5.8, medium 1 has $\varepsilon_{r1} = 9$, $\mu_{r1} = 1$ and medium 2 has $\varepsilon_{r2} = 16$, $\mu_{r2} = 4$. Write time-domain expressions for the fields if the incident electric field is $\mathbf{E}^i = 5\cos(\omega t - 2\pi z)\mathbf{a}_y$. Observe that the incident electric field vector is in the y direction. Determine the average power transmitted through a 4-m^2 area of the surface.

5.5.4. With reference to Fig. 5.8, medium 1 has $\varepsilon_{r1} = 9$, $\mu_{r1} = 4$ and medium 2 has $\varepsilon_{r2} = 1$, $\mu_{r2} = 16$. Write time-domain expressions for the fields if the incident magnetic field is

$\mathbf{H}^i = 0.1 \cos(\omega t - 8\pi z)\mathbf{a}_x$. Observe that the incident electric field vector is in the y direction. Determine the average power transmitted through a 3-m^2 area of the surface.

$$\left[\mathbf{E}^i = -25.13 \cos(4\pi \times 10^8 t - 8\pi z)\mathbf{a}_y, \; \mathbf{E}^r = -17.95 \cos(4\pi \times 10^8 t + 8\pi z)\mathbf{a}_y, \right.$$

$$\mathbf{E}^t = -43.08 \cos\left(4\pi \times 10^8 t - \frac{16\pi}{3}z\right)\mathbf{a}_y, \; \mathbf{H}^i = 0.1 \cos(4\pi \times 10^8 t - 8\pi z)\mathbf{a}_x,$$

$$\left. \mathbf{H}^r = -\frac{5}{7} \times 0.1 \cos(4\pi \times 10^8 t + 8\pi z)\mathbf{a}_x, \; \mathbf{H}^t = \frac{2}{7} \times 0.1 \cos\left(4\pi \times 10^8 t - \frac{16\pi}{3}z\right)\mathbf{a}_x, 1.85 \text{ W} \right]$$

5.5.5. With reference to Fig. 5.9, medium 1 is free space and medium 2 is lossy with $\varepsilon_r = 4$, $\mu_r = 1$, and $\sigma = 10^3$ S/m. The incident electric field is given by $\mathbf{E}^i = 10 \cos(6\pi \times 10^6 t - 0.063z)\mathbf{a}_x$. Write complete time-domain expressions for the fields. Determine the average power dissipated in a volume of material in the second medium consisting of a 2-m^2 area of the surface and 1 mm deep.

5.5.6. With reference to Fig. 5.9, medium 1 is free space and medium 2 is stainless steel with $\varepsilon_r = 1$, $\mu_r = 500$, and $\sigma = 0.02$ S/m. The incident electric field is given by $\mathbf{E}^i = 100 \cos(2\pi \times 10^9 t - 20.94z)\mathbf{a}_x$. Write complete time-domain expressions for the fields. Determine the average power dissipated in a volume of material in the stainless steel consisting of a 2-m^2 area of the surface and one skin depth deep.

$$[\mathbf{E}^i = 100 \cos(2\pi \times 10^9 t - 20.94z)\mathbf{a}_x, \; \mathbf{E}^r = 91 \cos(2\pi \times 10^9 t + 20.94z + 0.91°)\mathbf{a}_x,$$

$$\mathbf{H}^i = 0.265 \cos(2\pi \times 10^9 t - 20.94z)\mathbf{a}_y, \; \mathbf{H}^r = -0.241 \cos(2\pi \times 10^9 t + 20.94z + 0.91°)\mathbf{a}_y,$$

$$\mathbf{E}^t = 191 e^{-83z} \cos(2\pi \times 10^9 t - 475.62z + 0.43°)\mathbf{a}_x, \text{ and}$$

$$\mathbf{H}^t = 2.34 \times 10^{-2} e^{-83z} \cos(2\pi \times 10^9 t - 475.62z - 9.47°)\mathbf{a}_y, 2.49 \text{ W}]$$ ~~3.8w~~

5.5.7. With reference to Fig. 5.9, medium 1 is lossless and has parameters of $\varepsilon_r = 9$, $\mu_r = 1$. Medium 2 is lossy with $\varepsilon_r = 1$, $\mu_r = 1$, and $\sigma = 20$ S/m. The incident electric field is given by $\mathbf{E}^i = 5 \cos(10\pi \times 10^8 t - 10\pi z)\mathbf{a}_x$. Write complete time-domain expressions for the fields. Determine the average power dissipated in a volume of material in the second medium consisting of a 1-cm^2 area of the surface and 10 mm deep.

5.5.8. Airplanes use radar altimeters to accurately determine their low-level altitude. If an airplane is flying over the ocean ($\varepsilon_r = 81$, $\mu_r = 1$, and $\sigma = 4$ S/m), determine the percent of the transmitted power that is reflected from the ocean surface and the percent of the transmitted power that is lost in the ocean if the radar frequency is 7 GHz. [64.2% and 35.8%]

5.5.9. A radio wave strikes the surface of a copper conductor normal to it. If the total electric field is zero at a distance of 1 m away from the conductor surface, determine the lowest possible frequency of the radio wave.

SECTION 5.6 SNELL'S LAWS

5.6.1. A light ray traveling in air is incident at an angle θ on a sheet of transparent material of thickness t having an index of refraction n as shown in Fig. P5.6.1. Show that the path the beam

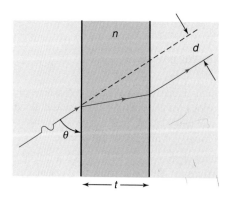

Figure P5.6.1 Problem 5.6.1.

takes as it exits the material is in the same direction as the incident beam. Determine the distance d that the ray will be displaced from its path on exiting the material.

$$\left[d = t\cos\theta - t\frac{\sin\theta\cos\theta}{\sqrt{n^2 - \cos^2\theta}} \right]$$

5.6.2. A glass isoceles prism is used to change the path of the light ray as shown in Fig. P5.6.2. If the index of refraction of glass is 1.5, determine the ratio of transmitted and incident power densities.

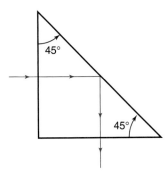

Figure P5.6.2 Problem 5.6.2.

5.6.3. A person fishing from a boat observes a fish feeding on the bottom of a shallow lake. The person's height is 6 ft, and the lake depth at this location is 10 ft. The fish appears to be a distance of 20 ft from the boat. Determine the true distance of the fish from the boat. [28.3 ft]

6

Transmission Lines

In this chapter we will investigate the use of two parallel conductors to guide a signal from a source to a load. These are called *transmission lines* and occur in many forms. A pair of wires (circular cross-sectional conductors) serve to transfer 60-Hz electric power from one point to another, carry low-frequency audio signals from the source (e.g., a CD player) to a load (e.g., a loudspeaker), and carry high-frequency signals from radio and TV transmitters to the antenna. The electronics in high-speed digital computers are mounted on printed circuit boards (PCBs). The PCB consists of a glass-epoxy board with rectangular cross section conductors (lands) etched on either side of the board. Pairs of these lands serve to carry the digital signals between the various electronic modules that are mounted on the board.

Although our earlier electric circuits courses indicated that the effect of these interconnection conductors can be ignored, this is not true in the case of high-speed digital computer interconnects and high-frequency analog circuits. For example, consider a pair of lands 10 cm in length (approximately 4 in.) etched on a glass-epoxy ($\varepsilon_r = 4.7$) board. The effective relative permittivity is approximately the average of the board relative permittivity and air or $\varepsilon_r' \cong 2.85$, giving a velocity of propagation of waves on these conductors of $v = v_o/\sqrt{\varepsilon_r'} = 1.777 \times 10^8$ m/s. Digital signals propagating along these lands will suffer a delay of $T = \mathscr{L}/v = 0.56$ ns in propagating from one end of the line to the other where $\mathscr{L}$ is the total length of the line. Digital computers require precise timing, and today a delay of 0.5 ns can be very significant. Also, if the signals they carry have frequency content such that the line is electrically long at those frequencies (line length not much less than wavelength), then the phase shift incurred in transiting the line will be significant and the effect of the line cannot be ignored. For example, suppose the lands are carrying a digital clock signal consisting of pulses at the basic repetition rate of 1 GHz. This signal is periodic and will contain, according to the Fourier series, frequency components at all multiples of this basic repetition rate, that is, 1 GHz, 2 GHz, 3 GHz, 4 GHz, . . . , etc. A length of 10-cm line will be 0.56λ in electrical length at the fundamental frequency component of 1 GHz. In other words, the line will be about one-half wavelength at 1 GHz. At the second harmonic of 2 GHz the line will be about one wavelength long. At the higher frequency components, the line will be even longer, electrically. Clearly the effect of the line cannot be ignored in this situation.

Our purpose in this chapter is to examine the effect of interconnect lines, whether they be parallel wires or interconnect lands on PCBs, on the transmission of a signal from one end to the other. In particular, as clock speeds of digital computers continue to increase, seemingly without bound, the effects of interconnect lands will continue to increase in importance. If we are to design high-speed digital circuits, we must consider the interconnects in that design in order for the circuit to correctly process the data. Additionally, radio transmission with wireless devices such as cell phones rely on very high frequency carriers in the GHz range. Designing circuits that reliably carry these signals from one point in the phone to another again requires that we consider the effect of the interconnects in the physical design. Some 20 years ago we could ignore the effects of these interconnect lines because either the clock frequencies were very low or the radio transmission frequencies were in the MHz range. Today and in the future, we can no longer ignore the effects of interconnect lines.

In addition, the transmission line can radiate the signal it is carrying to other neighboring lines, causing interference. This is referred to as *crosstalk*. Typically the higher the frequency of the signal being carried by the line, the larger the radiated emissions and hence the crosstalk. With increasing frequencies of analog signals and increasing repetition rates of digital signals, crosstalk is an ever-increasing problem that must be considered in the design or else the design will not function properly.

Chapter Learning Objectives

After completing the contents of this chapter, you should be able to

- ▷ understand the concepts of waves propagating on transmission lines, reflection of waves at terminations, and time delay of propagation,

- ▷ compute the time history of terminal voltages and currents on transmission lines for pulse waveform excitation of the line,

- ▷ use the SPICE transmission-line model to compute the time history of terminal voltages and currents on transmission lines for pulse waveform excitation of the line,

- ▷ compute the terminal voltages and currents as well as the input impedance of transmission lines for single-frequency, sinusoidal excitation of the line,

- ▷ use the SPICE transmission-line model to compute the terminal voltages and currents of transmission lines for single-frequency, sinusoidal excitation of the line,

- ▷ use the Smith chart to compute the input impedance of a transmission line for single-frequency, sinusoidal excitation of the line,

- ▷ create a lumped-circuit approximate model of a transmission line,

- ▷ understand the effect of losses on the voltage and current on a lossy transmission line, and

- ▷ cite and explain several engineering applications of the principles of this chapter.

▷ 6.1 THE TRANSMISSION-LINE EQUATIONS

Consider a two-conductor transmission line shown in Fig. 6.1 where the conductors are parallel to the z axis. If we apply a voltage V between the two conductors as in Fig. 6.1a, charge will be deposited on the conductors, resulting in an electric field, $\mathbf{E}^t$, lying in the transverse or xy plane. Since the two conductors separate charge, this suggests that the line has a capacitance per unit of length, c F/m. Now suppose we apply a current I passing to the right in the upper conductor and "returning" on the lower conductor as shown in Fig. 6.1b. This current will cause a magnetic field, $\mathbf{H}^t$, that also lies in the transverse or xy plane. This magnetic field passes through the loop between the two conductors and suggests that the line has an inductance per unit of length, l H/m. This suggests that the line can be modeled as a *distributed-parameter* circuit consisting of

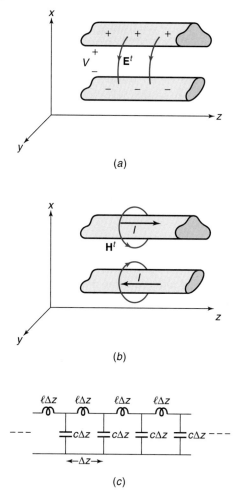

Figure 6.1 The two-conductor transmission line. (a) Electric field about a transmission line caused by the voltage between the two conductors. (b) Magnetic field about a transmission line caused by the current on the conductors. (c) Representing a transmission line as a distributed parameter circuit consisting of cells of per-unit-length inductance, l, and per-unit-length capacitance, c.

a sequence of inductors and capacitors as shown in Fig. 6.1c. Note that the total inductance and capacitance in a length Δz of the line is the per-unit-length value multiplied by the length of that section, $l\Delta z$ and $c\Delta z$.

Transmission lines have, in addition to inductance and capacitance, losses. The conductors have a finite, nonzero resistance, and the medium surrounding the conductors has loss in its dielectric (except for free space insulation). Usually these represent second-order effects and can be neglected. At frequencies in the GHz range the resistance of the conductors may become significant due to skin effect. We will defer consideration of losses until Section 6.6.

There is an important point about the line that can be observed from this equivalent circuit. If a pulse is applied to the left end of the line, it will charge the first capacitance and energize the first inductance. As the pulse moves down the line to the right, it will discharge the first capacitor and deenergize the first inductor and then charge and energize the next capacitor and inductor, and so forth. Hence waves of voltage and current (and their associated transverse electric and magnetic fields) will move down the line with a velocity v. It takes a certain time to energize and de-energize these elements, so that it will take a finite, nonzero time for the waves to transit the line. This will result in a *time delay* for a line of total length $\mathcal{L}$ of $T = \mathcal{L}/v$.

In the remainder of this chapter we will investigate this model more quantitatively from the differential equations relating the voltage and current at various points on the line.

6.1.1 Types of Transmission Lines

Figure 6.2 depicts various types of lines composed of wires (conductors of circular cross section). Figure 6.2a shows a two-wire line, Fig. 6.2b shows a wire above a ground plane, and Fig. 6.2c shows a coaxial cable where the inner wire is on the axis of an overall

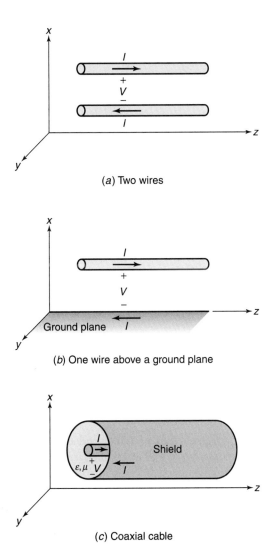

(a) Two wires

(b) One wire above a ground plane

(c) Coaxial cable

Figure 6.2 Illustration of wire-type transmission lines. (a) A two-wire line. (b) One wire above a ground plane. (c) A coaxial cable.

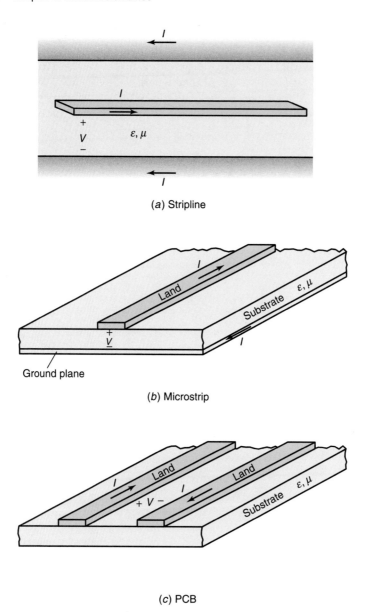

(a) Stripline

(b) Microstrip

(c) PCB

Figure 6.3 Transmission lines composed of rectangular cross section conductors. (a) A stripline. (b) A microstrip line. (c) A printed circuit board (PCB).

cylindrical shield. Figure 6.3 depicts various types of integrated circuit and printed circuit board (PCB) structures. Figure 6.3a depicts what is called a *stripline*. A rectangular cross section conductor (land) is sandwiched between two ground planes. This commonly occurs in PCBs that have "innerplanes." Figure 6.3b depicts a printed circuit board or substrate having a ground plane on one side and a land on the other. This structure is also common on PCBs as well as microwave circuits and is called a *microstrip*. Figure 6.3c depicts two lands on one side of a substrate. This is commonly found on the outer layers of PCBs, and the lands interconnect the electronic modules that are mounted on the top and bottom of the board.

6.1.2 The Transmission-Line Equations

Consider a Δz section of the line shown in Fig. 6.4. The line voltage and current are functions of time t and position z. Writing Kirchhoff's voltage law around the outside loop gives

$$V(z + \Delta z, t) - V(z,t) = -l\Delta z \frac{\partial I(z,t)}{\partial t}$$

Dividing both sides by Δz and taking the limit as $\Delta z \to 0$ gives

$$\left. \frac{V(z + \Delta z, t) - V(z,t)}{\Delta z} \right|_{\lim \Delta z \to 0} = \frac{\partial V(z,t)}{\partial z}$$

and we obtain the first transmission-line equation:

$$\boxed{\frac{\partial V(z,t)}{\partial z} = -l\frac{\partial I(z,t)}{\partial t}} \tag{6.1a}$$

Similarly, writing Kirchhoff's current law at the upper node of the capacitor gives

$$I(z + \Delta z, t) - I(z,t) = -c\Delta z \frac{\partial V(z + \Delta z, t)}{\partial t}$$

Dividing both sides by Δz and taking the limit as $\Delta z \to 0$ gives

$$\left. \frac{I(z + \Delta z, t) - I(z,t)}{\Delta z} \right|_{\lim \Delta z \to 0} = \frac{\partial I(z,t)}{\partial z}$$

and we obtain the second transmission-line equation:

$$\boxed{\frac{\partial I(z,t)}{\partial z} = -c\frac{\partial V(z,t)}{\partial t}} \tag{6.1b}$$

Equations (6.1a) and (6.1b) are called the *transmission-line equations*. Observe that they are coupled in that each equation involves both V and I. We can uncouple these equations by, for example, differentiating (6.1a) with respect to z to give

$$\frac{\partial^2 V(z,t)}{\partial z^2} = -l\frac{\partial^2 I(z,t)}{\partial t \partial z}$$

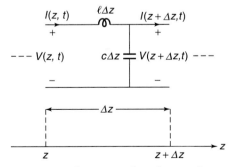

Figure 6.4 The per-unit-length equivalent circuit of a transmission line.

and differentiating (6.1b) with respect to t to give

$$\frac{\partial^2 I(z,t)}{\partial z \partial t} = -c\frac{\partial^2 V(z,t)}{\partial t^2}$$

Substituting the second into the first gives the first uncoupled equation as

$$\frac{\partial^2 V(z,t)}{\partial z^2} = lc\frac{\partial^2 V(z,t)}{\partial t^2} \qquad (6.2a)$$

Differentiating (6.1b) with respect to z and differentiating (6.1a) with respect to t and substituting gives the second uncoupled equation as

$$\frac{\partial^2 I(z,t)}{\partial z^2} = lc\frac{\partial^2 I(z,t)}{\partial t^2} \qquad (6.2b)$$

We will solve these equations in this chapter in order to understand the wave propagation on the line.

6.1.3 The Per-Unit-Length Parameters

The above transmission-line equations contain the per-unit-length parameters of capacitance c (F/m) and inductance l (H/m). All of the structural information, such as type of conductor, wire radii, and wire separation, that distinguish one type of line from another are contained in these two parameters. For wire lines, these parameters were derived in Chapter 3. First consider a two-wire line composed of wires of radii r_w that are separated by distance s as shown in Fig. 6.5a. The exact expressions for the per-unit-length capacitance and inductance are

$$c = \frac{\pi \varepsilon_o}{\ln\left[\dfrac{s}{2r_w} + \sqrt{\left(\dfrac{s}{2r_w}\right)^2 - 1}\right]} \qquad \text{F/m} \qquad (6.3a)$$

and

$$l = \frac{\mu_o}{\pi}\ln\left[\frac{s}{2r_w} + \sqrt{\left(\frac{s}{2r_w}\right)^2 - 1}\right] \qquad \text{H/m} \qquad (6.3b)$$

We have assumed that the surrounding medium is free space. If the wires are separated sufficiently so that the charge and current are uniformly distributed around the wire peripheries (proximity effect is not pronounced, which is satisfied for $s/r_w > 5$), then these expressions reduce to

$$c \cong \frac{\pi \varepsilon_o}{\ln\left[\dfrac{s}{r_w}\right]} \qquad \text{F/m} \qquad (6.4a)$$

and

$$l \cong \frac{\mu_o}{\pi}\ln\left[\frac{s}{r_w}\right] \qquad \text{H/m} \qquad (6.4b)$$

The approximate expressions for wide separations in (6.4) were derived directly in Chapter 3.

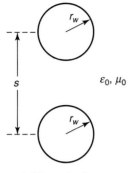

(a) Two-wire line

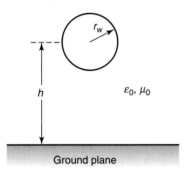

(b) Wire above ground

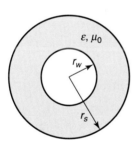

(c) Coaxial cable

Figure 6.5 Cross-sectional dimensions of wire-type lines. (a) A two-wire line. (b) One wire above a ground plane. (c) A coaxial cable.

For the case of one wire of radius r_w at a height h above a ground plane in Fig. 6.5b, the exact expressions for capacitance and inductance can be obtained from the two-wire result using the method of images to give

$$c = \frac{2\pi\varepsilon_o}{\ln\left[\dfrac{h}{r_w} + \sqrt{\left(\dfrac{h}{r_w}\right)^2 - 1}\right]} \quad \text{F/m} \tag{6.5a}$$

and

$$l = \frac{\mu_o}{2\pi}\ln\left[\frac{h}{r_w} + \sqrt{\left(\frac{h}{r_w}\right)^2 - 1}\right] \quad \text{H/m} \tag{6.5b}$$

If the wire is sufficiently elevated above the ground plane so that proximity effect is not a factor (a ratio of $h/r_w > 2.5$ is sufficient), these approximate to

$$c \cong \frac{2\pi\varepsilon_o}{\ln\left[\dfrac{2h}{r_w}\right]} \qquad \text{F/m} \tag{6.6a}$$

and

$$l \cong \frac{\mu_o}{2\pi}\ln\left[\frac{2h}{r_w}\right] \qquad \text{H/m} \tag{6.6b}$$

These approximate expressions in (6.6) were also derived in Chapter 3.

For the coaxial cable shown in Fig. 6.5c having a wire of radius r_w centered on the axis of a shield of inner radius r_s, the exact values are (because of symmetry, proximity effect is not a factor)

$$c = \frac{2\pi\varepsilon_r\varepsilon_o}{\ln\left[\dfrac{r_s}{r_w}\right]} \qquad \text{F/m} \tag{6.7a}$$

and

$$l = \frac{\mu_o}{2\pi}\ln\left[\frac{r_s}{r_w}\right] \qquad \text{H/m} \tag{6.7b}$$

In the case of the coaxial cable, the dielectric interior to the shield has a relative permittivity of ε_r. These exact expressions were derived in Chapter 3.

Dimensions of transmission lines such as wire radius, r_w, separation, s, height above ground, h, shield inner radius, r_s, are commonly given in the English system of units in *mils* where 1 mil = 1/1000 in. To convert dimensions in mils to meters, multiply by 2.54×10^{-5}.

Observe that all three structures are immersed in a homogeneous medium. It can be shown that for a homogeneous medium

$$\boxed{\begin{aligned} lc &= \mu_o\varepsilon_o\varepsilon_r \\ &= \frac{\varepsilon_r}{v_o^2} \qquad \text{homogeneous medium} \end{aligned}} \tag{6.8}$$

where the velocity of propagation in free space is $v_o = 1/\sqrt{\mu_o\varepsilon_o} \cong 3 \times 10^8$ m/s. For the case of two wires in free space, (6.3) and (6.4) give $lc = \mu_o\varepsilon_o$; for the case of one wire above a ground plane in free space, (6.5) and (6.6) also give $lc = \mu_o\varepsilon_o$; and for the case of a coaxial cable where the interior dielectric has relative permittivity ε_r, (6.7) gives $lc = \mu_o\varepsilon_o\varepsilon_r$. For lines in a homogeneous medium, the relation in (6.8) allows us to determine l in terms of c or vice versa. We will later find that voltage and current waves propagate on the line with velocity $v = 1/\sqrt{lc}$. If the medium is homogeneous, then the waves travel at the speed of light in that medium:

$$\boxed{\begin{aligned} v &= \frac{1}{\sqrt{lc}} \\ &= \frac{1}{\sqrt{\mu_o\varepsilon_o\varepsilon_r}} \\ &= \frac{v_o}{\sqrt{\varepsilon_r}} \qquad \text{homogeneous medium} \end{aligned}} \tag{6.9}$$

QUICK REVIEW EXERCISE 6.1

Determine the exact and approximate values for the per-unit-length capacitance and inductance of a two-wire line (typical of ribbon cables used to interconnect electronic components) whose wires have radii of 7.5 mils (0.19 mm) and are separated by 50 mils (1.27 mm).

ANSWERS   Exact, 14.8 pF/m and 0.75 μH/m; Approximate, 14.6 pF/m and 0.759 μH/m.   ◀

QUICK REVIEW EXERCISE 6.2

Determine the exact and approximate values for the per-unit-length capacitance and inductance of one wire of radius 16 mils (0.406 mm) at a height of 100 mils (2.54 mm) above a ground plane.

ANSWERS   Exact, 22.1 pF/m and 0.504 μH/m; Approximate, 22.0 pF/m and 0.505 μH/m.   ◀

QUICK REVIEW EXERCISE 6.3

Determine the per-unit-length capacitance and inductance of a coaxial cable (RG-58U) where the inner wire has radius 16 mils (0.406 mm) and the shield has an inner radius of 58 mils (1.47 mm). The dielectric is polyethylene having a relative permittivity of 2.3.

ANSWERS   99.2 pF/m and 0.258 μH/m.   ◀

For lines composed of rectangular cross section conductors as illustrated in Fig. 6.3, the parameters cannot be obtained in closed form as was the case for wires. Approximate expressions are given in C.R. Paul, *Introduction to Electromagnetic Compatibility*, John Wiley Interscience, 1992. First, consider the stripline shown in cross section in Fig. 6.6a. The center strip (land) is of width w and is midway between two planes that are separated a distance s. The space between the planes and surrounding the strip is filled with a dielectric with relative permittivity ε_r. Assuming a zero thickness strip, $t = 0$, the per-unit-length inductance is

$$l = \frac{30\pi}{v_o} \frac{1}{\left[\dfrac{w_e}{s} + 0.441\right]} \qquad \text{H/m} \tag{6.10a}$$

and the effective width of the center conductor is

$$\frac{w_e}{s} = \begin{cases} \dfrac{w}{s} & \dfrac{w}{s} \geq 0.35 \\[2mm] \dfrac{w}{s} - \left(0.35 - \dfrac{w}{s}\right)^2 & \dfrac{w}{s} \leq 0.35 \end{cases} \tag{6.10b}$$

The per-unit-length capacitance can be found from the inductance using the relation in (6.8) since the surrounding medium is homogeneous:

$$\begin{aligned} c &= \frac{\varepsilon_r}{l v_o^2} \\[2mm] &= \frac{\varepsilon_r}{30\pi v_o}\left[\frac{w_e}{s} + 0.441\right] \qquad \text{F/m} \end{aligned} \tag{6.10c}$$

The microstrip line shown in Fig. 6.6b has a land of width w placed on a board of thickness h having a ground plane on the opposite side. The board has a relative

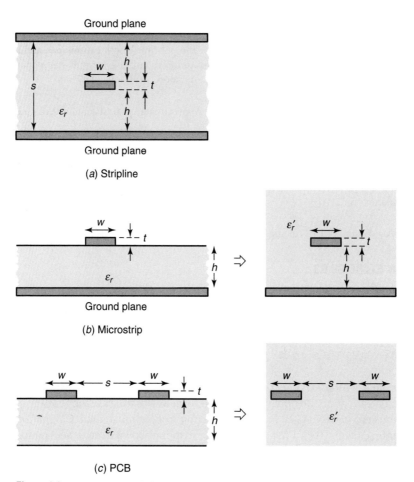

Figure 6.6 Cross-sectional dimensions of lines composed of rectangular cross section conductors. (a) A stripline. (b) A microstrip line. (c) A printed circuit board (PCB).

permittivity ε_r. Assuming the land thickness is zero ($t = 0$) the per-unit-length inductance is

$$
l = \begin{cases}
\dfrac{60}{v_o} \ln\left[\dfrac{8h}{w} + \dfrac{w}{4h}\right] & \text{H/m} \quad \dfrac{w}{h} \le 1 \\[2ex]
\dfrac{120\pi}{v_o}\left[\dfrac{w}{h} + 1.393 + 0.667 \ln\left(\dfrac{w}{h} + 1.444\right)\right]^{-1} & \text{H/m} \quad \dfrac{w}{h} \ge 1
\end{cases}
\tag{6.11a}
$$

The effective relative permittivity is

$$
\varepsilon'_r = \frac{\varepsilon_r + 1}{2} + \frac{\varepsilon_r - 1}{2}\frac{1}{\sqrt{1 + 10\dfrac{h}{w}}}
\tag{6.11b}
$$

This effective relative permittivity accounts for the fact that the electric field lines are partly in air and partly in the substrate dielectric. If this inhomogeneous medium (air and the dielectric) is replaced with a homogeneous one having an effective relative permittivity of ε_r' as shown in Fig. 6.6b, all properties of the line remain unchanged. But the homogeneous medium problem is much easier to analyze. Hence the per-unit-length capacitance can be found from the relation in (6.8) as

$$c = \frac{\varepsilon_r'}{lv_o^2}$$

$$= \begin{cases} \dfrac{\varepsilon_r'}{60v_o \ln\left[\dfrac{8h}{w} + \dfrac{w}{4h}\right]} & \text{F/m} & \dfrac{w}{h} \leq 1 \\[4mm] \dfrac{\varepsilon_r'}{120\pi v_o}\left[\dfrac{w}{h} + 1.393 + 0.667 \ln\left(\dfrac{w}{h} + 1.444\right)\right] & \text{F/m} & \dfrac{w}{h} \geq 1 \end{cases} \tag{6.11c}$$

The PCB shown in Fig. 6.6c has two lands of width w placed on one side of a board and separated edge-to-edge a distance s. The board has thickness h and a relative permittivity ε_r. Assuming the land thickness is zero ($t = 0$), the per-unit-length inductance is

$$l = \begin{cases} \dfrac{120}{v_o} \ln\left(2\dfrac{1 + \sqrt{k}}{1 - \sqrt{k}}\right) & \text{H/m} & \dfrac{1}{\sqrt{2}} \leq k \leq 1 \\[4mm] \dfrac{377\pi}{v_o \ln\left(2\dfrac{1 + \sqrt{k'}}{1 - \sqrt{k'}}\right)} & \text{H/m} & 0 \leq k \leq \dfrac{1}{\sqrt{2}} \end{cases} \tag{6.12a}$$

where k is

$$k = \frac{s}{s + 2w} \tag{6.12b}$$

and $k' = \sqrt{1 - k^2}$. The effective relative permittivity is

$$\varepsilon_r' = \frac{\varepsilon_r + 1}{2}\left\{\tanh\left[0.775 \ln\left(\frac{h}{w}\right) + 1.75\right] \right.$$
$$\left. + \frac{kw}{h}[0.04 - 0.7k + 0.01(1 - 0.1\varepsilon_r)(0.25 + k)]\right\} \tag{6.12c}$$

which again accounts for the fact that the electric field lines are partly in air and partly in the substrate dielectric. If this inhomogeneous medium (air and the dielectric) is replaced with a homogeneous one having an effective relative permittivity of ε_r' as shown in Fig. 6.6c, all properties of the line remain unchanged. Hence the per-unit-length capacitance can be found from the relation in (6.8) as

$$c = \frac{\varepsilon_r'}{lv_o^2}$$

$$= \begin{cases} \dfrac{\varepsilon_r'}{120v_o \ln\left(2\dfrac{1 + \sqrt{k}}{1 - \sqrt{k}}\right)} & \text{F/m} & \dfrac{1}{\sqrt{2}} \leq k \leq 1 \\[4mm] \dfrac{\varepsilon_r' \ln\left(2\dfrac{1 + \sqrt{k'}}{1 - \sqrt{k'}}\right)}{377\pi v_o} & \text{F/m} & 0 \leq k \leq \dfrac{1}{\sqrt{2}} \end{cases} \tag{6.12d}$$

▶ **QUICK REVIEW EXERCISE 6.4**

Determine the per-unit-length capacitance and inductance of a stripline with dimensions $s = 20$ mils (0.508 mm), $w = 5$ mils (0.127 mm), and $\varepsilon_r = 4.7$.

ANSWERS   113.2 pF/m and 0.461 μH/m.   ◀

▶ **QUICK REVIEW EXERCISE 6.5**

Determine the per-unit-length capacitance and inductance of a microstrip line with dimensions $h = 50$ mils (1.27 mm), $w = 5$ mils (0.127 mm), and $\varepsilon_r = 4.7$.

ANSWERS   38.46 pF/m and 0.877 μH/m. The effective relative permittivity is $\varepsilon'_r = 3.034$.   ◀

▶ **QUICK REVIEW EXERCISE 6.6**

Determine the per-unit-length capacitance and inductance of a PCB with dimensions $s = 15$ mils (0.381 mm), $w = 15$ mils (0.381 mm), $h = 62$ mils (1.575 mm), and $\varepsilon_r = 4.7$.

ANSWERS   38.53 pF/m and 0.804 μH/m. The effective relative permittivity is $\varepsilon'_r = 2.787$.   ◀

▶ 6.2 TIME-DOMAIN EXCITATION OF TRANSMISSION LINES

We will now examine a transmission line connecting a source to a load as illustrated in Fig. 6.7. The source consists of an open-circuit voltage source, $V_S(t)$, and source resistance, R_S, and the load is represented by a resistance R_L. The line will have a total length $\mathcal{L}$. The source voltage can have an arbitrary waveform.

6.2.1   The General Solution

The second-order, uncoupled transmission-line equations are given in (6.2). It is a simple matter to show that their solutions are

$$
V(z,t) = \underbrace{V^+\!\left(t - \frac{z}{v}\right)}_{\substack{\text{forward } (+z) \\ \text{traveling wave}}} + \underbrace{V^-\!\left(t + \frac{z}{v}\right)}_{\substack{\text{backward } (-z) \\ \text{traveling wave}}}
$$

(6.13a)

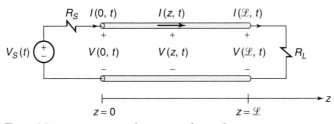

Figure 6.7   Terminations of a two-conductor line.

and

$$I(z,t) = \underbrace{\frac{V^+\left(t - \dfrac{z}{v}\right)}{Z_C}}_{\substack{\text{forward } (+z) \\ \text{traveling wave}}} - \underbrace{\frac{V^-\left(t + \dfrac{z}{v}\right)}{Z_C}}_{\substack{\text{backward } (-z) \\ \text{traveling wave}}}$$

(6.13b)

where the *characteristic impedance* is

$$Z_C = \sqrt{\frac{l}{c}} \quad \Omega$$

(6.14)

and the velocity of propagation is

$$v = \frac{1}{\sqrt{lc}} \quad \text{m/s}$$

(6.15)

The voltage and current solutions in (6.13) consist of the sum and difference of forward-traveling (in the $+z$ direction) and backward-traveling (in the $-z$ direction) waves. The forward-traveling wave is given by the function $V^+(t - z/v)$. It is easy to see that this is traveling in the $+z$ direction because the argument of the function, $t - z/v$, must be constant in order to track movement of a point on the waveform. Hence as t increases, z must also increase, and the wave travels in the $+z$ direction to accomplish this. Similarly, the function $V^-(t + z/v)$ represents a wave traveling in the $-z$ direction because, in order to keep the argument constant, when t increases, z must decrease in order to keep the argument constant. The precise form of the functions V^+ and V^- depend on the waveform of the source voltage, $V_S(t)$. However, t, z, and v can only appear in these as $t - z/v$ or $t + z/v$.

The characteristic impedance, Z_C, is a real (not complex) number. Hence it would be more properly called the characteristic *resistance*. The word "impedance" is a frequency-domain (phasor) term but here the source voltage waveform is not necessarily a single-frequency sinusoid but may have an arbitrary waveform. However, it has become an industry standard to refer to Z_C as the characteristic *impedance*, as we will continue to do here.

It is important to observe that because of the important minus sign between the two waves in the current expression in (6.13b) that $I(z,t) \neq V(z,t)/Z_C$. In other words, the input resistance seen looking into one end of the line is not equal to the characteristic impedance. The characteristic impedance Z_C only relates the voltage and current in the forward-traveling wave and the voltage and current in the backward-traveling wave.

▷ **QUICK REVIEW EXERCISE 6.7**

Determine the characteristic impedances and velocities of propagation of the wire-type lines in Quick Review Exercise Problems 6.1, 6.2, and 6.3.

ANSWERS $225.1\ \Omega$, 3×10^8 m/s; $151\ \Omega$, 3×10^8 m/s, $51\ \Omega$; 1.98×10^8 m/s. ◀

▷ **QUICK REVIEW EXERCISE 6.8**

Determine the characteristic impedances and velocities of propagation of the rectangular cross section lines in Quick Review Exercise Problems 6.4, 6.5, and 6.6.

ANSWERS 63.8 Ω, 1.38×10^8 m/s; 151 Ω, 1.72×10^8 m/s; 144.45 Ω, 1.8×10^8 m/s. ◀

6.2.2 Wave Tracing and the Reflection Coefficients

Now consider the termination at the load, $z = \mathcal{L}$. The total voltage and current at the load are

$$V(\mathcal{L},t) = V^+(t - T) + V^-(t + T) \tag{6.16a}$$

and

$$I(\mathcal{L},t) = \frac{V^+(t - T)}{Z_C} - \frac{V^-(t + T)}{Z_C} \tag{6.16b}$$

where the one-way time delay is

$$\boxed{T = \frac{\mathcal{L}}{v}} \quad \text{s} \tag{6.17}$$

At the load, Ohm's law dictates that

$$\frac{V(\mathcal{L},t)}{I(\mathcal{L},t)} = R_L \tag{6.18}$$

But this cannot be satisfied with only a forward-traveling wave because

$$\frac{V^+(t - T)}{I^+(t - T)} = Z_C \tag{6.19}$$

Hence there must exist both forward- and backward-traveling waves in order to satisfy (6.18).

If $R_L = Z_C$ we say that the line is *matched* at the load. In this case there will only be forward-traveling waves on the line as evidenced by (6.19). In the case of a mismatched line, $R_L \neq Z_C$, we define a (voltage) reflection coefficient at the load as the ratio of the backward-traveling (reflected) voltage wave to the forward-traveling (incident) voltage wave as

$$\Gamma_L = \frac{V^-(t + T)}{V^+(t - T)} \tag{6.20}$$

so that the voltage and current expressions at the load become

$$V(\mathcal{L},t) = V^+(t - T)[1 + \Gamma_L] \tag{6.21a}$$

and

$$I(\mathcal{L},t) = \frac{V^+(t - T)}{Z_C}[1 - \Gamma_L] \tag{6.21b}$$

Taking the ratio of (6.21a) and (6.21b) must yield

$$\frac{V(\mathcal{L},t)}{I(\mathcal{L},t)} = R_L$$
$$= Z_C \frac{[1 + \Gamma_L]}{[1 - \Gamma_L]} \tag{6.22}$$

Solving (6.22) for the reflection coefficient we obtain the load (voltage) reflection coefficient as

$$\boxed{\Gamma_L = \frac{R_L - Z_C}{R_L + Z_C}}$$

(6.23)

Because of the minus sign in the current expression in (6.13b), the reflection coefficient for current is the negative of the voltage reflection coefficient.

It is interesting to compare this to the case of a uniform plane wave incident normal to a boundary discussed in Section 5.5 of Chapter 5. (See Fig. 5.8.) The portion of the incident electric field wave that is reflected is

$$\Gamma = \frac{\eta_2 - \eta_1}{\eta_2 + \eta_1}$$

where $\eta_1 = \sqrt{\mu_1/\varepsilon_1}$ is the intrinsic impedance of the left medium (containing the incident electric field) and $\eta_2 = \sqrt{\mu_2/\varepsilon_2}$ is the intrinsic impedance of the right medium containing the transmitted electric field. This allows us to make a correspondence between the two similar problems if we make the following correspondence:

$$V \Leftrightarrow E$$
$$I \Leftrightarrow H$$
$$Z_C \Leftrightarrow \eta_1$$
$$R_L \Leftrightarrow \eta_2$$

as indicated in Fig. 6.8. Hence what we learned about uniform plane waves can, in many cases, be carried over to transmission lines in an analogous fashion.

The preceding has shown that at a mismatched load, the incoming (forward-traveling) voltage wave is partially reflected and sent back to the source. This is illustrated in Fig. 6.9.

The wave initially sent out from the source is a forward-traveling wave and has not reached the mismatched load to produce a reflection. Hence until the wave reaches the

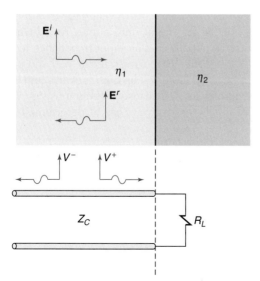

Figure 6.8 The analogy of a transmission-line termination and normal incidence of uniform plane waves on a plane, material boundary.

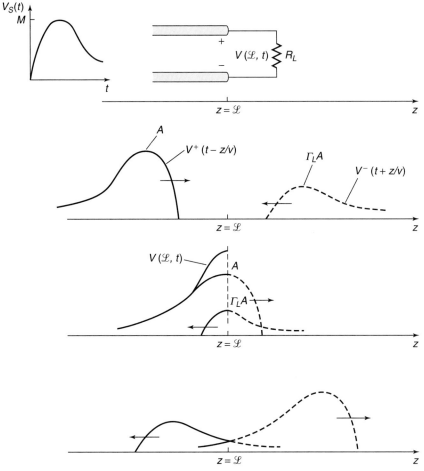

Figure 6.9 Illustration of the reflection of voltage waves at the load of a transmission line.

load and causes a reflected wave, there is only a forward-traveling wave on the line. But this reflected wave does not appear at the line input until after a round-trip delay of $2T$. Thus, according to (6.13), the source initially sees an input impedance looking into the line of Z_C as shown in Fig. 6.10. After the wave arrives at the load and produces a reflected wave, this is no longer true because there will be both forward- and backward-traveling waves on the line and the ratio is not Z_C because of the minus sign in the current expression in (6.13). Hence the initially sent out wave can be determined by voltage division as shown in Fig. 6.10 as

$$V(0,t) = \frac{Z_C}{R_S + Z_C} V_S(t) \qquad t < 2T \tag{6.24a}$$

Similarly, the initial current wave sent out by the source is

$$I(0,t) = \frac{V_S(t)}{R_S + Z_C} \qquad t < 2T \tag{6.24b}$$

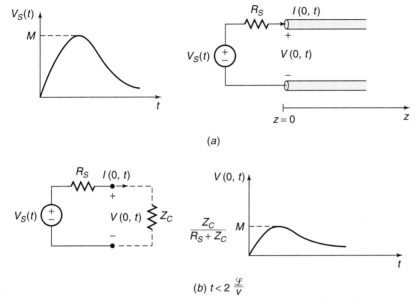

Figure 6.10 Illustration of determining the initially sent out pulse at the source.

The wave reflected at the load will, after a travel delay of $T = \mathcal{L}/v$, be incident on the source. We define in a similar manner the source (voltage) reflection coefficient as

$$\Gamma_S = \frac{R_S - Z_C}{R_S + Z_C} \tag{6.25}$$

Thus a portion of the incoming voltage wave (which was reflected by the load) will be reflected back toward the load. The reflection coefficient for the current wave is, because of the minus sign in the current expression in (6.13b), the negative of the voltage reflection coefficient. This process continues indefinitely.

This is conveniently shown by the "bounce" or lattice diagram in Fig. 6.11. From this we can write an expression for the voltage at $z = 0$ and at the load, $z = \mathcal{L}$, as

$$\begin{aligned} V(0,t) = \frac{Z_C}{R_S + Z_C} & [V_S(t) + (1 + \Gamma_S)\Gamma_L V_S(t - 2T) \\ & + (1 + \Gamma_S)(\Gamma_S\Gamma_L)\Gamma_L V_S(t - 4T) \\ & + (1 + \Gamma_S)(\Gamma_S\Gamma_L)^2\Gamma_L V_S(t - 6T) + \cdots] \end{aligned} \tag{6.26a}$$

and

$$\begin{aligned} V(\mathcal{L},t) = \frac{Z_C}{R_S + Z_C} & [(1 + \Gamma_L)V_S(t - T) + (1 + \Gamma_L)\Gamma_S\Gamma_L V_S(t - 3T) \\ & + (1 + \Gamma_L)(\Gamma_S\Gamma_L)^2 V_S(t - 5T) \\ & + (1 + \Gamma_L)(\Gamma_S\Gamma_L)^3 V_S(t - 7T) + \cdots] \end{aligned} \tag{6.26b}$$

So the total voltages are the sum of the source voltage waveforms scaled and delayed by multiples of the one-way time delay, T. Although the source and load voltage waveforms could be sketched from (6.26), it is much simpler to "trace the individual incident

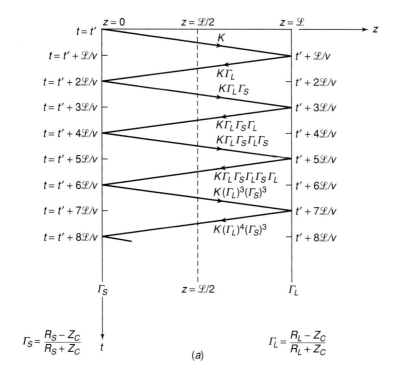

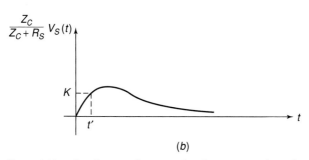

Figure 6.11 The "bounce diagram" for determining the voltages on the line at different instants of time.

and reflected waves" and at any time add all those present at that time. The following example illustrates this simple technique.

▶ **EXAMPLE 6.1**

Figure 6.12 shows a line wherein a 30-V battery is switched onto a line of length 400 m. The line has a characteristic impedance of 50 Ω and a velocity of propagation of 2×10^8 m/s or 200 m/μs. The source resistance is zero ($R_S = 0$) and the load resistance is 100 Ω($R_L = 100$ Ω). Sketch the current at the input to the line and the voltage at the load.

SOLUTION The source reflection coefficient is

$$\Gamma_S = \frac{0 - 50}{0 + 50}$$

$$= -1$$

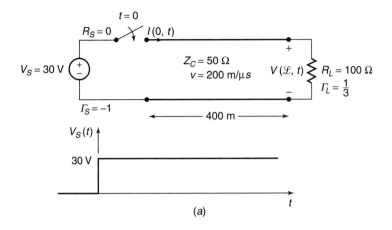

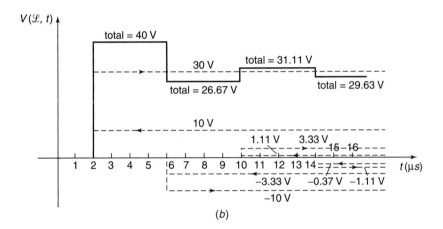

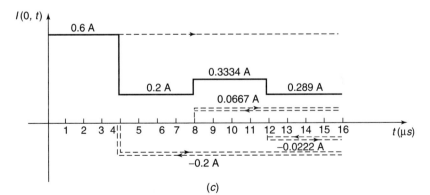

Figure 6.12 Example 6.1; determining the current at the input to the line and the load voltage. (a) The problem specification. (b) The load voltage. (c) The input current to the line.

and the load reflection coefficient is

$$\Gamma_L = \frac{100 - 50}{100 + 50}$$

$$= \frac{1}{3}$$

The one-way time delay is

$$T = \frac{\mathcal{L}}{v}$$

$$= 2 \ \mu s$$

First we sketch the load voltage, $V(\mathcal{L}, t)$. The initially sent out voltage is, since the source resistance is zero, 30 V. The incident and reflected voltages are sketched as dashed lines with an arrow added to indicate whether it is associated with a forward-traveling or a backward-traveling wave. After a time delay of 2 μs the initially sent out voltage of 30 V arrives at the load and a reflected voltage of $\frac{1}{3} \times 30 = 10$ V is sent back toward the source. This reflected voltage arrives at the source at 4 μs and a reflected voltage of $-1 \times 10 = -10$ V is sent back toward the load. When this, now incident, voltage arrives at the load a portion, $\frac{1}{3} \times -10 = -3.33$ V is sent back toward the source. These incident and reflected voltages are sketched in Fig. 6.12b. Adding the voltages present at any one time gives the total as shown by a solid line.

The current can be similarly sketched. The initially sent out current is 30 V/Z_C = 0.6 A. The current reflection coefficient at the load is the negative of the voltage reflection coefficient or $-\frac{1}{3}$. Similarly, the current reflection coefficient at the source is the negative of the voltage reflection coefficient or $+1$. Tracing these incident and reflected currents in the same manner as for the voltage produces the result shown in Fig. 6.12c.

And finally we should "sanity check" these results. After a sufficiently long time, the reflections will decay to zero and we would expect the line to have no effect. Consequently the results should converge after several round-trip delays to 30 V and 30 V/100 Ω = 0.3 A. The sketches indicate that this is indeed the case. ◀

▶ EXAMPLE 6.2

This example shows the effect of pulse width on the total voltages. Consider a line of length 0.2 m (7.9 in.) shown in Fig. 6.13a. The source voltage is a pulse of 20 V amplitude and 1 ns duration. The line has a characteristic impedance of 100 Ω and a velocity of propagation of 2×10^8 m/s. The source resistance is 300 Ω ($R_S = 300 \ \Omega$) and the load is open circuited ($R_L = \infty$). Sketch the voltage at the input to the line and at the load.

SOLUTION The source reflection coefficient is

$$\Gamma_S = \frac{300 - 100}{300 + 100}$$

$$= \frac{1}{2}$$

and the load reflection coefficient is

$$\Gamma_L = \frac{\infty - 100}{\infty + 100}$$

$$= 1$$

The one-way time delay is

$$T = \frac{\mathcal{L}}{v}$$

$$= 1 \ ns$$

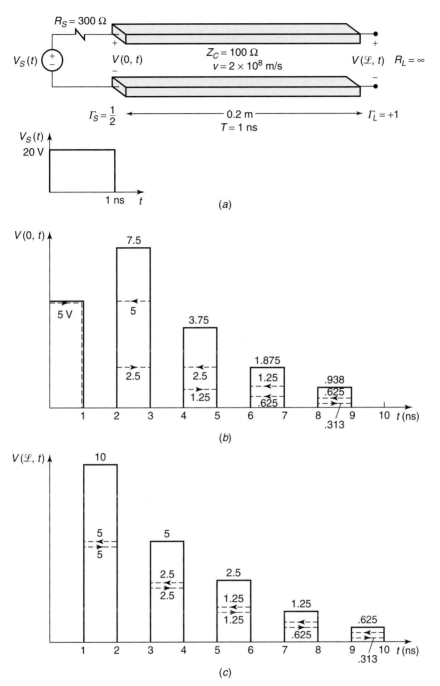

Figure 6.13 Example 6.2. Illustration of the effect of pulse width on the terminal voltages. (a) The problem specification. (b) The voltage at the input to the line. (c) The load voltage.

First we sketch the source voltage, $V(0,t)$. The initially sent out voltage is

$$\frac{100}{300 + 100} \times 20 = 5\,\text{V}$$

The incident and reflected voltages are again sketched in Fig. 6.13b with dashed lines with an arrow added to indicate whether they are associated with a forward-traveling or a backward-traveling wave. The incident pulse is sent to the load, arriving there after one time delay of 1 ns, where it is reflected as a pulse of 5 V because the load reflection coefficient is $\Gamma_L = 1$. This pulse reflected at the load arrives at the source after an additional 1-ns time delay. This incoming pulse is reflected as $\Gamma_S \times 5V = 2.5$ V, which arrives at the load after 1 ns, where it is reflected as 2.5 V, arriving at the source after a 1-ns delay. The process continues as shown. Adding all the incident and reflected pulses at the source, we obtain the total voltage drawn with a solid line. Clearly this total decays to zero, as it should in steady state.

Now we sketch the voltage at the load. After a time delay of 1 ns, the initially sent out voltage of 5 V arrives at the load and a reflected voltage of 5 V is sent back toward the source. This reflected voltage arrives at the source after 2 ns, and a reflected voltage of 2.5 V is sent back toward the load and arrives there at 3 ns. When this pulse reflected at the source arrives at the load, a reflected voltage of 2.5 V is sent back toward the source, which is reflected at the source as 1.25 V, arriving at the load at 5 ns. These incident and reflected voltages are sketched in Fig. 6.13c. Adding the voltages present at any one time gives the total as shown by a solid line. Clearly this load voltage is decaying to zero, as it should in steady state. ◀

▶ **EXAMPLE 6.3**

This example illustrates the effect of a pulse width that is greater than the round-trip delay. Consider the coaxial cable shown in Fig. 6.14a. The source voltage is a pulse of 100 V amplitude and 6 μs duration. The line is specified by its per-unit-length capacitance and inductance: $c = 100$ pF/m and $l = 0.25$ μH/m. This corresponds to the RG-58U coaxial cable whose per-unit-length parameters were computed in Quick Review Exercise Problem 6.3. The line has a characteristic impedance of

$$Z_C = \sqrt{\frac{l}{c}}$$
$$= 50\ \Omega$$

The velocity of propagation is

$$v = \frac{1}{\sqrt{lc}}$$
$$= 200\ \text{m/μs}$$

The source resistance is 150 Ω ($R_S = 150$ Ω) and the load resistance is a short circuit ($R_L = 0$ Ω). Sketch the voltage at the input to the line.

SOLUTION   The source reflection coefficient is

$$\Gamma_S = \frac{150 - 50}{150 + 50}$$
$$= \frac{1}{2}$$

and the load reflection coefficient is

$$\Gamma_L = \frac{0 - 50}{0 + 50}$$
$$= -1$$

The one-way time delay is

$$T = \frac{\mathcal{L}}{v}$$
$$= 2\ \text{μs}$$

The initially sent out voltage is

$$\frac{50}{150 + 50} \times 100 = 25\ \text{V}$$

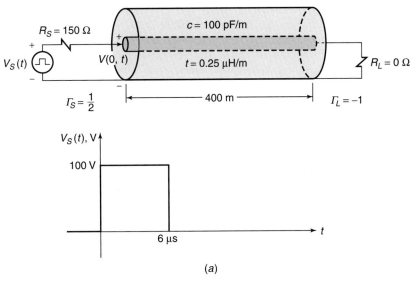

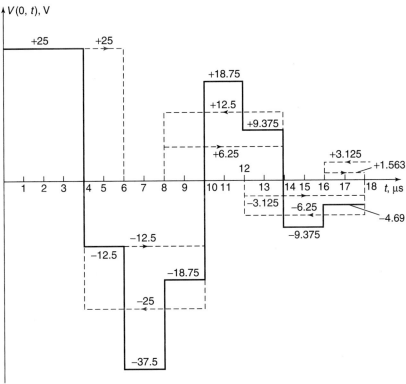

(b)

Figure 6.14 Example 6.3. Illustration of the effect of pulse width on the terminal voltages. (a) The problem specification. (b) The voltage at the input to the line.

The incident and reflected voltages are again sketched with dashed lines in Fig. 6.14b with an arrow added to indicate whether they are associated with a forward-traveling or a backward-traveling wave. The incident pulse is sent to the load, arriving there after one time delay of 2 μs, where it is reflected as a pulse of −25 V. This pulse reflected at the load arrives at the source after an additional 2-μs time delay. This incoming pulse is reflected as −12.5 V, which arrives at the load after 2 μs, where it is reflected as 12.5 V, arriving at the source after a 2-μs delay. The process continues as shown. Adding all the incident and reflected pulses at the source, we obtain the total voltage drawn with a solid line. Clearly this total decays to zero, as it should in steady state.

Observe that in this example the pulse width of 6 μs is three time delays. Hence the initially sent out pulse and the arriving pulse (which was the sent out pulse reflected at the load) overlap. This overlap creates a rather interesting and complicated waveform.   ◀

6.2.3   The SPICE Model

The solution to the transmission-line equations in (6.13) is repeated here:

$$V(z,t) = V^+\left(t - \frac{z}{v}\right) + V^-\left(t + \frac{z}{v}\right) \tag{6.27a}$$

$$Z_C I(z,t) = V^+\left(t - \frac{z}{v}\right) - V^-\left(t + \frac{z}{v}\right) \tag{6.27b}$$

In the equation for the current we have multiplied both sides by the characteristic impedance. Add and subtract these to give

$$V(z,t) + Z_C I(z,t) = 2V^+\left(t - \frac{z}{v}\right) \tag{6.28a}$$

$$V(z,t) - Z_C I(z,t) = 2V^-\left(t + \frac{z}{v}\right) \tag{6.28b}$$

Evaluating (6.28a) at the source and at the load gives

$$V(\mathcal{L},t) + Z_C I(\mathcal{L},t) = 2V^+(t - T) \tag{6.29a}$$

$$V(0,t - T) + Z_C I(0,t - T) = 2V^+\left(t - \frac{0}{v} - T\right) = 2V^+(t - T) \tag{6.29b}$$

Observe that we have delayed (6.29b) by one time delay, T, for reasons to become apparent. Eliminating V^+ by subtracting these, we obtain

$$\boxed{V(\mathcal{L},t) + Z_C I(\mathcal{L},t) = V(0,t - T) + Z_C I(0,t - T)} \tag{6.30a}$$

Performing a similar evaluation on (6.28b) yields

$$\boxed{V(0,t) - Z_C I(0,t) = V(\mathcal{L},t - T) - Z_C I(\mathcal{L},t - T)} \tag{6.30b}$$

Equations (6.30) allow us to model the line as a two port as shown in Fig. 6.15a. Each controlled voltage source is a function of the voltage and current at the other end of the line delayed by one time delay.

The SPICE circuit analysis program includes an exact model of a two-conductor, lossless transmission line which implements the model in Fig. 6.15a. For details on the use of SPICE see C.R. Paul, *Fundamentals of Electric Circuit Analysis*, John Wiley, 2001. The form of the SPICE coding is shown in Fig. 6.15b:

```
TXXX  N1  N2  N3  N4  Z0=    TD=
```

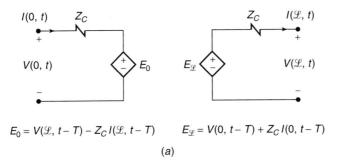

$$E_0 = V(\mathcal{L}, t - T) - Z_C I(\mathcal{L}, t - T) \qquad E_\mathcal{L} = V(0, t - T) + Z_C I(0, t - T)$$

(a)

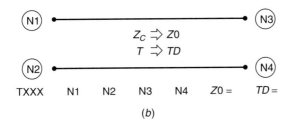

Figure 6.15 The SPICE (PSPICE) model of a transmission line. (a) An exact model of the line. (b) The SPICE coding.

(b)

The XXX is an arbitrary name for this line chosen by the user. SPICE refers to the characteristic impedance as Z0 and the time delay as TD.

▷ **EXAMPLE 6.4**

Use the SPICE (or the personal computer version, PSPICE) to solve the problem shown in Fig. 6.12 that was obtained in Example 6.1.

SOLUTION The SPICE (PSPICE) coding is shown in Fig. 6.16a:

```
EXAMPLE 6.4
VS 1 0 PWL(0 0 .01U 30)
T 1 0 2 0 Z0=50 TD=2U
RL 2 0 100
.TRAN .01U 20U 0 .01U
.PRINT TRAN V(2) I(VS)
*THE LOAD VOLTAGE IS V(2) AND
*THE INPUT CURRENT IS -I(VS)
.PROBE
.END
```

We have used the SPICE piecewise linear function (PWL) to specify the source voltage. This function specifies a piecewise linear graph of it as a sequence of straight lines between time points T1, T2, T3, . . . whose values are V1, V2, V3, . . . as

```
    VXXX N1 N2 PWL(T1 V1 T2 V2 T3 V3 . . . .)
```

Also, we have specified the battery voltage with a very small (.01 μs) rise time in order to specify it with the PWL function. We have used the .PROBE feature of PSPICE to provide plots of

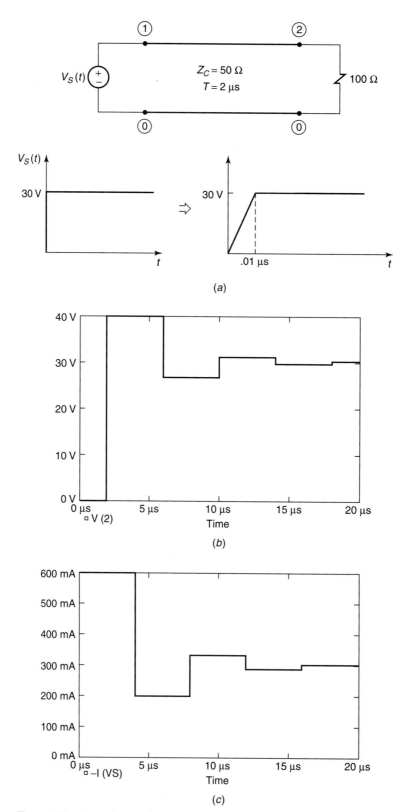

Figure 6.16 Example 6.4; the SPICE solution of the problem of Example 6.1. (a) The SPICE node labeling. (b) The SPICE solution for the line load voltage. (c) The SPICE solution for the input current to the line.

the load voltage and input current which are shown in Fig. 6.16b and Fig. 6.16c, respectively. Compare these to the corresponding plots obtained manually and shown in Fig. 6.12.

The format of the .TRAN line is

.TRAN [*print step*] [*final solution time*] [*print start*] [*maximum solution time step*]

The *print step* is the time interval that solutions are printed to a file if so requested in a .PRINT statement, and the *final solution time* is the final time for which a solution is desired. These first two parameters are required, and the remaining two are optional. All solutions start at $t = 0$ but the *print start* parameter delays the printing of the results to an output file until this time. Usually the *print start* parameter is set to zero. Specification of the remaining term, *maximum solution time step*, is often required in order to control the accuracy and resolution of the solution. SPICE (PSPICE) solves the equations of the transmission line and associated termination circuits by discretizing the time interval into increments Δt. These are solved in a "bootstrapping" fashion by updating the results at the next time step with those from the previous intervals. The *maximum solution time step* parameter in the .TRAN line sets that maximum discretization time step. When the circuit contains a transmission line, the line voltages and currents will be changing in intervals of time on the order of the one-way time delay, T, as we have seen. In order to not miss any such important variations, the *maximum solution time step* must be considerably less than this one-way delay. The SPICE program developed in the 1960s automatically set the maximum discretization time step to be one-half of the smallest line delay when the circuit contained transmission lines. In many problems the voltages and currents will be varying in time intervals much smaller than this. For example, the source voltage waveform may be specified as having a rise/fall time in order to specify it with the PWL function. This rise/fall time may be (and usually is) much smaller than the line one-way delay. Hence the *maximum solution time step* should be set on the order of the smallest time variation. ◂

▷ EXAMPLE 6.5

Use the SPICE (or the personal computer version, PSPICE) to solve the problem shown in Fig. 6.13 that was obtained in Example 6.2.

SOLUTION The SPICE (PSPICE) coding is shown in Fig. 6.17a:

```
EXAMPLE 6.5
VS 1 0 PWL(0 0 0.01N 20 1N 20 1.01N 0)
RS 1 2 300
T 2 0 3 0 Z0=100 TD=1N
RL 3 0 1E8
.TRAN 0.01N 10N 0 0.01N
.PRINT TRAN V(2) V(3)
*THE LOAD VOLTAGE IS V(3) AND
*THE INPUT VOLTAGE IS V(2)
.PROBE
.END
```

We have again used the SPICE piecewise linear function to specify the source voltage. Also, we have specified the pulse with very small (.01 ns) rise and fall times in order to specify it with

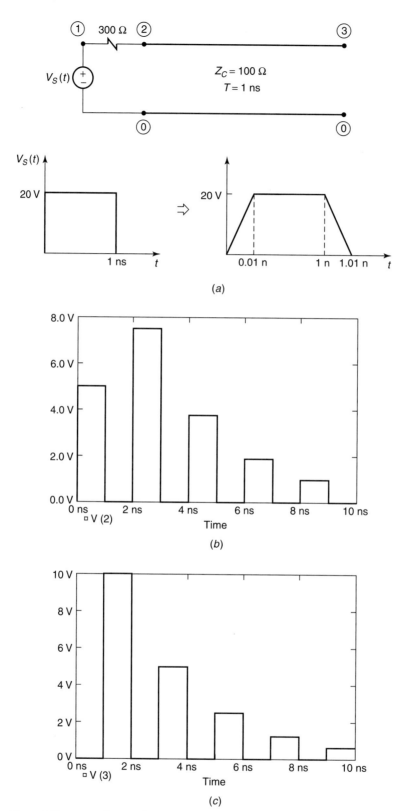

Figure 6.17 Example 6.5; the SPICE solution of the problem of Example 6.2. (a) The SPICE node labeling. (b) The SPICE solution for the input voltage to the line. (c) The SPICE solution for the line load voltage.

the PWL function. The open-circuit load is specified as a large (10^8-Ω) resistance. The input voltage and output voltage are shown in Fig. 6.17b and Fig. 6.17c, respectively. Compare these to the corresponding plots obtained manually and shown in Fig. 6.13. ◄

▷ **EXAMPLE 6.6**

Use the SPICE (or the personal computer version, PSPICE) to solve the problem shown in Fig. 6.14 that was obtained in Example 6.3.

SOLUTION   The SPICE (PSPICE) coding is shown in Fig. 6.18a:

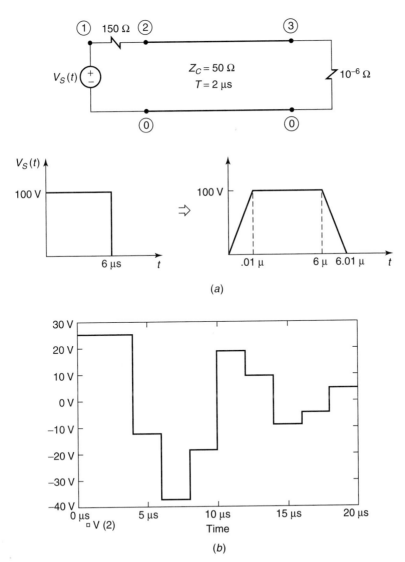

(a)

(b)

Figure 6.18   Example 6.6; the SPICE solution of the problem of Example 6.3. (a) The SPICE node labeling. (b) The SPICE solution for the input voltage to the line.

```
EXAMPLE 6.6
VS 1 0 PWL(0 0 .01U 100 6U 100 6.01U 0)
RS 1 2 150
T 2 0 3 0 Z0=50 TD=2U
RL 3 0 1E-6
.TRAN .01U 20U 0 .01U
.PRINT TRAN V(2) V(3)
*THE INPUT VOLTAGE IS V(2)
.PROBE
.END
```

Again we used the PWL function to specify the pulse. The short-circuit load is represented with a 1-µΩ resistor since SPICE does not allow for zero-ohm resistors. The input voltage to the line is shown in Fig. 6.18b. Compare this to the corresponding plot obtained manually and shown in Fig. 6.14b. ◀

▶ 6.3 SINUSOIDAL (PHASOR) EXCITATION OF TRANSMISSION LINES

In this section we will investigate the solution for the transmission-line voltage and current when the source voltage is a single-frequency sinusoid, that is, $V_S(t) = V_S \sin(\omega t + \phi)$ or $V_S(t) = V_S \cos(\omega t + \phi)$, and the line is in steady state, that is, any transients occurring when the source was connected to the line have died out. The solution process is the same as was used to solve lumped circuits and is called the phasor method. The idea is to replace the source voltage with its complex exponential representation as

$$\left.\begin{array}{l} V_S(t) = V_S \sin(\omega t + \phi) \\ V_S(t) = V_S \cos(\omega t + \phi) \end{array}\right\} \Rightarrow \underbrace{V_S \angle \phi e^{j\omega t}}_{\hat{V}_S}$$

The transmission-line voltages and currents in the time domain are replaced by their phasor equivalents in the *frequency domain* as

$$V(z,t) \Rightarrow \underbrace{V \angle \theta_V e^{j\omega t}}_{\hat{V}(z)} \tag{6.31a}$$

and

$$I(z,t) \Rightarrow \underbrace{I \angle \theta_I e^{j\omega t}}_{\hat{I}(z)} \tag{6.31b}$$

The advantage of this method is that the time variable is removed and the voltage and current (phasor) depend only on position z. Differentiating with respect to time amounts to simply multiplying by $j\omega$:

$$\frac{\partial V(z,t)}{\partial t} \Rightarrow j\omega V(z,t) \tag{6.32a}$$

$$\frac{\partial I(z,t)}{\partial t} \Rightarrow j\omega I(z,t) \tag{6.32b}$$

Hence the transmission-line equations in (6.1) can be written in terms of the phasor voltage and current, $\hat{V}(z)$ and $\hat{I}(z)$, by substituting (6.31) and canceling the $e^{j\omega t}$ on both sides and become

$$\frac{d\hat{V}(z)}{dz} = -j\omega l\, \hat{I}(z) \tag{6.33a}$$

and

$$\frac{d\hat{I}(z)}{dz} = -j\omega c\hat{V}(z) \tag{6.33b}$$

The uncoupled, second-order transmission-line equations in (6.2) become

$$\frac{d^2\hat{V}(z)}{dz^2} = -\omega^2 lc\hat{V}(z) \tag{6.34a}$$

and

$$\frac{d^2\hat{I}(z)}{dz^2} = -\omega^2 lc\hat{I}(z) \tag{6.34b}$$

6.3.1 The General Solution

The solution to the second-order equations in (6.34) is well known:

$$\hat{V}(z) = \hat{V}^+ e^{-j\beta z} + \hat{V}^- e^{j\beta z} \tag{6.35a}$$

and

$$\hat{I}(z) = \frac{\hat{V}^+}{Z_C} e^{-j\beta z} - \frac{\hat{V}^-}{Z_C} e^{j\beta z} \tag{6.35b}$$

where the characteristic impedance is, as before, $Z_C = \sqrt{l/c}$ and the phase constant is

$$\beta = \omega\sqrt{lc}$$
$$= \frac{\omega}{v} \quad \text{radians/m} \tag{6.36}$$

and the velocity of propagation is as before, $v = 1/\sqrt{lc}$. The quantities $\hat{V}^+$ and $\hat{V}^-$ are, as yet, undetermined constants. These will be determined by the specific load and source parameters. The phase constant β represents a phase shift as the waves propagate along the line.

The source and load configuration are as shown in Fig. 6.19, but here we may include inductors and capacitors in those impedances, $\hat{Z}_S$ and $\hat{Z}_L$. Once we incorporate the source and load into the general solution in (6.35), we can return to the time domain in the usual fashion by (1) multiplying the phasor result by $e^{j\omega t}$ and (2) taking the real or imaginary part of the result as was done in electric circuit analysis. Assuming the source voltage is a cosine, the time-domain result becomes

$$V(z,t) = V^+ \cos(\omega t - \beta z + \theta^+) + V^- \cos(\omega t + \beta z + \theta^-) \tag{6.37a}$$

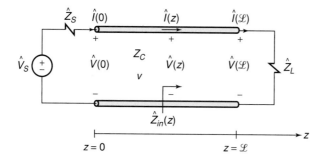

Figure 6.19 The specification of the terminations for sinusoidal steady-state (phasor) excitation of the line.

and

$$I(z, t) = \frac{V^+}{Z_C}\cos(\omega t - \beta z + \theta^+) - \frac{V^-}{Z_C}\cos(\omega t + \beta z + \theta^-) \tag{6.37b}$$

where the undetermined constants are, in general, complex as $\hat{V}^{\pm} = V^{\pm}\angle\theta^{\pm}$. Observe from (6.36) that since $\beta = \omega/v$, we may write

$$(\omega t - \beta z) \Leftrightarrow \omega(t - z/v) \tag{6.38}$$

This shows that *the phase shift term $(\omega t - \beta z)$ is equivalent to a time delay $(t - z/v)$.*

6.3.2 The Reflection Coefficient and Input Impedance

In order to incorporate the source and load into the general solution, we again define a voltage reflection coefficient as the ratio of the (phasor) voltages in the backward-traveling and forward-traveling waves:

$$\begin{aligned}\hat{\Gamma}(z) &= \frac{\hat{V}^- e^{j\beta z}}{\hat{V}^+ e^{-j\beta z}} \\ &= \frac{\hat{V}^-}{\hat{V}^+}e^{j2\beta z}\end{aligned} \tag{6.39}$$

This is a general reflection coefficient at any z along the line. Evaluating this at the load, $z = \mathscr{L}$, gives

$$\hat{\Gamma}_L = \frac{\hat{V}^-}{\hat{V}^+}e^{j2\beta\mathscr{L}} \tag{6.40}$$

where the load reflection coefficient is as before:

$$\hat{\Gamma}_L = \frac{\hat{Z}_L - Z_C}{\hat{Z}_L + Z_C} \tag{6.41}$$

Substituting (6.40) into (6.39) gives an expression for the reflection coefficient at any point along the line in terms of the load reflection coefficient, which can be directly calculated via (6.41):

$$\hat{\Gamma}(z) = \hat{\Gamma}_L e^{j2\beta(z - \mathscr{L})} \tag{6.42}$$

The general solution in (6.35) for the phasor voltage and current along the line can be written in terms of the reflection coefficient at that point by substituting (6.39):

$$\hat{V}(z) = \hat{V}^+ e^{-j\beta z}[1 + \hat{\Gamma}(z)] \tag{6.43a}$$

and

$$\hat{I}(z) = \frac{\hat{V}^+}{Z_C} e^{-j\beta z} [1 - \hat{\Gamma}(z)] \tag{6.43b}$$

Substituting the explicit relation for the reflection coefficient in terms of the load reflection coefficient given in (6.42) gives

$$\hat{V}(z) = \hat{V}^+ e^{-j\beta z} [1 + \hat{\Gamma}_L e^{j2\beta(z - \mathscr{L})}] \tag{6.44a}$$

and

$$\hat{I}(z) = \frac{\hat{V}^+}{Z_C} e^{-j\beta z} [1 - \hat{\Gamma}_L e^{j2\beta(z - \mathscr{L})}] \tag{6.44b}$$

An important parameter that we will need to calculate is the input impedance to the line at any point on the line. The input impedance is the ratio of the total voltage and current at that point as shown in Fig. 6.19:

$$
\begin{aligned}
\hat{Z}_{in}(z) &= \frac{\hat{V}(z)}{\hat{I}(z)} \\
&= Z_C \frac{[1 + \hat{\Gamma}(z)]}{[1 - \hat{\Gamma}(z)]}
\end{aligned}
\tag{6.45}
$$

where we have substituted (6.43). This important relation will be used many times. The process for determining the input impedance is to (1) compute the load reflection coefficient from (6.41), (2) compute the reflection coefficient at the desired point from (6.42), and (3) then substitute that result into (6.45).

An important position to calculate the input impedance is at the input to the entire line, $z = 0$. The reflection coefficient at the input to the line is obtained from (6.42):

$$\hat{\Gamma}(0) = \hat{\Gamma}_L e^{-j2\beta \mathscr{L}} \tag{6.46}$$

Hence the input impedance to a line of length $\mathscr{L}$ is

$$
\begin{aligned}
\hat{Z}_{in} &= Z_C \frac{[1 + \hat{\Gamma}(0)]}{[1 - \hat{\Gamma}(0)]} \\
&= Z_C \frac{[1 + \hat{\Gamma}_L e^{-j2\beta \mathscr{L}}]}{[1 - \hat{\Gamma}_L e^{-j2\beta \mathscr{L}}]}
\end{aligned}
\tag{6.47}
$$

and we will reserve the special notation $\hat{Z}_{in} \equiv \hat{Z}_{in}(0)$.

6.3.3 The Solution for the Terminal Voltages and Currents

We will be interested in determining the current and voltage at the input and output of the line. Evaluating (6.43) at $z = 0$ and $z = \mathscr{L}$ gives

$$\hat{V}(0) = \hat{V}^+ [1 + \hat{\Gamma}(0)] \tag{6.48a}$$

$$\hat{I}(0) = \frac{\hat{V}^+}{Z_C} [1 - \hat{\Gamma}(0)] \tag{6.48b}$$

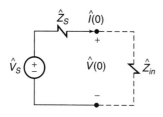

Figure 6.20 Determining the input voltage to the line by modeling the line input as its input impedance.

and

$$\hat{V}(\mathcal{L}) = \hat{V}^+ e^{-j\beta\mathcal{L}}[1 + \hat{\Gamma}_L] \qquad (6.49a)$$

$$\hat{I}(\mathcal{L}) = \frac{\hat{V}^+}{Z_C} e^{-j\beta\mathcal{L}}[1 - \hat{\Gamma}_L] \qquad (6.49b)$$

If we can determine the undetermined constant $\hat{V}^+$, then we can determine the source and load voltages and currents by substituting into (6.48) and (6.49).

In order to determine this undetermined constant, a convenient position to evaluate it is at the input to the line, $z = 0$. At the line input, the line can be replaced by the input impedance to the line as shown in Fig. 6.20. The input voltage to the line can be determined by voltage division as

$$\hat{V}(0) = \frac{\hat{Z}_{in}}{\hat{Z}_S + \hat{Z}_{in}} \hat{V}_S \qquad (6.50)$$

Hence the undetermined constant is determined from (6.48a) as

$$\hat{V}^+ = \frac{\hat{V}(0)}{[1 + \hat{\Gamma}(0)]} \qquad (6.51)$$

The process in determining the input and output voltages and currents to the line is

1 determine the load reflection coefficient, $\hat{\Gamma}_L$, from (6.41),
2 determine the input reflection coefficient, $\hat{\Gamma}(0)$, from (6.46),
3 determine the input impedance to the line, $\hat{Z}_{in}$, from (6.47),
4 determine the input voltage, $\hat{V}(0)$, from (6.50),
5 determine the undetermined constant, $\hat{V}^+$, from (6.51), and
6 substitute this into (6.48) and (6.49) to determine the source and load voltages and currents.

The following example illustrates this process.

▶ **EXAMPLE 6.7**

A line 2.7 m in length is excited by a 100-MHz source as shown in Fig. 6.21. Determine the source and load voltages.

SOLUTION   First we compute the load reflection coefficient:

$$\hat{\Gamma}_L = \frac{\hat{Z}_L - Z_C}{\hat{Z}_L + Z_C}$$

$$= \frac{50 + j200}{150 + j200}$$

$$= 0.82 \angle 22.83°$$

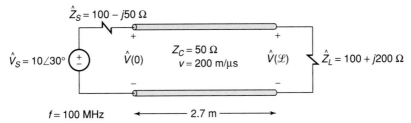

$\hat{Z}_S = 100 - j50\ \Omega$

$\hat{V}_S = 10\angle 30°$

$\hat{V}(0)$

$Z_C = 50\ \Omega$
$v = 200\ \text{m/}\mu\text{s}$

$\hat{V}(\mathcal{L})$

$\hat{Z}_L = 100 + j200\ \Omega$

$f = 100\ \text{MHz}$ ⟵ 2.7 m ⟶

Figure 6.21 Example 6.7.

Next we compute the reflection coefficient at the input to the line from (6.46):

$$\hat{\Gamma}(0) = \hat{\Gamma}_L e^{-j2\beta\mathcal{L}}$$

But this requires that we compute the phase shift $e^{-j2\beta\mathcal{L}} = 1\angle -2\beta\mathcal{L}$. This can be written as

$$-2\beta\mathcal{L} = -4\pi f\frac{\mathcal{L}}{v}$$

$$= -4\pi\frac{\mathcal{L}}{\lambda}$$

At the excitation frequency of 100 MHz, the line length of 2.7 m is 1.35λ. Hence

$$-2\beta\mathcal{L} = -4\pi\frac{\mathcal{L}}{\lambda}$$

$$= -5.4\pi \qquad \text{radians}$$

$$= -972 \qquad \text{degrees}$$

Hence

$$\hat{\Gamma}(0) = \hat{\Gamma}_L e^{-j2\beta\mathcal{L}}$$

$$= 0.82\angle -949.17°$$

and

$$\hat{Z}_{in} = Z_C\frac{[1 + \hat{\Gamma}(0)]}{[1 - \hat{\Gamma}(0)]}$$

$$= 23.35\angle 75.62°$$

The input voltage to the line is

$$\hat{V}(0) = \frac{\hat{Z}_{in}}{\hat{Z}_S + \hat{Z}_{in}}\hat{V}_S$$

$$= 2.14\angle 120.13°$$

and the undetermined constant is

$$\hat{V}^+ = \frac{\hat{V}(0)}{[1 + \hat{\Gamma}(0)]}$$

$$= 2.75\angle 66.58°$$

The voltages are

$$\hat{V}(0) = \hat{V}^+[1 + \hat{\Gamma}(0)]$$

$$= 2.14\angle 120.13°$$

$$\hat{V}(\mathcal{L}) = \hat{V}^+ e^{-j\beta\mathcal{L}}[1 + \hat{\Gamma}_L]$$

$$= 4.93\angle -409.12°$$

Hence the time-domain voltages are

$$V(0,t) = 2.14 \cos(6.28 \times 10^8 t + 120.13°)$$
$$V(\mathcal{L},t) = 4.93 \cos(6.28 \times 10^8 t - 49.12°)$$

In the load voltage we have added 360° to give an equivalent phase angle of $-409.12° \Rightarrow -49.12°$.

◀

▶ QUICK REVIEW EXERCISE 6.9

Determine the source and load voltages for a line whose length is 10 m and has a characteristic impedance of $Z_C = 50\ \Omega$ and a velocity of propagation of $v = 200$ m/μs. The source voltage is $\hat{V}_S = 100\angle 0°$ V at 26 MHz and $\hat{Z}_S = 50\ \Omega$, $\hat{Z}_L = 100 + j50\ \Omega$.

ANSWERS $\hat{V}(0) = 28.28\angle 7.48°$ V and $\hat{V}(\mathcal{L}) = 70.71\angle -99.84°$ V. ◀

6.3.4 The SPICE Solution

SPICE (or PSPICE) can be used to solve these phasor problems. There are three changes from the time-domain use earlier. First, the specification of a phasor voltage (or current) source is

```
VS N1 N2 AC mag phase
```

Second, the .TRAN card needs to be changed to

```
.AC DEC 1 f f
```

where f is the frequency of the source. Lastly the output is obtained as

```
.PRINT AC VM(NX) VP(NX)
```

where VM is the magnitude and VP is the phase of this node voltage.

▶ EXAMPLE 6.8

Solve the problem of Example 6.7 using SPICE.

SOLUTION First, we must synthesize equivalent circuits consisting of R, L, C elements that represent the phasor source and load impedances, $\hat{Z}_S = 100 - j50$ and $\hat{Z}_L = 100 + j200$, at the source frequency of 100 MHz. Figure 6.22a shows that these can be represented by a 100-Ω resistor in series with a 31.8-pF capacitor and a 100-Ω resistor in series with a 0.318-μH inductor, respectively. The SPICE circuit is shown in Fig. 6.22b and the coding is

```
EXAMPLE 6.8
VS 1 0 AC 10 30
RS 1 2 100
CS 2 3 31.8P
T 3 0 4 0 Z0=50 TD=13.5N
RL 4 5 100
LL 5 0 0.318U
.AC DEC 1 1E8 1E8
.PRINT AC VM(3) VP(3) VM(4) VP(4)
.END
```

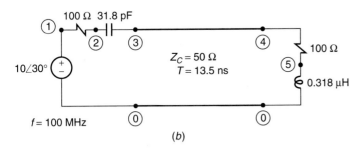

Figure 6.22 The SPICE solution of the problem of Example 6.7. (a) Modeling complex impedances for inclusion into the SPICE program. (b) The SPICE modeling of the problem.

The results are

$$\hat{V}(0) = \text{VM}(3)\angle\text{VP}(3)$$
$$= 2.136\angle 120.1°$$

and

$$\hat{V}(\mathcal{L}) = \text{VM}(4)\angle\text{VP}(4)$$
$$= 4.926\angle -49.1°$$

These results are in excellent agreement with those obtained by hand in Example 6.7. ◀

▷ QUICK REVIEW EXERCISE 6.10

Solve the problem in Quick Review Exercise Problem 6.9 using SPICE.

ANSWERS $\hat{V}(0) = 28.28\angle 7.473°$ and $\hat{V}(\mathcal{L}) = 70.71\angle -99.87°$. ◀

6.3.5 Voltage and Current as a Function of Position on the Line

We have been primarily interested in determining the voltages and currents at the endpoints of the line. In this section we will investigate and provide plots of the magnitudes of the phasor line voltage and current at various points along the line. Recall the expressions for the voltage and current at various positions along the line in (6.44), which are repeated here

$$\hat{V}(z) = \hat{V}^+ e^{-j\beta z}\big[1 + \hat{\Gamma}_L e^{j2\beta(z-\mathcal{L})}\big] \tag{6.44a}$$

and

$$\hat{I}(z) = \frac{\hat{V}^+}{Z_C} e^{-j\beta z}\big[1 - \hat{\Gamma}_L e^{j2\beta(z-\mathcal{L})}\big] \tag{6.44b}$$

We will plot the *magnitude* of these for distances $d = \mathcal{L} - z$ away from the load. Taking the magnitudes of (6.44) gives

$$|\hat{V}(d = \mathcal{L} - z)| = |\hat{V}^+|\big|1 + \hat{\Gamma}_L e^{-j2\beta d}\big| \tag{6.52a}$$

and

$$|\hat{I}(d = \mathcal{L} - z)| = \frac{|\hat{V}^+|}{Z_C}|1 - \hat{\Gamma}_L e^{-j2\beta d}| \tag{6.52b}$$

There are three important cases of special interest that we will investigate: (1) the load is a short circuit, $\hat{Z}_L = 0$, (2) the load is an open circuit, $\hat{Z}_L = \infty$, and (3) the load is matched, $\hat{Z}_L = Z_C$.

For the case where the load is a short circuit, $\hat{Z}_L = 0$, the load reflection coefficient is $\hat{\Gamma}_L = -1$. Equations (6.52) reduce to

$$\begin{aligned}
|\hat{V}(d)| &= |\hat{V}^+||1 - e^{-j2\beta d}| \\
&= |\hat{V}^+|\underbrace{|e^{-j\beta d}|}_{1}\underbrace{|e^{j\beta d} - e^{-j\beta d}|}_{|2j\sin(\beta d)|} \\
&\propto |\sin(\beta d)| \\
&= \left|\sin\left(2\pi\frac{d}{\lambda}\right)\right|
\end{aligned} \tag{6.53a}$$

and

$$\begin{aligned}
|\hat{I}(d)| &= \frac{|\hat{V}^+|}{Z_C}|1 + e^{-j2\beta d}| \\
&= \frac{|\hat{V}^+|}{Z_C}\underbrace{|e^{-j\beta d}|}_{1}\underbrace{|e^{j\beta d} + e^{-j\beta d}|}_{|2\cos(\beta d)|} \\
&\propto |\cos(\beta d)| \\
&= \left|\cos\left(2\pi\frac{d}{\lambda}\right)\right|
\end{aligned} \tag{6.53b}$$

and we have written the result in terms of the distance from the load, d, in terms of wavelengths. These are plotted in Fig. 6.23b. Observe that the voltage is zero at the load and at distances from the load which are multiples of a half wavelength. The current is a maximum at the load and is zero at distances from the load that are odd multiples of a quarter wavelength. Further observe that corresponding points are separated by one-half wavelength. We will find this to be an important property for all other loads.

For the case where the load is an open circuit, $\hat{Z}_L = \infty$, the load reflection coefficient is $\hat{\Gamma}_L = 1$. Equations (6.52) reduce to

$$\begin{aligned}
|\hat{V}(d)| &= |\hat{V}^+||1 + e^{-j2\beta d}| \\
&\propto |\cos(\beta d)| \\
&= \left|\cos\left(2\pi\frac{d}{\lambda}\right)\right|
\end{aligned} \tag{6.54a}$$

and

$$\begin{aligned}
|\hat{I}(d)| &= \frac{|\hat{V}^+|}{Z_C}|1 - e^{-j2\beta d}| \\
&\propto |\sin(\beta d)| \\
&= \left|\sin\left(2\pi\frac{d}{\lambda}\right)\right|
\end{aligned} \tag{6.54b}$$

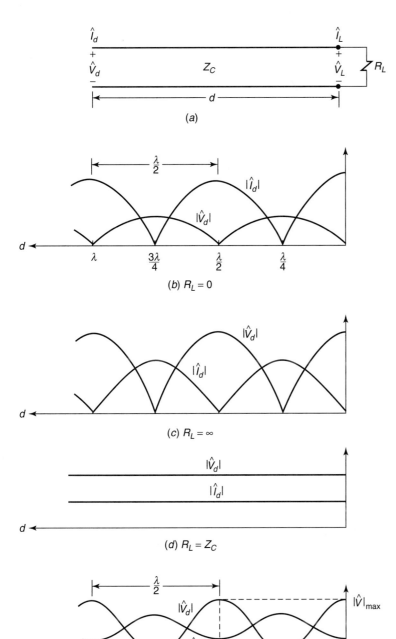

Figure 6.23 Variation of the magnitudes of the line voltage and current for various termination impedances. (a) The line specification. (b) A short-circuit load. (c) An open-circuit load. (d) A matched load. (e) General termination impedance.

and we have again written the result in terms of the distance from the load, d, in terms of wavelengths. These are plotted in Fig. 6.23c. Observe that the current is zero at the load and at distances from the load which are multiples of a half wavelength. The voltage is a maximum at the load and is zero at distances from the load that are odd multiples of a quarter wavelength. This is the reverse of the short-circuit load case. Further, again observe that corresponding points are separated by one-half wavelength.

And finally we investigate the case of a *matched* load, $\hat{Z}_L = Z_C$. For this case the load reflection coefficient is zero, $\hat{\Gamma}_L = 0$, so that the results in (6.52) show that the voltage and current are constant in magnitude along the line as shown in Fig. 6.23d. This is an important reason to match a line. Although the magnitudes of the voltage and current are constant along the line, they suffer a phase shift, $e^{-j\beta z}$, as they propagate along it.

The general case is sketched in Fig. 6.23e. The locations of the voltage and current maxima and minima are determined by the actual load impedance, but adjacent, corresponding points on each waveform are separated by one-half wavelength. This is an important and general result and can be proven from the general expressions in (6.52). In those expressions the only dependence on the distance from the load, d, is in the phase term $e^{-j2\beta d}$. Substituting $\beta = 2\pi/\lambda$ and distances that are multiples of a half wavelength, $d \Rightarrow d \pm n\,(\lambda/2)$, we obtain

$$e^{-j2\beta\left(d\pm\frac{n\lambda}{2}\right)} = e^{-j2\beta d}\underbrace{e^{\mp j2n\pi}}_{+1} \tag{6.55}$$

Hence we obtain the important result:

Corresponding points on the magnitude of the line voltage and current are separated by one-half wavelength in distance.

Similarly we can show that

the input impedance to the line replicates for multiples of a half wavelength.

This is illustrated in Fig. 6.24 and can be demonstrated from the expression for the input impedance in (6.45). The dependence on distance from the load is contained in the expression for the reflection coefficient at that point given in (6.42):

$$\begin{aligned} \hat{\Gamma}(z) &= \hat{\Gamma}_L e^{j2\beta(z-\mathscr{L})} \\ &= \hat{\Gamma}_L e^{-j2\beta d} \end{aligned} \tag{6.42}$$

Because of (6.55) we see that $e^{-j2\beta d}$ replicates for distances that are separated by one-half wavelength and hence the reflection coefficient replicates also in this fashion.

6.3.6 Matching and VSWR

We have seen in Fig. 6.23 how the magnitudes of the voltage and current vary with position on the line. In the matched case shown in Fig. 6.23d the magnitudes of the voltage and current do not vary with position, but in the cases of a short-circuit or an open-circuit load (severely mismatched), the magnitudes of the voltage and current vary drastically with position achieving minima of zero. Observe that

a maximum and the adjacent minimum are separated by one-quarter wavelength.

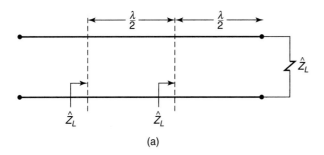

(a)

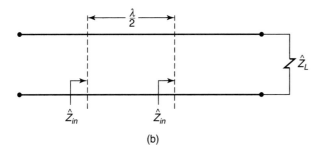

(b)

Figure 6.24 Important properties of input impedance. (a) At distances from the load that are multiples of a half wavelength, $Z_{in} = Z_L$. (b) The input impedance replicates for distances that are multiples of a half wavelength.

This radical variation of the voltage and current along the line for a mismatched load is very undesirable since it can cause spots along the line where large voltages can occur that may damage insulation or produce load voltages and currents that are radically different from the input voltage to the line. Ideally we would like the line to have no effect on the signal transmission, as will occur for the matched load. In the matched load case, there will only be a phase shift or time delay.

Although we ideally always would like to match the line, there are many practical cases where we cannot exactly match the line. If $\hat{Z}_L \neq Z_C$ the line is mismatched but how do we quantitatively measure this degree of "mismatch"? The answer to this is the voltage standing wave ratio or VSWR. The VSWR is the ratio of the maximum voltage on the line to the minimum voltage on the line. Figure 6.23 shows that a maximum and a minimum will be separated by one-quarter of a wavelength. Nevertheless the VSWR is

$$\text{VSWR} = \frac{|\hat{V}|_{\max}}{|\hat{V}|_{\min}} \tag{6.56}$$

Observe that for the extreme cases of a short-circuit load $\hat{Z}_L = 0$ or an open-circuit load $\hat{Z}_L = \infty$, the minimum voltage is zero and hence the VSWR is infinite:

$$\text{VSWR} = \begin{cases} \infty & \hat{Z}_L = 0 \\ \infty & \hat{Z}_L = \infty \end{cases} \tag{6.57}$$

but the VSWR is unity for a matched load:

$$\text{VSWR} = 1 \qquad \hat{Z}_L = Z_C \tag{6.58}$$

The VSWR must therefore lie between these two bounds:

$$1 \leq \text{VSWR} < \infty \tag{6.59}$$

Industry considers a line to be "matched" if VSWR < 2

We can derive a quantitative result for the VSWR from (6.52a), which shows that the voltage at some distance d is proportional to $|\hat{V}(d)| \propto |1 + \hat{\Gamma}_L e^{-j2\beta d}|$. The maximum will be proportional to $|\hat{V}|_{max} \propto 1 + |\hat{\Gamma}_L|$ and the minimum will be proportional to $|\hat{V}|_{min} \propto 1 - |\hat{\Gamma}_L|$. Hence the VSWR is

$$\boxed{\text{VSWR} = \frac{1 + |\hat{\Gamma}_L|}{1 - |\hat{\Gamma}_L|}} \qquad (6.60)$$

► **QUICK REVIEW EXERCISE 6.11**

Determine the VSWR for the problem in Example 6.7.

ANSWER $\quad \text{VSWR} = \dfrac{1 + 0.82}{1 - 0.82} = 10.11.$ ◄

► **QUICK REVIEW EXERCISE 6.12**

Determine the VSWR for the problem in Quick Review Exercise Problem 6.9.

ANSWER $\quad \text{VSWR} = \dfrac{1 + 0.4472}{1 - 0.4472} = 2.62.$ ◄

6.3.7 Power Flow on the Line

We now investigate the power flow on the line. The forward- and backward-traveling voltage and current waves carry power. The average power flowing to the right $(+z)$ shown in Fig. 6.25a can be determined in terms of the total voltage and current

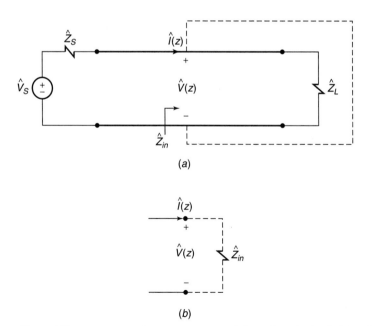

(a)

(b)

Figure 6.25 Computing the average power delivered to the load. (a) and (b) Viewing a section of the line and the load as a lumped impedance represented by the input impedance.

at that point in the same fashion as for a lumped circuit as shown in Fig. 6.25b, so that

$$P_{AV} = \frac{1}{2}\text{Re}\{\hat{V}(z)\hat{I}^*(z)\} \tag{6.61}$$

where ° denotes the conjugate. Substituting the general expressions for voltage and current given in (6.44) gives,

$$
\begin{aligned}
P_{AV} &= \frac{1}{2}\text{Re}\{\hat{V}(z)\hat{I}^*(z)\} \\
&= \frac{1}{2}\text{Re}\left\{\hat{V}^+ e^{-j\beta z}[1 + \hat{\Gamma}_L e^{j2\beta(z-\mathcal{L})}]\frac{\hat{V}^{+*}}{Z_C}e^{j\beta z}[1 - \hat{\Gamma}_L^* e^{-j2\beta(z-\mathcal{L})}]\right\} \\
&= \frac{1}{2}\text{Re}\left\{\frac{\hat{V}^+\hat{V}^{+*}}{Z_C}[1 + \underbrace{(\hat{\Gamma}_L e^{j2\beta(z-\mathcal{L})} - \hat{\Gamma}_L^* e^{-j2\beta(z-\mathcal{L})})}_{\text{imaginary}} - \hat{\Gamma}_L\hat{\Gamma}_L^*]\right\} \\
&= \frac{|\hat{V}^+|^2}{2Z_C}[1 - |\hat{\Gamma}_L|^2]
\end{aligned}
\tag{6.62}
$$

and we have used the complex algebra result that the product of a complex number and its conjugate is the magnitude squared of that complex number. This result confirms that if the load is a short circuit, $\hat{\Gamma}_L = -1$, or an open circuit, $\hat{\Gamma}_L = 1$, then the *net power flow in the +z direction is zero.* Clearly an open circuit or a short circuit can absorb no power. Essentially, then, the power in the forward-traveling wave is equal to the power in the backward-traveling wave and all the incident power is reflected.

This can be confirmed by separately computing the average power flow in the individual waves. The power flowing to the right in the forward-traveling wave is

$$
\begin{aligned}
P_{AV}^+ &= \frac{1}{2}\text{Re}\left\{(\hat{V}^+ e^{-j\beta z})\left(\frac{\hat{V}^{+*}}{Z_C}e^{j\beta z}\right)\right\} \\
&= \frac{|\hat{V}^+|^2}{2Z_C}
\end{aligned}
\tag{6.63a}
$$

and the power flowing to the right in the backward-traveling wave is

$$
\begin{aligned}
P_{AV}^- &= \frac{1}{2}\text{Re}\left\{(\hat{V}^- e^{j\beta z})\left(-\frac{\hat{V}^{-*}}{Z_C}e^{-j\beta z}\right)\right\} \\
&= -\frac{|\hat{V}^-|^2}{2Z_C} \\
&= -\frac{|\hat{V}^+|^2}{2Z_C}|\hat{\Gamma}_L|^2
\end{aligned}
\tag{6.63b}
$$

The sum of (6.63a) and (6.63b) gives (6.62). The ratio of reflected power to incident power is $|\hat{\Gamma}_L|^2$. For a matched load, $\hat{Z}_L = Z_C$, $\hat{\Gamma}_L = 0$, and all the power in the forward-traveling wave is absorbed in the (matched) load.

▷ **QUICK REVIEW EXERCISE 6.13**

Determine the ratio of reflected power to incident power for the problem of Example 6.7 as a percentage.

ANSWER   67.24%.   ◀

▷ **QUICK REVIEW EXERCISE 6.14**

Determine the ratio of reflected power to incident power for Quick Review Exercise Problem 6.9 as a percentage.

ANSWER   20%.   ◀

▶ ## 6.4   THE SMITH CHART

The Smith chart is a *nomograph* whose primary use is to compute the input impedance to a transmission line[1]. In addition, it has several other computational advantages. Although it can be used to analyze lossy lines, we will investigate its use for lossless lines. The key to the construction and understanding of the Smith chart is the basic relation between the input impedance to a line of length $\mathscr{L}$ in terms of the reflection coefficient at that point on the line:

$$\hat{Z}_{in} = Z_C \frac{[1 + \hat{\Gamma}_{in}]}{[1 - \hat{\Gamma}_{in}]} \tag{6.64a}$$

and the relation between the load reflection coefficient and the reflection coefficient at the input to the line:

$$\begin{aligned} \hat{\Gamma}_{in} &= \hat{\Gamma}_L e^{-j2\beta\mathscr{L}} \\ &= \hat{\Gamma}_L e^{-j4\pi\frac{\mathscr{L}}{\lambda}} \end{aligned} \tag{6.64b}$$

and we have written the reflection coefficient in terms of the load reflection coefficient given in (6.41) and wavelength. The first step is to write the input impedance *normalized* to the load impedance:

$$\begin{aligned} \hat{z}_{in} &= \frac{\hat{Z}_{in}}{Z_C} \\ &= \frac{[1 + \hat{\Gamma}_{in}]}{[1 - \hat{\Gamma}_{in}]} \\ &= r + jx \end{aligned} \tag{6.65}$$

and we have written this complex *normalized impedance* in terms of a real and an imaginary form. Next, write the reflection coefficient at the input in terms of a real and an imaginary form:

$$\begin{aligned} \hat{\Gamma}_{in} &= \hat{\Gamma}_L e^{-j4\pi\frac{\mathscr{L}}{\lambda}} \\ &= p + jq \end{aligned} \tag{6.66}$$

Substituting the expression for the reflection coefficient in (6.66) into the expression for the (normalized) input impedance in (6.65) gives:

$$\begin{aligned} \hat{z}_{in} &= r + jx \\ &= \frac{1 + p + jq}{1 - p - jq} \end{aligned} \tag{6.67}$$

---

[1]P.H. Smith, "Transmission Line Calculator," *Electronics*, January 1939; and P.H. Smith, "An Improved Transmission Line Calculator," *Electronics*, January 1944.

Equating the real and imaginary parts of this gives two equations:

$$\left(p - \frac{r}{r+1}\right)^2 + q^2 = \frac{1}{(r+1)^2} \qquad \left(\begin{array}{l} \text{Circles of radius } \dfrac{1}{r+1} \\[2mm] \text{centered at } p = \dfrac{r}{r+1} \text{ and } q = 0 \end{array}\right) \qquad (6.68a)$$

and

$$(p-1)^2 + \left(q - \frac{1}{x}\right)^2 = \frac{1}{x^2} \qquad \left(\begin{array}{l} \text{Circles of radius } \dfrac{1}{x} \\[2mm] \text{centered at } p = 1 \text{ and } q = \dfrac{1}{x} \end{array}\right) \qquad (6.68b)$$

Figure 6.26 shows Equation (6.68a) drawn in the p,q plane, which is represented by circles of radius $1/(r + 1)$ centered at $p = r/(r + 1)$ and $q = 0$. Similarly, Fig. 6.27 shows Equation (6.68b) drawn in the p,q plane, which is represented by circles of radius $1/x$ centered at $p = 1$, $q = 1/x$. These two graphs relate the real and imaginary parts of the reflection coefficient at a point (p,q) to the real and imaginary parts of the normalized input impedance (r,x) at that point. Hence, if we know the reflection coefficient at a

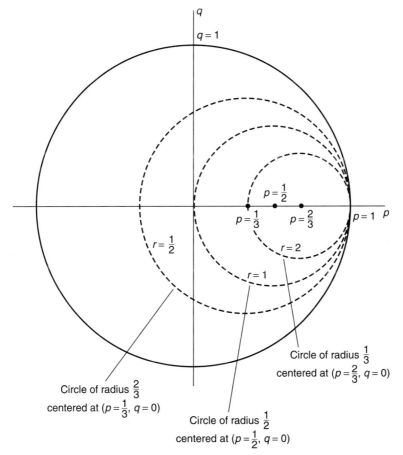

Figure 6.26 Graphical representation of Equation (6.68a) in terms of the real and imaginary parts of the reflection coefficient, $\hat{\Gamma} = p + jq$.

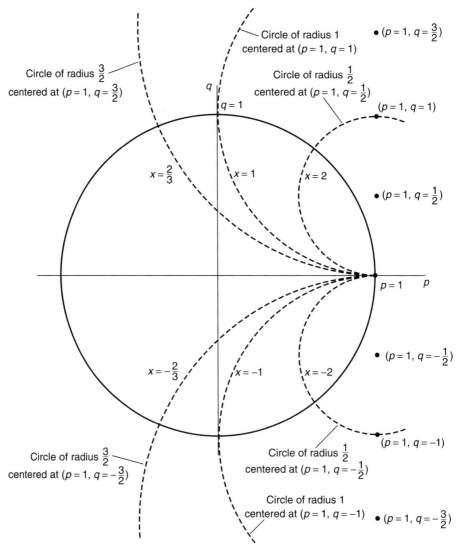

Figure 6.27 Graphical representation of Equation (6.68b) in terms of the real and imaginary parts of the reflection coefficient, $\hat{\Gamma} = p + jq$.

point, we can look up the real and imaginary parts of the normalized impedance at that point on these charts and vice versa. If we overlay Fig. 6.26 and Fig. 6.27, we obtain the Smith chart shown in Fig. 6.28. Although not normally shown on the Smith chart, we have drawn the vertical and horizontal axes of the real part of the reflection coefficient, p, and the imaginary part of the reflection coefficient, q, to remind the reader that the reflection coefficient is in the background. Figure 6.29 shows this relation between the real and imaginary parts of the normalized input impedance to the line and the real and imaginary parts of the reflection coefficient *at that point*. If we plot the normalized input impedance on the Smith chart, we implicitly have the magnitude of the reflection coefficient, $|\hat{\Gamma}_{in}| = \sqrt{p^2 + q^2}$, as the distance from the center of the chart to the plotted normalized input impedance, $r + jx$. The outer rim of the chart represents the maximum magnitude of the reflection coefficient of unity, $|\hat{\Gamma}|_{max} = 1$. The angle of

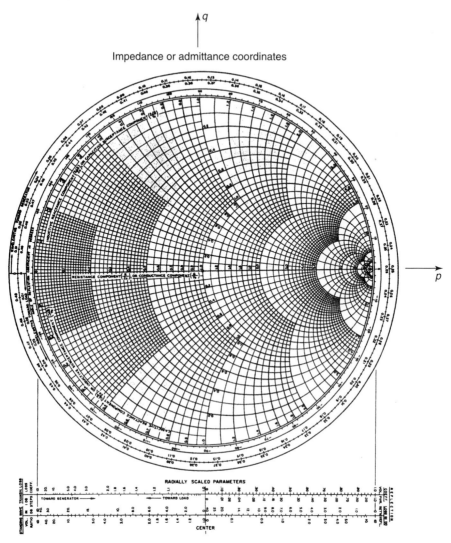

q

Impedance or admittance coordinates

p

RADIALLY SCALED PARAMETERS

Figure 6.28 The Smith chart.

the reflection coefficient, $\theta_{\hat{\Gamma}_{in}} = \angle -2\beta\mathscr{L} = \angle -4\pi(\mathscr{L}/\lambda)$, is measured (as is convention for complex numbers) counterclockwise from the positive real axis, p, as shown in Fig. 6.29.

The use of the Smith chart in determining input impedance to the line is illustrated in Fig. 6.30. First we plot the normalized load impedance, $\hat{z}_L = \hat{Z}_L/Z_C = r_L + jx_L$, on the Smith chart. Then, according to (6.64b), we rotate this for a constant radius (using a compass) and angle of $\theta_{\hat{\Gamma}_{in}} = \angle -2\beta\mathscr{L} = \angle -4\pi(\mathscr{L}/\lambda)$ (in this case clockwise because of the minus sign). At that point we simply read off the real and imaginary parts of the normalized input impedance, $\hat{z}_{in} = \hat{Z}_{in}/Z_C = r_{in} + jx_{in}$, and then unnormalize to get the input impedance by multiplying by the characteristic impedance, $\hat{Z}_{in} = \hat{z}_{in}Z_C$. Conversely, if we knew the input impedance (perhaps by measurement), we could plot its normalized value, rotate an angle of $2\beta\mathscr{L}$ (in this case counterclockwise), and read off the normalized load impedance. The direction one must rotate, toward the load or toward the source, is indicated on the outer perimeter of the Smith chart, as are the

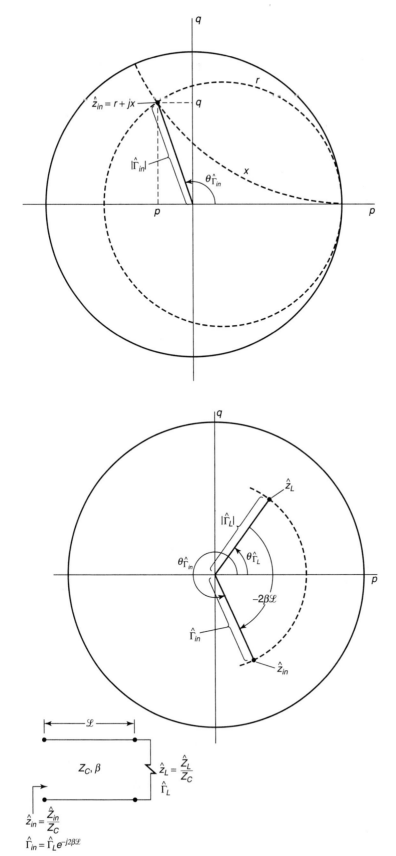

Figure 6.29 The relationship between the normalized input impedance to the line and the reflection coefficient at that point as viewed on the Smith chart.

Figure 6.30 Using the Smith chart to determine the normalized input impedance from the normalized load impedance and vice versa.

angles in degrees, $2\beta\mathcal{L}$, or distance in wavelengths, $4\pi(\mathcal{L}/\lambda)$. We will refer to these in a shorthand fashion as TG (toward generator) or TL (toward load).

► **EXAMPLE 6.9**

Consider a coaxial cable whose interior dielectric is polyethylene ($\varepsilon_r = 2.25$) and is of length 10 m. The line is driven by a source whose frequency is 34 MHz. The characteristic impedance is $Z_C = 50 \ \Omega$, and the line is terminated in a load impedance of $\hat{Z}_L = (50 + j100) \ \Omega$. Determine the input impedance to the line.

SOLUTION The velocity of propagation on the line is

$$v = \frac{v_o}{\sqrt{\varepsilon_r}}$$

$$= \frac{3 \times 10^8}{\sqrt{2.25}}$$

$$= 2 \times 10^8 \qquad \text{m/s}$$

A wavelength at the operating frequency is

$$\lambda = \frac{v}{f}$$

$$= \frac{2 \times 10^8}{34 \times 10^6}$$

$$= 5.882 \qquad \text{m}$$

Therefore, the length of the line in terms of wavelength is

$$\mathcal{L} = 10 \ \text{m} \times \frac{1}{5.882 \, \text{m}/\lambda}$$

$$= 1.70 \ \lambda$$

The normalized load impedance is

$$\hat{z}_L = \frac{50 + j100}{50}$$

$$= 1 + j2$$

Plot this on the Smith chart as shown in Fig. 6.31. Now rotate 1.7λ toward the generator (TG) to give a normalized input impedance of

$$\hat{z}_{in} = 0.29 - j0.82$$

Unnormalizing this gives the input impedance to the line as

$$\hat{Z}_{in} = \hat{z}_{in}Z_C$$

$$= 14.5 - j41 \qquad \Omega$$

The exact results are $\hat{Z}_{in} = (14.52 - j40.52) \ \Omega$.

Observe that one full rotation around the Smith chart occurs for 0.5λ. This shows that *the input impedance replicates for lengths of line that are multiples of a half wavelength,* as we showed previously. Hence, in rotating the full electrical length of the line, 1.7λ, we rotate three full revolutions (1.5λ) plus an additional 0.2λ.

Additionally, we can determine the magnitude of the reflection coefficient by measuring the distance from the center of the chart to the plotted point (using a compass) and transferring this length to one of the bottom scales that is marked REFLECTION COEFF. VOL., giving

Impedance or admittance coordinates

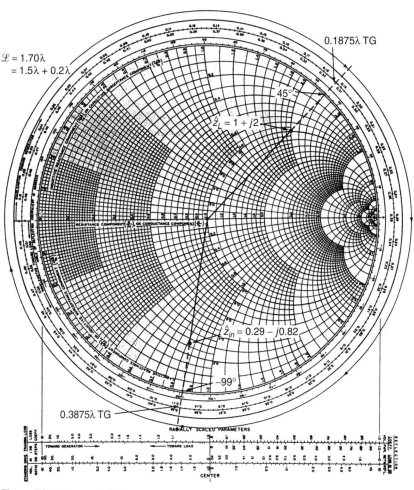

Figure 6.31 Example 6.9.

$|\hat{\Gamma}_L| = |\hat{\Gamma}_{in}| = 0.71$. This shows that *the magnitude of the reflection coefficient is the same at all points on the line; only the angle of the reflection coefficient varies at points along the line.* The angle of the load reflection coefficient is read off one of the outer scales around the periphery that is marked ANGLE OF REFLECTION COEFFICIENT IN DEGREES as 45 degrees. Hence the load reflection coefficient is

$$\hat{\Gamma}_L = 0.71\angle 45°$$

The reflection coefficient at the input to the line is similarly read off at that point as

$$\hat{\Gamma}_{in} = 0.71\angle -99°$$

The VSWR can also be read off the chart by measuring (with a compass) the magnitude of the reflection coefficient and transferring that to one of the bottom scales that is marked STANDING WAVE VOL. RATIO, giving VSWR = 5.8, and the exact value is VSWR = 5.83. ◀

Moving in the correct direction around the Smith chart in determining input impedance from the load impedance (TG) or the load impedance from the input impedance (TL) is very important.

▶ **EXAMPLE 6.10**

The measured input impedance to a line is $\hat{Z}_{in} = (20 - j40)\,\Omega$ and the load impedance is $\hat{Z}_L = (20 + j40)\,\Omega$. Determine the length of the line in wavelengths if its characteristic impedance is $\hat{Z}_C = 100\,\Omega$.

SOLUTION Plot the normalized input and load impedances on the chart:

$$\hat{z}_{in} = 0.2 - j0.4$$
$$\hat{z}_L = 0.2 + j0.4$$

as shown in Fig. 6.32. The length of the line in wavelengths is the circumferential distance between the two plotted points. But there are two circumferential distances between the two points. Which one is correct? The simple way of determining the correct distance is to simply start at the plotted load impedance and go toward the input impedance. This requires movement TOWARD THE GENERATOR or a clockwise movement from $\hat{z}_L(0.062\lambda, \mathrm{TG})$ to $\hat{z}_{in}(0.435\lambda, \mathrm{TG})$, giving the line length as

$$\mathscr{L} = 0.435\lambda - 0.062\lambda$$
$$= 0.373\lambda$$

Impedance or admittance coordinates

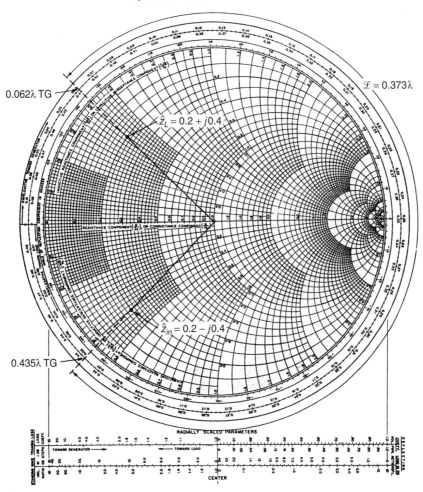

Figure 6.32 Example 6.10.

We could alternatively start at the plotted input impedance and go toward the load impedance. This would require movement TOWARD THE LOAD or a counterclockwise movement from $\hat{z}_{in}(0.062\lambda,\text{TL})$ to $\hat{z}_L(0.438\lambda,\text{TL})$, giving the same result. Note that this electrical length is only unique within a multiple of a half wavelength, since the input impedance replicates for line lengths that are a multiple of a half wavelength. So the answer could be $\mathcal{L} = 0.373\lambda + 0.5\lambda = 0.873\lambda$ or $\mathcal{L} = 0.373\lambda + 1.0\lambda = 1.373\lambda$, etc. Knowing the velocity of propagation on the line and the frequency of operation, we can determine the physical length from

$$\mathcal{L}(\text{m}) = \mathcal{L}(\lambda) \times \left(\lambda = \frac{v}{f}\, \text{m} \right)$$

For example, if the line is being operated at a frequency of 30 MHz, and the velocity of propagation is 250 m/μs, then the wavelength is $\lambda = 8.333$ m and the physical length is $\mathcal{L} = 3.11$ m. The VSWR can be read off as VSWR $= 5.8$. ◀

▶ **QUICK REVIEW EXERCISE 6.15**

A 100-Ω line has a measured input impedance of $(22 + j0)\,\Omega$ and a load impedance of $(150 - j200)\,\Omega$. Determine the shortest length of the line.

ANSWER 0.198λ. ◀

▶ **6.5 LUMPED-CIRCUIT APPROXIMATE MODELS OF TRANSMISSION LINES**

If a transmission line is electrically short at the frequency of operation, that is, much shorter than a wavelength, then the distributed nature of the line becomes less important. In this case we may model the line, as an approximation, as a lumped circuit and avoid directly solving the transmission-line equations. When is a line short enough, electrically, to use a lumped-circuit approximation of it? There is no fixed criterion for this, but we will assume that the line is electrically short if its length is less than $\frac{1}{10}$ of a wavelength at the frequency of the source, that is, $\mathcal{L} < \frac{1}{10}\lambda$.

When a line is electrically short, there are many possible lumped-circuit models. One that is useful is the lumped-Pi model shown in Fig. 6.33. This is so named because of its structural resemblance to the symbol π. The total line inductance is the product of the per-unit-length inductance multiplied by the total line length. This is lumped as an inductance of $l\mathcal{L}$ H and included as one inductance as shown in Fig. 6.33. The total capacitance of the line is the product of the per-unit-length capacitance multiplied by the total line length or $c\mathcal{L}$ F. This is split and lumped as capacitances on either side of the inductance as shown in Fig. 6.33. The capacitance is split and placed on either side of the inductance so as to give the model a reciprocal property, that is, interchanging the two ends of the line should give the same result.

The per-unit-length inductance and capacitance can be found from the line characteristic impedance, $Z_C = \sqrt{l/c}$, and velocity of propagation, $v = 1/\sqrt{lc}$, as

$$l = \frac{Z_C}{v} \tag{6.69a}$$

and

$$c = \frac{1}{vZ_C} \tag{6.69b}$$

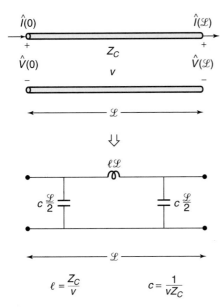

$$\ell = \frac{Z_C}{v} \qquad\qquad c = \frac{1}{vZ_C}$$

Figure 6.33 Representing an electrically short line with a lumped circuit.

▷ **EXAMPLE 6.11**

Consider the transmission line shown in Fig. 6.34a. Since the frequency of operation is 100 MHz and the velocity of propagation is 3×10^8 m/s, a wavelength is 3 m. Hence the total line length of 0.3 m is $(1/10)\lambda$. Determine the load voltage and compare to the exact value.

SOLUTION The per-unit-length parameters are computed from (6.69) as $l = 0.167\ \mu\text{H/m}$ and $c = 66.7$ pF/m. The total parameters are 50 nH and 20 pF. The SPICE circuit is shown in

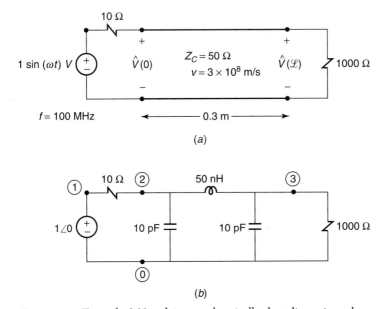

Figure 6.34 Example 6.11; solving an electrically short line using a lumped-circuit model. (a) The problem parameters. (b) The lumped-Pi model.

Fig. 6.34b, and the SPICE code is

```
EXAMPLE 6.11
VS 1 0 AC 1 0
RS 1 2 10
CS 2 0 10P
L 2 3 50N
CL 3 0 10P
RL 3 0 1000
.AC DEC 1 100MEG 100MEG
.PRINT AC VM(2) VP(2) VM(3) VP(3)
.END
```

The computed value of the load voltage is $V(\mathcal{L}) = 1.214\angle-10.2°$ V, whereas the exact value is $V(\mathcal{L}) = 1.205\angle-10.2°$ V. ◀

▷ 6.6 LOSSY LINES

We have been assuming that the transmission line is lossless in all of our previous work. We will now briefly consider the effects of losses. Losses come from two mechanisms. The resistance of the line conductors causes a per-unit-length resistance of r Ω/m, and losses in the surrounding medium cause a per-unit-length conductance of g S/m. The per-unit-length equivalent circuit for a Δz section of the line is shown in Fig. 6.35a. We can derive the transmission-line equations in a manner similar to the lossless case and obtain

$$\frac{\partial V(z,t)}{\partial z} = -rI(z,t) - l\frac{\partial I(z,t)}{\partial t} \tag{6.70a}$$

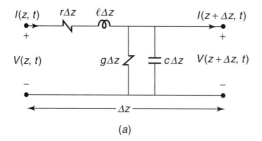

(a)

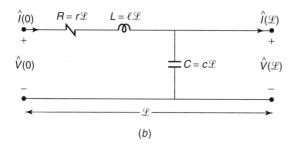

(b)

Figure 6.35 A lossy line. (a) The per-unit-length equivalent circuit. (b) An electrically short, lossy line.

and

$$\frac{\partial I(z,t)}{\partial z} = -gV(z,t) - c\frac{\partial V(z,t)}{\partial t} \tag{6.70b}$$

In the frequency (phasor) domain these become $(\partial/\partial t \Rightarrow j\omega)$

$$\boxed{\frac{d\hat{V}(z)}{dz} = -\hat{z}\hat{I}(z)} \tag{6.71a}$$

and

$$\boxed{\frac{d\hat{I}(z)}{dz} = -\hat{y}\hat{V}(z)} \tag{6.71b}$$

where the per-unit-length impedance and admittance are

$$\boxed{\hat{z} = (r + j\omega l) \qquad \Omega/\text{m}} \tag{6.72a}$$

and

$$\boxed{\hat{y} = (g + j\omega c) \qquad \text{S}/\text{m}} \tag{6.72b}$$

The coupled first-order transmission-line equations in (6.71) can be converted to uncoupled, second-order form by differentiating one with respect to z and substituting the other to yield

$$\boxed{\frac{d^2\hat{V}(z)}{dz^2} = \hat{z}\hat{y}\,\hat{V}(z)} \tag{6.73a}$$

and

$$\boxed{\frac{d^2\hat{I}(z)}{dz^2} = \hat{z}\hat{y}\,\hat{I}(z)} \tag{6.73b}$$

The general solution is very similar to the lossless case that was obtained previously and becomes

$$\boxed{\hat{V}(z) = \hat{V}^+ e^{-\alpha z}e^{-j\beta z} + \hat{V}^- e^{\alpha z}e^{j\beta z}} \tag{6.74a}$$

and

$$\boxed{\hat{I}(z) = \frac{\hat{V}^+}{\hat{Z}_C}e^{-\alpha z}e^{-j\beta z} - \frac{\hat{V}^-}{\hat{Z}_C}e^{\alpha z}e^{j\beta z}} \tag{6.74b}$$

where the characteristic impedance is now complex and defined by

$$\boxed{\begin{aligned} \hat{Z}_C &= \sqrt{\frac{\hat{z}}{\hat{y}}} \\ &= Z_C\angle\theta_{Z_C} \qquad \Omega \end{aligned}} \tag{6.75}$$

and the propagation constant is

$$\hat{\gamma} = \sqrt{\hat{z}\hat{y}}$$
$$= \alpha + j\beta \tag{6.76}$$

The real part of the propagation constant, α, is said to be the *attenuation constant*, whereas the imaginary part, β, is again said to be the *phase constant* with units of rad/m. In the time domain, the phasor solutions in (6.74) become

$$V(z,t) = \underbrace{V^+ e^{-\alpha z}}_{\text{amplitude}} \cos(\omega t - \beta z + \theta^+) + \underbrace{V^- e^{\alpha z}}_{\text{amplitude}} \cos(\omega t + \beta z + \theta^-) \tag{6.77a}$$

and

$$I(z,t) = \underbrace{\frac{V^+}{Z_C} e^{-\alpha z}}_{\text{amplitude}} \cos(\omega t - \beta z - \theta_{Z_C} + \theta^+) - \underbrace{\frac{V^-}{Z_C} e^{\alpha z}}_{\text{amplitude}} \cos(\omega t + \beta z - \theta_{Z_C} + \theta^-) \tag{6.77b}$$

and the undetermined constants are, in general, complex as $\hat{V}^\pm = V^\pm \angle \theta^\pm$. These are again in the form of forward-traveling and backward-traveling waves. The primary differences between the lossless line and the lossy line are that (1) the amplitude of the waves are attenuated, that is, decrease with increasing distance along the line according to the factor $e^{\pm \alpha z}$, and (2) the current lags behind the voltage because of the phase angle of the characteristic impedance, θ_{Z_C}, which, in the case of a lossy line, is not zero.

The parallels between the lossy transmission line and uniform plane waves propagating in a lossy medium are striking. The reader should compare the results and forms of the equations to those for a uniform plane wave traveling in a lossy medium given in Section 5.2 of Chapter 5.

The reflection coefficient and input impedance for a lossy line can be obtained in a fashion virtually identical to that used for the lossless line. The reflection coefficient at a point on the line is the ratio of the backward-traveling wave to the forward-traveling wave. From (6.74a), this is

$$\hat{\Gamma}(z) = \hat{\Gamma}_L e^{2\alpha(z - \mathcal{L})} e^{2j\beta(z - \mathcal{L})} \tag{6.78}$$

Thus the line voltages and currents in (6.74) can be written in terms of the reflection coefficient as

$$\hat{V}(z) = \hat{V}^+ e^{-\alpha z} e^{-j\beta z}(1 + \hat{\Gamma}(z)) \tag{6.79a}$$

and

$$\hat{I}(z) = \frac{\hat{V}^+}{\hat{Z}_C} e^{-\alpha z} e^{-j\beta z}(1 - \hat{\Gamma}(z)) \tag{6.79b}$$

The input impedance to the line is the ratio of these:

$$\hat{Z}_{in} = \hat{Z}_C \frac{[1 + \hat{\Gamma}(0)]}{[1 - \hat{\Gamma}(0)]}$$
$$= \hat{Z}_C \frac{[1 + \hat{\Gamma}_L e^{-2\alpha\mathcal{L}} e^{-j2\beta\mathcal{L}}]}{[1 - \hat{\Gamma}_L e^{-2\alpha\mathcal{L}} e^{-j2\beta\mathcal{L}}]} \tag{6.80}$$

Note in (6.78) that for the lossy line, the magnitude of the reflection coefficient, $|\hat{\Gamma}(z)| = |\hat{\Gamma}_L| e^{-2\alpha(\mathcal{L}-z)}$, is no longer constant along the line but decreases with distance, unlike the lossless line case.

The velocity of propagation of the waves on the line is

$$v = \frac{\omega}{\beta}$$
$$\neq \frac{1}{\sqrt{lc}}$$

(6.81)

The velocity of propagation is the ratio of ω and β but for a lossy line is not simply $1/\sqrt{lc}$ as was the case for a lossless line because $\beta \neq \omega\sqrt{lc}$.

▷ **EXAMPLE 6.12**

A RG-58U coaxial cable has a per-unit-length inductance of 0.3 μH/m, capacitance of 100 pF/m, resistance of 1.3 Ω/m, and negligible dielectric loss. Determine the attenuation and phase constants, the characteristic impedance, and the velocity of propagation of waves on the line at 100 MHz. Compare these to the lossless case.

SOLUTION   The per-unit-length impedance and admittance are

$$\hat{z} = r + j\omega l$$
$$= 1.3 + j188.5$$
$$= 188.5 \angle 89.6° \quad \Omega$$

and

$$\hat{y} = j\omega c$$
$$= j6.28 \times 10^{-2}$$
$$= 6.28 \times 10^{-2} \angle 90°$$

The propagation constant is

$$\hat{\gamma} = \sqrt{(r + j\omega l)(j\omega c)}$$
$$= \underbrace{1.187 \times 10^{-2}}_{\alpha} + \underbrace{j3.441}_{\beta}$$

and the characteristic impedance is

$$\hat{Z}_C = \sqrt{\frac{\hat{z}}{\hat{y}}}$$
$$= 54.77 \angle -0.198° \quad \Omega$$

The velocity of propagation of waves on the line is

$$v = \frac{\omega}{\beta}$$
$$= 1.83 \times 10^8 \quad \text{m/s}$$

For the lossless line we obtain $\alpha = 0$, $\beta = 3.441$, $Z_C = 54.77$ Ω, and $v = 1.83 \times 10^8$ m/s. Evidently this is a "low loss" cable at 100 MHz since the phase constant, characteristic impedance, and velocity of propagation with and without losses are virtually identical.   ◀

6.6.1   Modeling Lossy Lines at Low Frequencies

We have generally assumed transmission-line losses to be negligible and have characterized the line solely in terms of capacitance and inductance. At very high frequencies (in the GHz range) the conductor resistance increases due to skin effect and losses become important. Line losses such as conductor resistance also become significant when the line is operated at low frequencies. Lossy lines that are operated at low frequencies behave less like a transmission line and more like an RC circuit. For example, consider a lossy line that is electrically short. We may lump the per-unit-length resistance, r, inductance, l, and capacitance, c, into three lumped elements representing the entire line as shown in Fig. 6.35b. The total element value is the product of the per-unit-length value and the line length: $R = r\mathscr{L}$, $L = l\mathscr{L}$, and $C = c\mathscr{L}$. Observe that the series impedance is $\hat{Z} = R + j\omega L$. At low frequencies, the total resistance dominates the inductive reactance term so that the line appears as an RC circuit. For example, consider a parallel-wire household telephone transmission line having a per-unit-length resistance of $0.2\ \Omega/\text{m}$, a per-unit-length inductance of $1\ \mu\text{H/m}$, and a per-unit-length capacitance of $100\ \text{pF/m}$. In the absence of losses, the characteristic impedance would be $100\ \Omega$. The line carries audio frequencies whose highest frequency component is on the order of 10 kHz. At 10 kHz, the total series impedance is $\hat{Z} = R + j\omega L = 0.2\mathscr{L} + j0.063\mathscr{L}\ \Omega \cong R$. Hence, the inductive reactance is negligible, and the line looks like an RC circuit at these low frequencies. Hence at low frequencies, the conductor resistance cannot be neglected and the line cannot be considered lossless.

▶ 6.7 ENGINEERING APPLICATIONS

In this section we will examine some important applications of the principles of this chapter. The ever-increasing speeds of digital computers and frequencies of communications links require that interconnection conductors such as lands on a printed circuit board be modeled as transmission lines and the effect of the line be calculated in order to determine whether a digital or analog device design will operate as intended.

6.7.1   High-Speed Digital Interconnects and Signal Integrity

The clock speeds of digital devices today are in the GHz range and will continue to increase as the demand for increased computational speed continues. In addition, there are numerous analog devices that also operate in the GHz range. The electronic modules in these devices are commonly mounted on a glass-epoxy (commonly known as FR4) printed circuit board (PCB) whose thicknesses ranges from 40 mils (1.016 mm) to 70 mils (1.778 mm). Rectangular cross section conductors or lands interconnect these modules. Recall that the wavelength in free space at 1 GHz is 30 cm. Wave propagation along the lands of PCBs causes the wavelengths to further decrease. For example, on a glass-epoxy PCB, the electric fields lie partly in the board and partly in the surrounding air. Hence the effective relative permittivity is approximately the average of the two relative permittivities. For FR4, the relative permittivity is $\varepsilon_r = 4.7$, giving an approximate relative permittivity of $\varepsilon_r' \cong (1 + 4.7)/2 = 2.85$. Hence the velocity of wave propagation along the lands is reduced from that of free space to $v = v_o/\sqrt{\varepsilon_r'} = 1.77 \times 10^8$ m/s so that the wavelength at 1 GHz reduces to 18 cm.

Hence an analog circuit on a PCB operated at 1 GHz is electrically large if its largest dimension exceeds $\frac{1}{10}$ of a wavelength or 1.8 cm. Typical circuit dimensions of these high-frequency circuits easily exceed this dimension. Hence the interconnect lands must be modeled as distributed-parameter circuits with the transmission-line model, and lumped-circuit models no longer apply. In addition, the propagation delay along these lands is becoming increasingly important in high-speed digital devices. For example, a pair of lands on a PCB of length $\mathscr{L}$ will have a time delay in transmitting a digital pulse

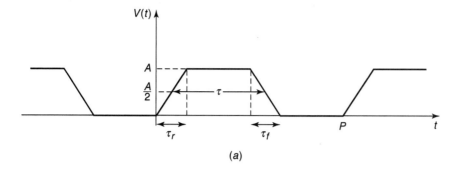

(a)

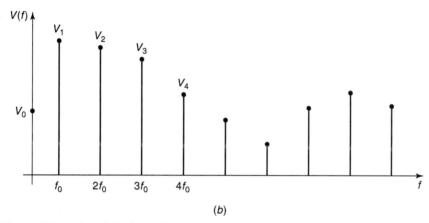

(b)

Figure 6.36 A digital clock signal (a) in the time domain and (b) in the frequency domain.

from one end to the other of

$$T = \frac{\mathcal{L}}{v} \quad \text{s}$$

(6.82)

For example, a pair of lands on a PCB of length 10 cm (approximately 4 in.) will suffer a delay of approximately 0.5 ns or 500 ps. This delay is becoming an increasingly important factor in the design of high-speed digital devices. In this section we will investigate more fully the effect of these interconnect lands in high-speed digital and high-frequency analog devices.

Clock signals in digital devices consist of trapezoidal pulses with a period P and a repetition rate or repetition frequency of $f_o = 1/P$ Hz as shown in Fig. 6.36a. These trapezoidal pulses have an amplitude of A and rise and fall times of τ_r and τ_f, respectively. The pulse width, τ, is defined between the 50% points, $A/2$. Digital data waveforms have a similar shape. Because these clock signals are periodic, we can view them with the Fourier series as being composed of sinusoidal components whose frequencies (harmonics) are at multiples of the fundamental frequency $f_o = 1/P$, that is, $f_o, 2f_o, 3f_o, 4f_o, \cdots$ as shown in Fig. 6.36b. Hence the pulse train can be written as[2]

$$V(t) = V_0 + V_1 \cos(\omega_o t + \theta_1) + V_2 \cos(2\omega_o t + \theta_2) + V_3 \cos(3\omega_o t + \theta_3) + \cdots \quad (6.83)$$

[2]C.R. Paul, *Introduction to Electromagnetic Compatibility*, John Wiley Interscience, 1992.

where $\omega_o = 2\pi f_o$. The amplitudes and phase angle of these sinusoidal components are given for the trapezoidal pulse train in Fig. 6.36a by[3]

$$V_0 = \frac{1}{P} \int_0^P V(t)dt$$

$$= A\frac{\tau}{P} \quad \text{(average value)}$$

$$(6.84a)$$

and

$$V_n = 2A\frac{\tau}{P}\left|\frac{\sin(n\pi\tau/P)}{n\pi\tau/P}\right|\,\left|\frac{\sin(n\pi\tau_r/P)}{n\pi\tau_r/P}\right| \qquad \tau_r = \tau_f \qquad (6.84b)$$

and

$$\theta_n = \pm n\pi\frac{\tau + \tau_r}{P} \qquad \tau_r = \tau_f \qquad (6.84c)$$

where the rise and fall times of the pulse are assumed equal, $\tau_r = \tau_f$. Typical digital clock signals have rise and fall times that are approximately equal. The amplitudes of these *spectral components* for a clock signal are shown in Fig. 6.36b. A measurement device known as a *spectrum analyzer* is commonly used to view the spectral components of a periodic waveform such as clock signals. (See Fig. 1.3c of Chapter 1.)

Bounds on these spectral components can be obtained as shown in Fig. 6.37a.[4] There are two break points in this Bode plot of the spectral amplitudes. The first level is 0 dB/decade up to the first break point of $1/\pi\tau$. Above this the amplitudes drop off at a rate of -20dB/decade up to the second break point of $1/\pi\tau_r$. Above this second break point, the amplitudes drop off at a rate of -40 dB/decade. The exact amplitudes and the bounds are compared in Fig. 6.37b for a 1-MHz clock signal that has 1-V amplitude, 20-ns rise/fall times, and a 50% duty cycle. The duty cycle is the ratio of the pulse width to the period:

$$D(\text{duty cycle}) = \frac{\tau}{P} \qquad (6.85)$$

A 50% duty cycle means that the pulse is on half the period and off the remainder of the period. This also is fairly typical of digital clock signals. At certain frequencies the bounds match the actual level very closely, whereas at a few points they exceed the actual level. However, the bounds are easier to work with, computationally. Observe that

the high-frequency spectral content is determined by the pulse rise/fall times.

The majority of the spectrum lies at frequencies below the second break point. To reduce the high-frequency spectral content, we should *increase* (make larger) the pulse rise/fall times, which moves this second break point lower in frequency, thereby reducing the

[3]C.R. Paul, *Introduction to Electromagnetic Compatibility*, John Wiley Interscience, 1992.

[4]C.R. Paul, *Introduction to Electromagnetic Compatibility*, John Wiley Interscience, 1992.

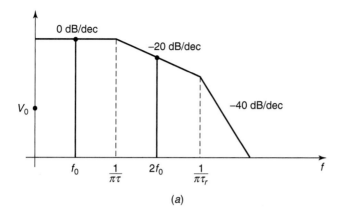

(a)

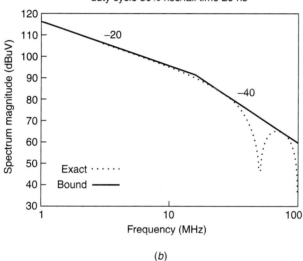

Spectrum of 1 V. 1 MHz trapezoidal wave
duty cycle 50% rise/fall time 20 ns

(b)

Figure 6.37 Asymptotic bounds on the spectrum of a digital clock signal. (a) The Bode plot. (b) Comparison of the bounds to the exact values for a 1-V, 1-MHz digital clock signal having a 50% duty cycle and 20-ns rise/fall times.

high-frequency spectral amplitudes further. To be conservative, we will assume that the majority of the spectrum lies below three times the second break point or

$$f_{\max} = 3\left(\frac{1}{\pi \tau_r}\right)$$
$$\cong \frac{1}{\tau_r}$$

(6.86)

For example, for a digital pulse having 1-ns rise/fall times (fairly typical, although future digital signals will have shorter rise/fall times in the hundreds of picoseconds), the spectrum is essentially contained below 1 GHz. Shorter rise/fall times in the picosecond range will have significant spectral components in the tens of GHz range. Hence, lands in digital devices in the future will become increasingly short, electrically. We will now investigate the effects of these interconnection lands in high-speed digital devices.

The quantity of interest in the effect of interconnect lands in digital devices is *signal integrity*. Consider a typical connection of two digital gates with a pair of lands as shown

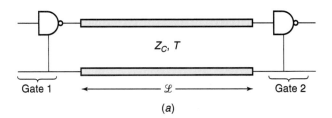

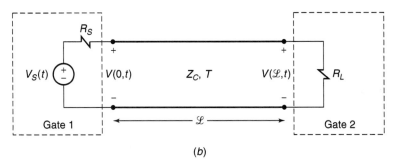

Figure 6.38 A typical digital application of transmission lines. (a) The line connects two digital gates. (b) Representing the terminal gates.

in Fig. 6.38a. This is modeled as shown in Fig. 6.38b as a source representing the output of Gate 1 and a load representing the input to Gate 2. Although we have used a resistor here to model the input to a gate, typically the inputs to devices such as CMOS (complementary metal oxide semiconductor) inverters are capacitive with values ranging from 1 pF to 10 pF. We will later model devices such as CMOS as capacitive at their input. For the moment, resistive inputs will serve to illustrate the essential problem of signal integrity. The two interconnect lands are modeled as a transmission line. Ideally we want the interconnect lands to have no effect. In other words, ideally we want

$$V(0,t) = V(\mathcal{L},t)$$
$$\cong \frac{R_L}{R_S + R_L} V_S(t) \qquad (6.87)$$

If the output voltage of the line does not approximately equal the input voltage to the line, then the second gate may respond incorrectly and we obtain logic errors. Hence digital designers must be concerned with the effects of the interconnect lands.

One of the important effects of the interconnect lands is to produce "ringing" at the output. This occurs because the line is not matched, that is, $R_L \neq Z_C$ and/or $R_S \neq Z_C$. To demonstrate this effect, consider the configuration shown in Fig. 6.39a. The line has a characteristic impedance of 50 Ω and a one-way time delay of T. The source is a 5-V pulse (zero rise time) and 10-Ω source impedance, and the load is an open circuit. This approximates to some degree the output of a CMOS gate (low source resistance) and the input to a CMOS gate (capacitive representing a high impedance). Wave tracing for the load voltage (output voltage of the line) is shown in Fig. 6.39b. The source reflection coefficient is $\Gamma_S = -\frac{2}{3}$, and the load reflection coefficient for the open circuit is $\Gamma_L = +1$. Observe that the load voltage oscillates about the desired 5-V level between levels of 8.33 V and 2.78 V, eventually settling down to 5 V. This

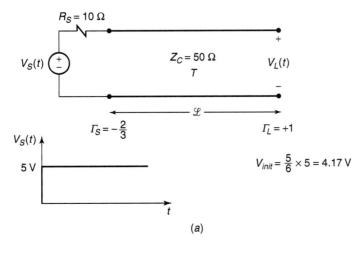

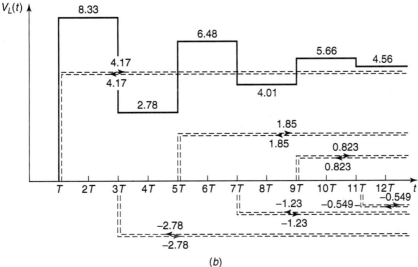

Figure 6.39 Illustration of the phenomenon of "ringing" in transmission lines that interconnect a low-impedance source and a high-impedance load. (a) An example to illustrate the phenomenon. (b) The results of the solution illustrating ringing at the terminals.

constitutes ringing and can cause the received voltage levels to lie outside the levels for a logic 0 and a logic 1 guaranteed by the manufacturer of the gates, thus creating logic errors.

As a practical example of interconnects, consider a CMOS inverter connected to another CMOS inverter by a pair of lands on a microstrip shown in Fig. 6.40a. The microstrip line consists of a land of width 100 mils above an FR4 ($\varepsilon_r = 4.7$) substrate of thickness 62 mils as shown in Fig. 6.40b. Using (6.11) we compute the per-unit-length inductance and capacitance as $l = 0.335\,\mu$H/m and $c = 117.5$ pH/m. The effective relative permittivity is computed from (6.11b) as $\varepsilon_r' = 3.54$. From these we compute the characteristic impedance as $Z_C = \sqrt{l/c} = 53.4\,\Omega$. The velocity of propagation is $v = v_o/\sqrt{\varepsilon_r'} = 1.59 \times 10^8$ m/s. The total line length is 20 cm, giving a one-way delay of $T = \mathcal{L}/v = 1.255$ ns. The source (output of gate 1) is represented by a 2.5-V, 25-MHz digital pulse train having rise/fall times of 2 ns and a 50% duty cycle. The source

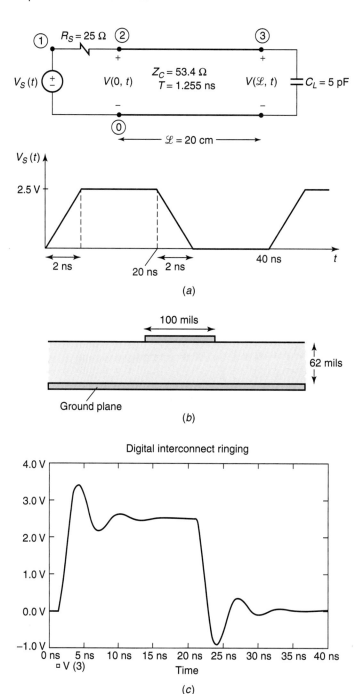

Figure 6.40 A typical signal integrity problem. (a) A transmission line interconnecting two CMOS gates. (b) The PCB dimensions. (c) The voltage at the line output (the input to the load CMOS gate) showing the ringing that may cause logic errors.

impedance is 25 Ω, representing a typical output resistance of a CMOS inverter. The load is represented as a 5-pF capacitance, simulating the input to a CMOS inverter. We will simulate this using SPICE to determine the voltage at the input to the line, $V(0,t)$, and the output voltage of the line, $V(\mathcal{L},t)$. The nodes are labeled on the diagram in

Fig. 6.40a. The SPICE (PSPICE) code is

```
EXAMPLE
VS 1 0 PWL(0 0 2N 2.5 20N 2.5 22N 0 40N 0)
RS 1 2 25
T 2 0 3 0 Z0=53.4 TD=1.255N
CL 3 0 5P
.TRAN 0.04N 40N 0 0.04N
.PROBE
.END
```

The plotted voltage at the output of the line (the input to the second gate) shown in Fig. 6.40c clearly shows the ringing due to the mismatch of the line. Experimental results for this problem were obtained. A photograph of the actual board is shown in Fig. 6.41a. The BNC connectors at each end allowed connection of coaxial cables to the pulse generator and the oscilloscope. We will show the measured and predicted voltages at the input to the line, $V(0,t)$. Leaving the output BNC connector open circuited simulates the input to a CMOS inverter since the BNC connector has a capacitance of about 5 pF. Figure 6.41b shows the SPICE prediction of the input voltage to the line, $V(0,t)$, and Fig. 6.41c shows the experimental result. The SPICE predictions match the experimental results very well.

Suppose a pair of lands on a PCB are mismatched at the load and/or the source. The ringing caused by the mismatch can cause errors, so what can the designer do to

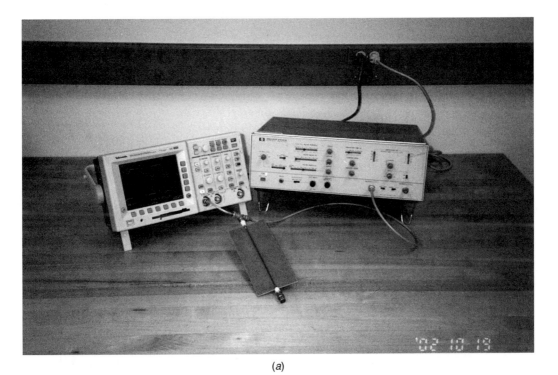

(a)

Figure 6.41 An experiment verifying the accuracy of the transmission-line model for digital circuits. (a) A photograph of the experiment. (b) The SPICE prediction of the voltage at the input to the line. (c) The measured voltage at the input to the line.

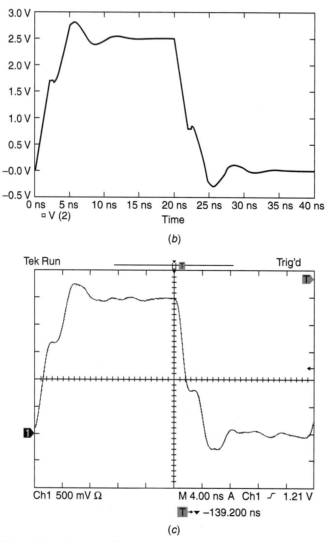

Figure 6.41 (*Continued*)

remedy this? There are two common methods that are used to match a line: series matching and parallel matching, shown in Fig. 6.42. The *series matching* shown in Fig. 6.42a places a resistor R in series with the output of the source (and at the input to the line). The value of this resistor is chosen so that $R_S + R = Z_C$ and the line is matched at the source. The line is not matched at the load. Initially a voltage wave is sent down the line whose value is one-half the source voltage because of the voltage division of the characteristic impedance of the line and the net source resistance, $R_S + R = Z_C$. For example, for a 5-V logic pulse, a 2.5-V pulse is sent down the line. If the load is an open circuit or other high impedance, the load reflection coefficient is unity, so that the incoming wave is completely reflected. The sum of this incident wave and the reflected wave then gives the total voltage at the load equal to twice the pulse sent down the line. Hence the load voltage is 5 V as desired (after a one-way delay incurred by the line). The *parallel match* shown in Fig. 6.42b seeks to match the line at

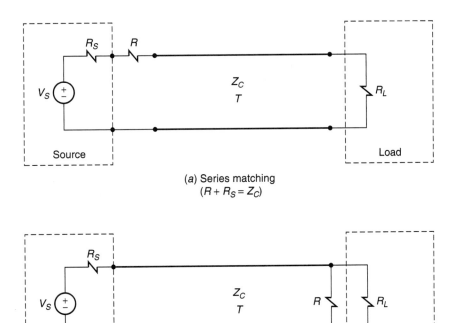

(a) Series matching
$(R + R_S = Z_C)$

(b) Parallel matching
$(R \| R_L = Z_C)$

Figure 6.42 Various schemes to match transmission lines. (a) The series match. (b) The parallel match.

the load by placing a resistance in parallel with the load such that their parallel combination is the characteristic impedance: $R \| R_L = Z_C$. In the case of a high-impedance load, the value of R would simply be chosen to be the characteristic impedance. Unlike the series match, the parallel match is matched at the load, so there will be no reflection at the load. However, the parallel match may suffer from an important deficiency. The initally sent out wave will have a value of

$$V_{initial} = \frac{Z_C}{Z_C + R_S} V_S$$

Because there is no reflection at the load for the parallel match, this is the total voltage at the load. Unlike the series match, there is no reflection to bring the load voltage up to its desired value. For example, if $R_S = 20\ \Omega$, $Z_C = 50\ \Omega$, and the source voltage level is 5 V, the initially sent out wave has a value of 3.57 V, which is also the load voltage. Hence logic errors may occur for the parallel match.

As an example of the effectiveness of these matching schemes, consider two CMOS inverters connected by a 50-Ω line that has a one-way delay of 0.2 ns as shown in Fig. 6.43a. The input of the CMOS inverter at the load is represented by a capacitance whose value is 5 pF. The output of the CMOS inverter at the input to the line is represented by a voltage source whose value transitions between 0 V and 5 V, and its source resistance is 20 Ω. This line is highly mismatched at both the source and the load. For a source voltage that has a magnitude of 5 V, a frequency of 100 MHz, a 50% duty cycle,

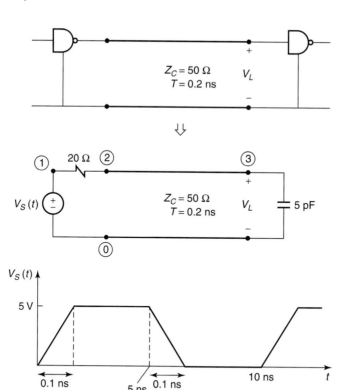

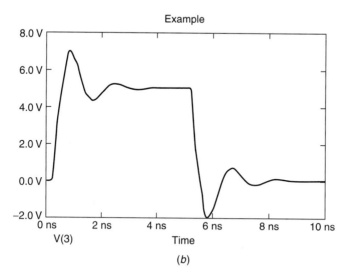

(b)

Figure 6.43 Using SPICE to investigate the effectiveness of matching schemes. (a) The problem specification. (b) SPICE predictions of the load voltage.

and rise/fall times of 0.1 ns, the PSPICE program is

```
EXAMPLE
VS  1  0  PWL(0  0  0.1N  5  5N  5  5.1N  0  10N  0)
RS  1  2  20
T  2  0  3  0  Z0=50  TD=0.2N
CL  3  0  5P
```

```
.TRAN 0.01N 10N 0 0.01N

.PROBE

.END
```

Figure 6.43b shows the PSPICE simulation of the load voltage showing the characteristic ringing caused by the mismatches.

Figure 6.44a shows the series match where a 30-Ω resistor is placed in series with the output of the source. The PSPICE program is

```
EXAMPLE

VS 1 0 PWL(0 0 0.1N 5 5N 5 5.1N 0 10N 0)

RS 1 2 20

R 2 3 30

T 3 0 4 0 Z0=50 TD=0.2N

CL 4 0 5P

.TRAN 0.01N 10N 0 0.01N

.PROBE

.END
```

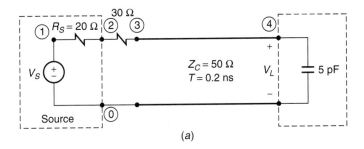

(a)

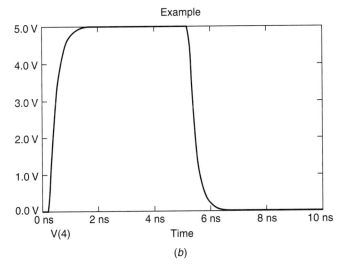

(b)

Figure 6.44 Applying the series match scheme to the problem of Fig. 6.43. (a) Inserting a resistor in series with the line input to match. (b) SPICE predictions of the load voltage showing that the ringing is eliminated.

The load voltage is shown in Fig. 6.44b, showing that the load voltage smoothly rises to the desired 5-V level.

Figure 6.45a shows the parallel match where a 50-Ω resistor is placed in parallel with the load. The PSPICE program is

```
EXAMPLE
VS 1 0 PWL(0 0 0.1N 5 5N 5 5.1N 0 10N 0)
RS 1 2 20
T 2 0 3 0 Z0=50 TD=0.2N
CL 3 0 5P
R 3 0 50
.TRAN 0.01N 10N 0 0.01N
.PROBE
.END
```

The load voltage is shown in Fig. 6.45b.

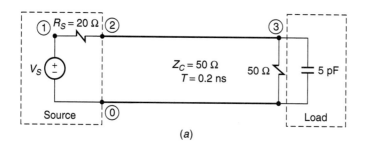

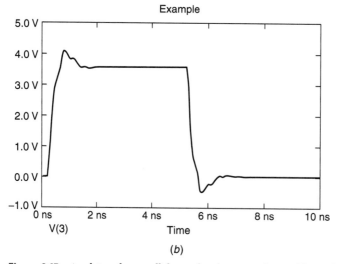

Figure 6.45 Applying the parallel match scheme to the problem of Fig. 6.43. (a) Inserting a resistor in parallel with the load to match. (b) SPICE predictions of the load voltage showing that the ringing is not eliminated.

Observe two important points. First, for the parallel match, the line is not completely matched at the load. At the lower-frequency components of the source waveform, the 5-pF capacitance has a large impedance at the lower frequencies of the source, so that 50 Ω in parallel with the capacitor is approximately 50 Ω. At the higher, frequency components of the source waveform, the impedance of the capacitor is much smaller, so that the parallel combination of 50 Ω and the 5-pF capacitor is dominated by the impedance of the capacitor. Hence at these higher, frequency components, the line is not matched. According to (6.86), the maximum spectral content of the pulse is $f_{max} = 1/\tau_r = 10$ GHz. The 50-Ω resistor in parallel with the 5-pF capacitor is approximately 50 Ω below a frequency of $f = 1/(2\pi RC) = 637$ MHz. Hence there are significant frequency components of the source waveform where the parallel match does not match the line. Another important point to observe is that the steady-state level is

$$V_{\text{steady state}} = \frac{R}{R + R_S} V_S$$

$$= \frac{50}{50 + 20} 5$$

$$= 3.57 \quad V$$

Thus, for the parallel match, the load voltage falls significantly below the desired 5-V level, and logic errors may occur.

6.7.2 Construction of Microwave Circuit Components Using Transmission Lines

Transmission lines that are electrically long display very interesting behavior that cannot be produced with lumped-circuit elements. For example, consider a transmission line that is one-quarter wavelength long ($\mathcal{L} = \lambda/4$) at the excitation frequency and is terminated in a short circuit as shown in Fig. 6.46a. The load reflection

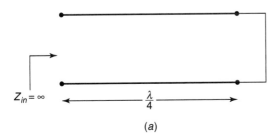

(a)

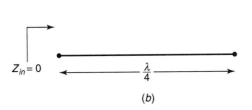

(b)

Figure 6.46 Effect of a quarter-wavelength line. (a) The input to a quarter-wavelength line that has a short-circuit load appears to be an open circuit. (b) The input to a quarter-wavelength line that has an open-circuit load appears to be a short circuit.

coefficient is

$$\hat{\Gamma}_L = \frac{0 - Z_C}{0 + Z_C}$$
$$= -1$$

The reflection coefficient at the line input is obtained from (6.46). The phase term in the reflection coefficient of $e^{-j2\beta\mathscr{L}}$ becomes, using $\beta = 2\pi/\lambda$,

$$e^{-j2\beta\mathscr{L}} = e^{-j4\pi\frac{\mathscr{L}}{\lambda}}$$

For a line length of one-quarter wavelength the phase term becomes

$$e^{-j2\beta\mathscr{L}} = e^{-j\pi}$$
$$= -1 \qquad \mathscr{L} = \frac{1}{4}\lambda$$

Thus the reflection coefficient at the input to the line becomes

$$\hat{\Gamma}_{in} = \hat{\Gamma}_L e^{-j2\beta\mathscr{L}}$$
$$= (-1)(-1)$$
$$= 1$$

Hence the input impedance becomes

$$\hat{Z}_{in} = Z_C \frac{1 + 1}{1 - 1}$$

$$= \infty \qquad \begin{cases} \hat{Z}_L = \text{short circuit} \\ \mathscr{L} = \dfrac{\lambda}{4} \end{cases}$$

Therefore, *a quarter-wavelength line with a short-circuit load looks like an open circuit at its input terminals.*

Conversely, suppose the quarter-wavelength line is terminated in an open circuit as shown in Fig. 6.46b. The load reflection coefficient in this case is $\hat{\Gamma}_L = 1$ and the reflection coefficient at the input is

$$\hat{\Gamma}_{in} = \hat{\Gamma}_L e^{-j2\beta\mathscr{L}}$$
$$= (+1)(-1)$$
$$= -1$$

Hence the input impedance is

$$\hat{Z}_{in} = Z_C \frac{1 - 1}{1 + 1}$$

$$= 0 \qquad \begin{cases} \hat{Z}_L = \text{open circuit} \\ \mathscr{L} = \dfrac{\lambda}{4} \end{cases}$$

Therefore, *a quarter-wavelength line with an open-circuit load looks like a short circuit at its input terminals.*

Transmission lines can be used to construct circuit elements such as capacitors and inductors that will work reliability at microwave frequencies (>1 GHz). Why can't we simply use lumped-element capacitors and inductors at microwave frequencies? The answer was given in Section 3.13.4 of Chapter 3. The attachment leads have inductance and capacitance (a parallel pair of wires constitute the attachment leads of lumped elements and are hence transmission lines). This lead inductance and lead capacitance will resonate with the lumped element to give undesired behavior above this resonant frequency. See Fig. 3.53 of Chapter 3. The attachment leads for a capacitor cause the capacitor to behave like an inductor above the resonant frequency $f = 1/(2\pi \sqrt{L_{lead}C})$. For example, a 1000-pF capacitor that has parallel-wire attachment leads that are $\frac{1}{2}$ in. long and separated by $\frac{1}{4}$ in. ($L_{lead} \cong 14$ nH) will have a resonant frequency of 43 MHz. Above this resonant frequency, the impedance of the capacitor (as seen looking into the terminals of the attachment leads) increases with increasing frequency and hence behaves like an inductor.

At microwave frequencies, it is difficult to construct lumped capacitors and inductors as well as many other lumped-circuit elements because of this attachment lead problem. We can use short-circuited (or open-circuited) lengths of transmission lines to simulate a lumped-circuit element such as a capacitor. For example, the input impedance to a line having a short-circuit load ($\hat{\Gamma}_L = -1$) is obtained from (6.47):

$$\hat{Z}_{in} = Z_C \frac{[1 - e^{-j2\beta\mathscr{L}}]}{[1 + e^{-j2\beta\mathscr{L}}]} \qquad \begin{cases} \text{short-circuit load} \\ \hat{Z}_L = 0 \end{cases}$$

But this can be written as

$$\begin{aligned}
\hat{Z}_{in} &= Z_C \frac{[1 - e^{-j2\beta\mathscr{L}}]}{[1 + e^{-j2\beta\mathscr{L}}]} \\
&= Z_C \frac{e^{-j\beta\mathscr{L}}[e^{j\beta\mathscr{L}} - e^{-j\beta\mathscr{L}}]}{e^{-j\beta\mathscr{L}}[e^{j\beta\mathscr{L}} + e^{-j\beta\mathscr{L}}]} \\
&= jZ_C \tan\left(2\pi\frac{\mathscr{L}}{\lambda}\right) \qquad \begin{cases} \text{short-circuit load} \\ \hat{Z}_L = 0 \end{cases}
\end{aligned} \tag{6.88}$$

This impedance is plotted, as a function of wavelength, in Fig. 6.47a. Observe that for line lengths up to $\lambda/4$, the input impedance is positive, indicating an inductive reactance, whereas, for line lengths between $\lambda/4$ and $\lambda/2$, the impedance is negative, indicating a capacitive reactance.

For example, suppose we wish to construct a 10-pF capacitor at a frequency of 1 GHz using a short-circuited length of transmission line. The impedance of a 10-pF capacitor at 1 GHz is $-j(1/\omega C) = -j15.92 \, \Omega$. Setting this equal to (6.88) we solve for

$$\tan\left(2\pi\frac{\mathscr{L}}{\lambda}\right) = \frac{-15.92}{Z_C}$$

If we use a length of 50-Ω line terminated in a short circuit, we obtain from this that $\mathscr{L}/\lambda = -0.049$ or $\mathscr{L}/\lambda = -0.049 + 0.5 = 0.451$. If the line is air filled ($\lambda|_{1 \text{ GHz}} = 30$ cm), then $\mathscr{L} = 13.53$ cm.

It should be kept in mind that although we can construct inductors or capacitors with short-circuited transmission lines, they are valid only at one frequency. Hence these elements are very much narrowband elements.

6.7.3 Antenna Feed Lines

Antennas have, as we will see in the next chapter, an input impedance, $\hat{Z}_{ant}$. The antenna must inevitably be connected to the source with a connection line (a transmission line)

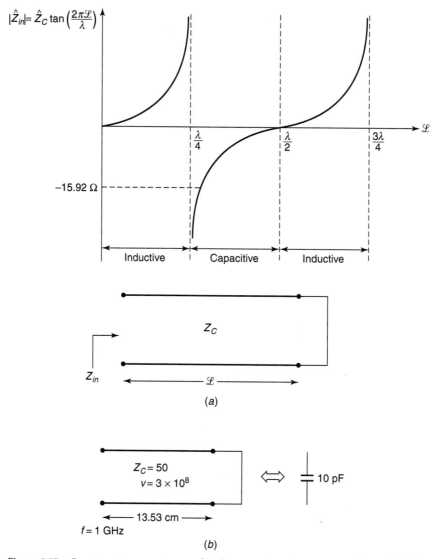

Figure 6.47 Constructing capacitors and inductors using short-circuited lines. (a) The input impedance to a short-circuited line whose length is given in wavelengths. (b) Constructing a 10-pF capacitor for use at 1 GHz using a short-circuited transmission line.

as shown in Fig. 6.48a. If the input impedance to the antenna is different from the characteristic impedance of the connection line, power will be reflected back to the source and not radiated as we saw in Section 6.3.7. How do we provide a match between the transmission line and the antenna to prevent this? There are many ways of doing this. One way is to use a *quarter-wave transformer* as shown in Fig. 6.48b. A quarter-wavelength section of a line having characteristic impedance Z'_C has an input impedance of

$$\hat{Z}'_{in} = Z'_C \frac{[1 - \hat{\Gamma}'_L]}{[1 + \hat{\Gamma}'_L]}$$

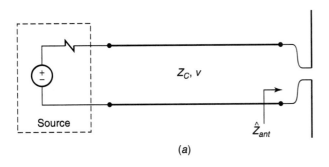

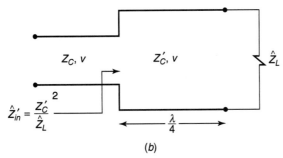

Figure 6.48 Matching antenna feed lines with the quarter-wave transformer. (a) Illustration of the effect of the feed line. (b) The quarter-wave transformer.

and we have used the result for a quarter-wavelength line derived earlier:

$$
\begin{aligned}
e^{-j2\beta\mathscr{L}} &= e^{-j4\pi\frac{\mathscr{L}}{\lambda}} \\
&= e^{-j\pi} \\
&= -1
\end{aligned}
$$

If we want this to equal the characteristic impedance of the line to which it is attached, $\hat{Z}'_{in} = Z_C$, we must have

$$
\begin{aligned}
\hat{Z}'_{in} &= Z_C \\
&= Z'_C \underbrace{\frac{[1 - \hat{\Gamma}'_L]}{[1 + \hat{\Gamma}'_L]}}_{\dfrac{Z'_C}{\hat{Z}_L}}
\end{aligned}
$$

giving

$$
Z'_C = \sqrt{Z_C \hat{Z}_L} \tag{6.89}
$$

For example, in order to match a 300-Ω load to a 75-Ω line we need to insert a quarter-wavelength line between them having a characteristic impedance of 150 Ω. Although the VSWR on the matching section is not unity, it is a lossless line. Hence all the power delivered by the source is delivered to the antenna input.

This is also a narrowband device. If the excitation frequency is changed, the length of the matching section is no longer $\frac{\lambda}{4}$.

6.7.4 Crosstalk between Transmission Lines

Crosstalk occurs when a signal (current or voltage) on one pair of conductors couples to an adjacent pair of conductors, causing an (unintended) reception of that signal at the terminals of the second pair of conductors. This is illustrated in Fig. 6.49a. A pair of parallel conductors called the generator circuit connects a source represented by a source voltage, $V_S(t)$, and its source impedance, R_S, to a load represented by R_L. Another pair of parallel conductors may be parallel and adjacent to the generator line.

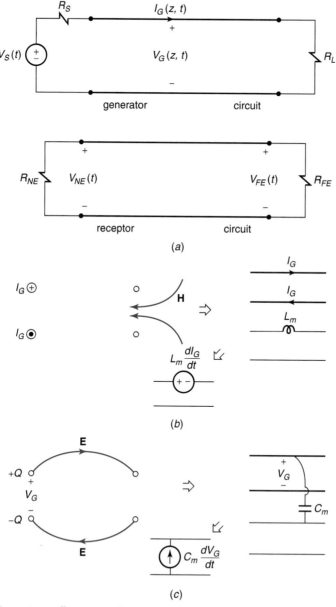

Figure 6.49 Illustration of crosstalk between transmission lines. (a) The problem specification. (b) Inductive coupling via the magnetic field and mutual inductance. (c) Capacitive coupling via the electric field and mutual capacitance.

These conductors connect two devices represented by R_{NE} and R_{FE} and are called the receptor circuit. The subscripts NE and FE represent "near end" and "far end," respectively, and are common in industry. We wish to determine the near-end and far-end coupled crosstalk voltages, $V_{NE}(t)$ and $V_{FE}(t)$. The current on the generator line, $I_G(z,t)$, produces a magnetic field that threads the loop between the two receptor conductors as shown in Fig. 6.49b. This creates a mutual inductance, L_m, between the two circuits. Similarly, the voltage between the two conductors of the generator circuit, $V_G(z,t)$, causes electric field lines, some of which terminate on the conductors of the receptor circuit as shown in Fig. 6.49c. This results in a mutual capacitance, C_m, between the two circuits. Hence the receptor circuit can be represented by two sources as shown in Fig. 6.50a. The terminal voltages of the receptor circuit are obtained from that circuit using superposition as

$$V_{NE}(t) = \underbrace{\frac{R_{NE}}{R_{NE} + R_{FE}} L_m \frac{dI_G}{dt}}_{\text{Inductive Coupling}} + \underbrace{\frac{R_{NE}R_{FE}}{R_{NE} + R_{FE}} C_m \frac{dV_G}{dt}}_{\text{Capacitive Coupling}}$$

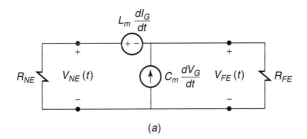

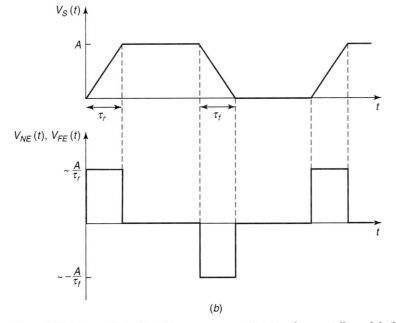

Figure 6.50 Crosstalk in digital interconnects. (a) A simple crosstalk model. (b) The time-domain crosstalk from a digital clock signal.

and

$$V_{FE}(t) = -\underbrace{\frac{R_{FE}}{R_{NE} + R_{FE}} L_m \frac{dI_G}{dt}}_{\text{inductive coupling}} + \underbrace{\frac{R_{NE}R_{FE}}{R_{NE} + R_{FE}} C_m \frac{dV_G}{dt}}_{\text{capacitive coupling}}$$

If the line is electrically short at the highest significant frequency component of $V_S(t)$, then the generator line voltage and current are essentially constant along the line (the line doesn't matter) and can be computed approximately from

$$V_G(t) \cong \frac{R_L}{R_S + R_L} V_S(t)$$

and

$$I_G(t) \cong \frac{1}{R_S + R_L} V_S(t)$$

Substituting these into the previous result gives

$$V_{NE}(t) = \left[\underbrace{\frac{R_{NE}}{R_{NE} + R_{FE}} L_m \frac{1}{R_S + R_L}}_{\text{inductive coupling}} + \underbrace{\frac{R_{NE}R_{FE}}{R_{NE} + R_{FE}} C_m \frac{R_L}{R_S + R_L}}_{\text{capacitive coupling}} \right] \frac{dV_S(t)}{dt} \quad (6.90a)$$

and

$$V_{FE}(t) = \left[-\underbrace{\frac{R_{FE}}{R_{NE} + R_{FE}} L_m \frac{1}{R_S + R_L}}_{\text{inductive coupling}} + \underbrace{\frac{R_{NE}R_{FE}}{R_{NE} + R_{FE}} C_m \frac{R_L}{R_S + R_L}}_{\text{capacitive coupling}} \right] \frac{dV_S(t)}{dt} \quad (6.90b)$$

This indicates that the crosstalk varies as the derivative of the source voltage, $dV_S(t)/dt$ (referred to as the "slew rate" of the voltage). Figure 6.50b shows the crosstalk when the generator circuit is carrying a digital clock signal. The crosstalk voltages, $V_{NE}(t)$ and $V_{FE}(t)$, are pulses occurring during the transition (rise/fall times) of the source voltage. Their amplitudes are proportional to the instantaneous slope of $V_S(t)$, which is the ratio of the amplitude to the rise/fall time. Hence the magnitude of the crosstalk is not only proportional to the mutual inductance and capacitance between the two circuits but is also proportional to the rise/fall times of the source. This indicates that one way of reducing crosstalk in digital circuits is to slow (make longer) the pulse rise/fall times. We saw previously that slowing (increasing) the pulse rise/fall times also reduces the high-frequency spectral content. So this makes sense because, in general, the higher the frequency of a signal, the more easily it will couple to adjacent circuits producing crosstalk.

For example, consider a coupled microstrip line shown in Fig. 6.51a. An FR4 board of thickness 62 mils supports two lands whose widths are 100 mils and are separated edge to edge by 100 mils. A ground plane beneath the board serves as the reference conductor for the two circuits. The total length of the line is 20 cm. The source has an open-circuit voltage that is a 5-V, 1-MHz trapezoidal pulse train having 50-ns rise/fall times and a 50% duty cycle, and the source resistance of 50 Ω, $R_S = 50$ Ω, as shown in Fig. 6.51b. This drives the generator circuit, which is terminated in a 50-Ω load,

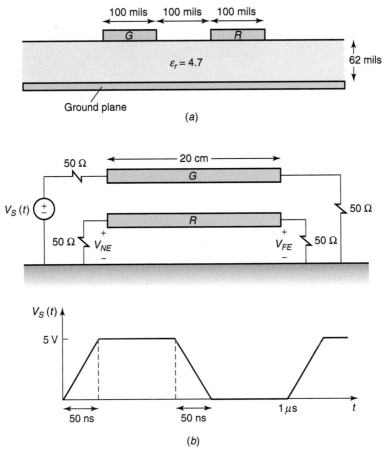

Figure 6.51 A typical crosstalk problem in a digital circuit. (a) A cross section of the PCB. (b) The problem specification.

$R_L = 50\ \Omega$. The receptor circuit is terminated in 50 Ω at both ends, $R_{NE} = R_{FE} = 50\ \Omega$. The total mutual inductance between the two circuits is $L_m = 7.44$ nH, and the total mutual capacitance between the two circuits is $C_m = 1$ pF. Substituting into (6.90a) gives the level of the near-end crosstalk pulse as

$$V_{NE} = \left[\frac{1}{2}L_m\frac{1}{100} + 25C_m\frac{1}{2}\right]\frac{5}{50 \times 10^{-9}}$$

$$= 4.96\ \text{mV}$$

Figure 6.52a shows a photograph of the actual board and the measurement setup for V_{NE}. Figure 6.52b shows the experimentally measured near-end crosstalk, V_{NE}. The source voltage waveform is also shown. Observe the similarity of the crosstalk waveform to Fig. 6.50b. The crosstalk occurs during the transition of the source waveform (during the rise and fall times). Figure 6.52c shows an expanded view of the crosstalk during the leading edge of the source pulse, and Fig. 6.52d shows an expanded view of the crosstalk pulse during the falling edge of the source pulse. The measured peak of the crosstalk pulses is 5 mV, which compares well with the prediction. In order for this

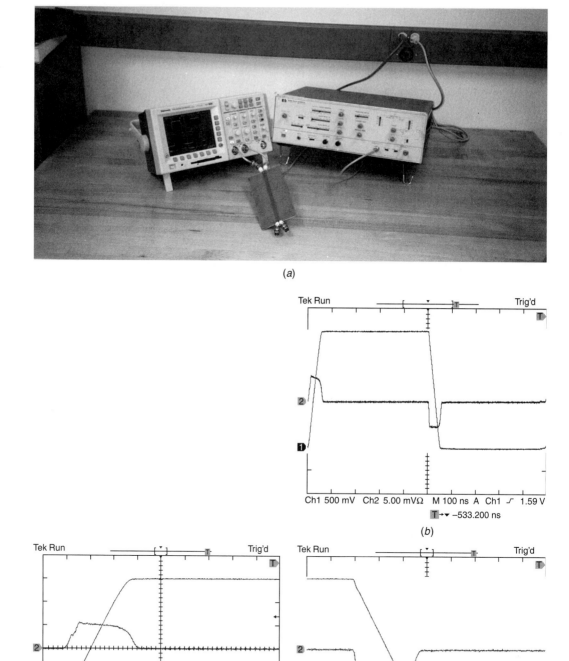

Figure 6.52 Experimental verification of the crosstalk problem of Fig. 6.51. (a) Photograph of the experiment. (b), (c), and (d) Oscilloscope photographs of the near-end crosstalk voltage.

simple model to be valid, the line length must be electrically short. According to (6.86), the highest significant frequency of the pulse is

$$
\begin{aligned}
f_{\max} &\cong \frac{1}{\tau_r} \\
&= \frac{1}{50 \times 10^{-9}} \\
&= 20 \text{ MHz}
\end{aligned}
$$

At this frequency the line length of 20 cm is $(1/45)\lambda$, where we have used the effective relative permittivity as the average of the board and air, $\varepsilon_r' = 2.85$, in computing the velocity of propagation. Hence the line is electrically short at the significant frequencies of the pulse, and this simple crosstalk model is applicable. Additional experimental results are compared to this simple model in C.R. Paul, *Introduction to Electromagnetic Compatibility*, John Wiley Interscience, 1992.

6.7.5 Use of Shielded Wires and Twisted Wire Pairs to Reduce Crosstalk

In the previous section we discussed the problem of crosstalk wherein a voltage and a current on one pair of conductors, the generator circuit, will cause electric and magnetic fields to interact with the conductors of a neighboring circuit, the receptor circuit, and induce noise or interference voltages at the terminations of the receptor circuit, V_{NE} and V_{FE}. (See Fig. 6.49 and Fig. 6.50.) Suppose those induced voltages are of sufficient magnitude and/or spectral content to cause interference with the devices represented by the terminations of the receptor circuit, R_{NE} and R_{FE}. There are several ways to reduce that crosstalk. One of the more obvious ways is to move the generator conductors and the receptor conductors farther apart to reduce the coupled electric and magnetic fields. Another way would be to route the conductors of the generator circuit perpendicular to the conductors of the receptor circuit. Finally, if we slow (increase) the pulse rise/fall times of the source voltage, this will reduce the high-frequency spectral content of it and hence reduce the crosstalk. When these measures are not feasible or have been used to the maximum extent possible and significant crosstalk still occurs, there are two remaining options that can be successful in reducing crosstalk: replacing the receptor conductors or the generator conductors with either shielded wires and/or twisted pairs of wires. The crosstalk is given in Equation (6.90) for electrically short lines and is the sum of (superposition of) two components: a capacitive coupling contribution due to the mutual capacitance between the two circuits, C_m, and an inductive coupling contribution due to the mutual inductance between the two circuits, L_m. Replacing the receptor wires with a shielded wire essentially reduces the crosstalk by reducing the capacitive coupling component. Replacing the receptor wires with a twisted wire pair essentially reduces the inductive coupling component. Using a shielded, twisted pair reduces both components.

First, we consider using shielded wires to reduce the crosstalk. Consider two wires above an infinite ground plane (the reference conductor for the voltages). One wire with the ground plane constitutes the generator circuit and the other wire with the ground plane constitutes the receptor circuit. Suppose we surround the receptor wire with a cylindrical shield, that is, replace it with a shielded wire as shown in Fig. 6.53a. The effect is to introduce capacitances between each pair as shown in Fig. 6.53b. Observe that some of the capacitances are absent. For example, the mutual capacitance between the generator wire and the receptor wire (which is now surrounded by the

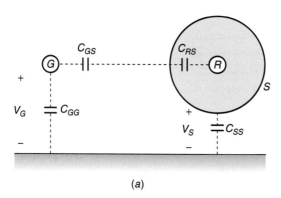

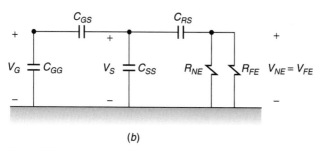

Figure 6.53 Use of a shielded wire to reduce crosstalk. (a) Illustration of the mutual capacitances. (b) The equivalent circuit.

shield) is zero, $C_{GR} = 0$. This is because the shield around the receptor wire essentially acts as a Faraday shield, discussed in Section 3.13.5 of Chapter 3. Similarly, the self capacitance between the receptor wire and the ground plane is also zero, $C_{RR} = 0$, for the same reason. Observe from this capacitive equivalent circuit that if the shield is "grounded" (connected to the ground plane, thereby rendering the shield voltage zero, $V_S = 0$), then the generator voltage can have no effect on the receptor voltages, thereby eliminating the capacitive coupling. However, this has no effect on the magnetic field or inductive coupling.

The magnetic field or inductive coupling can be reduced, essentially eliminated, with the use of a twisted wire pair. A pair of twisted wires is a bifilar helix, but we can approximate its effect by representing it as a set of abruptly alternating loops changing position by 180° with every half twist as shown in Fig. 6.54a. The magnetic field from the generator circuit will thread these loops and induce Faraday law voltage sources in each loop whose values are the time rate of change of the magnetic flux threading each loop. The polarity of these sources was determined in Chapter 4 so as to induce a current that would produce a secondary magnetic field that would tend to oppose the original magnetic field as shown in Fig. 6.54b. The key to understanding how this eliminates magnetic field or inductive coupling is to "untwist" the wires as shown in Fig. 6.54c. Observe that the Faraday law induced sources in adjacent loops have opposing polarities. Hence the inductive coupling to each adjacent half-twist is canceled out.

The best of both worlds is achieved using a shielded, twisted pair of wires. The shield eliminates (ideally) the capacitive coupling if it is grounded to at least one point, rendering the shield voltage zero, whereas the twisted pair internal to the shield eliminates (ideally) the inductive coupling.

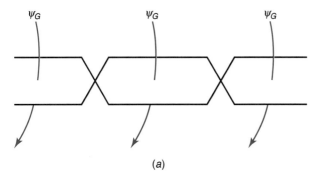

(a)

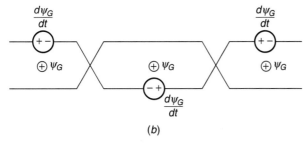

(b)

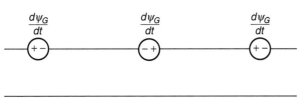

(c)

Figure 6.54 Use of twisted pairs of wires to reduce crosstalk. (a) Coupling of magnetic flux through the loops of the twisted pair. (b) Insertion of Faraday's law sources in each loop. (c) "Untwisting" the wires to show the cancellation of adjacent induced voltage sources.

▶ SUMMARY OF IMPORTANT CONCEPTS AND FORMULAE

1. **Waves on parallel-conductor transmission lines:** A pair of parallel conductors serves to guide voltage and current waves from one end to the other. The waves are traveling in the medium surrounding the conductors as electric and magnetic fields which constitute plane waves. The line can be visualized as a distributed-parameter circuit where per-unit-length inductances and capacitances form cells of infinitesimal length distributed along the line. A finite time is required to charge and discharge these elements as the wave propagates along the line, resulting in a time delay of propagation on the line of $T = \mathscr{L}/v$.

2. **The transmission-line equations:** The equations governing the voltage and current on the line are

$$\frac{\partial V(z,t)}{\partial z} = -l\frac{\partial I(z,t)}{\partial t}$$

$$\frac{\partial I(z,t)}{\partial z} = -c\frac{\partial V(z,t)}{\partial t}$$

whose solution is

$$V(z,t) = \underbrace{V^+\left(t - \frac{z}{v}\right)}_{\substack{\text{forward-} \\ \text{traveling } (+z) \text{ wave}}} + \underbrace{V^-\left(t + \frac{z}{v}\right)}_{\substack{\text{backward-} \\ \text{traveling } (-z) \text{ wave}}}$$

$$I(z,t) = \underbrace{\frac{V^+\left(t - \frac{z}{v}\right)}{Z_C}}_{\substack{\text{forward-} \\ \text{traveling } (+z) \text{ wave}}} - \underbrace{\frac{V^-\left(t + \frac{z}{v}\right)}{Z_C}}_{\substack{\text{backward-} \\ \text{traveling } (-z) \text{ wave}}}$$

where the characteristic impedance is $Z_C = \sqrt{l/c}$, and the velocity of propagation is $v = 1/\sqrt{lc}$. The time required for a wave to propagate from one end of the line to the other is $T = \mathcal{L}/v$ and $\mathcal{L}$ is the total line length.

3. **Reflection coefficients:** source end: $\Gamma_S = (R_S - Z_C)/(R_S + Z_C)$, load end $\Gamma_L = (R_L - Z_C)/(R_L + Z_C)$.

4. **The transmission-line equations for sinusoidal excitation:**

$$\frac{d\hat{V}(z)}{dz} = -j\omega l \, \hat{I}(z)$$

$$\frac{d\hat{I}(z)}{dz} = -j\omega c \, \hat{V}(z)$$

whose frequency-domain solution is

$$\hat{V}(z) = \hat{V}^+ e^{-j\beta z} + \hat{V}^- e^{j\beta z}$$

$$\hat{I}(z) = \frac{\hat{V}^+}{Z_C} e^{-j\beta z} - \frac{\hat{V}^-}{Z_C} e^{j\beta z}$$

where the characteristic impedance is, as before, $Z_C = \sqrt{l/c}$ and the phase constant is

$$\beta = \omega\sqrt{lc}$$

$$= \frac{\omega}{v} \quad \text{radians/m}$$

The corresponding time-domain solution is

$$V(z,t) = V^+\cos(\omega t - \beta z + \theta^+) + V^-\cos(\omega t + \beta z + \theta^-)$$

$$I(z,t) = \frac{V^+}{Z_C}\cos(\omega t - \beta z + \theta^+) - \frac{V^-}{Z_C}\cos(\omega t + \beta z + \theta^-)$$

5. **Reflection coefficient and input impedance:**

$$\hat{\Gamma}_L = \frac{\hat{Z}_L - Z_C}{\hat{Z}_L + Z_C}$$

$$\hat{\Gamma}(z) = \hat{\Gamma}_L e^{j2\beta(z - \mathcal{L})}$$

$$\hat{V}(z) = \hat{V}^+ e^{-j\beta z}[1 + \hat{\Gamma}_L e^{j2\beta(z - \mathcal{L})}]$$

$$\hat{I}(z) = \frac{\hat{V}^+}{Z_C} e^{-j\beta z}[1 - \hat{\Gamma}_L e^{j2\beta(z - \mathcal{L})}]$$

$$\hat{Z}_{in} = Z_C \frac{[1 + \hat{\Gamma}(0)]}{[1 - \hat{\Gamma}(0)]}$$

$$= Z_C \frac{[1 + \hat{\Gamma}_L e^{-j2\beta\mathcal{L}}]}{[1 - \hat{\Gamma}_L e^{-j2\beta\mathcal{L}}]}$$

6. **Properties of the voltage and current on the line:** Corresponding points on the magnitude of the line voltage (current) are separated by one-half wavelength in distance. The input impedance to the line replicates for multiples of a half wavelength. A maximum and the adjacent minimum are separated by one-quarter wavelength.

7. **Voltage standing wave ratio:** $\text{VSWR} = \dfrac{1 + |\hat{\Gamma}_L|}{1 - |\hat{\Gamma}_L|}$.

8. **Power flow on the line:** $P_{\text{AV}} = \dfrac{|\hat{V}^+|^2}{2Z_C}[1 - |\hat{\Gamma}_L|^2]$.

9. **Creating approximate lumped-circuit models of transmission lines:** If the line is electrically short, $\mathcal{L} < (1/10)\lambda$, a simple lumped-circuit model is accurate where $l = Z_C/v$ H/m and $c = 1/vZ_C$ F/m and the total inductance and capacitance are $L = l\mathcal{L}$ H and $C = c\mathcal{L}$ F.

10. **Digital clock signals:** have sinusoidal components at multiples of the repetition frequency. High-frequency content is determined by the rise/fall times of the pulse. The highest-frequency component of significance is approximately $f_{\text{max}} \cong 1/\tau_r$ where τ_r is the rise/fall time of the pulse.

► PROBLEMS

SECTION 6.1 THE TRANSMISSION-LINE EQUATIONS

6.1.1. The per-unit-length equivalent circuit for a Δz section of a transmission line shown in Fig. 6.4 is not the only configuration that will yield the transmission-line equations. Show that the three per-unit-length equivalent circuits in Fig. P6.1.1 also give the transmission-line equations

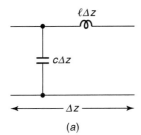

(a)

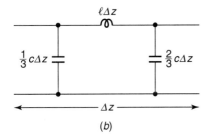

(b)

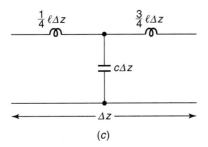

(c)

Figure P6.1.1 Problem 6.1.1.

in the limit as $\Delta z \to 0$. Observe in these circuits that the total capacitance is $c\Delta z$, and the total inductance is $l\Delta z$.

6.1.2. Two bare #20-gauge (radius = 16 mils) wires are separated center-to-center by 50 mils. Determine the exact and approximate values of the per-unit-length capacitance and inductance.

6.1.3. One bare #12-gauge (radius = 40 mils) wire is suspended at a height of 80 mils above its return path, which is a large ground plane. Determine the exact and approximate values of the per-unit-length capacitance and inductance. [Exact: 42.18 pF/m, 0.2634 μH/m, Approximate: 40.07 pF/m, 0.2773 μH/m. The ratio of wire height above ground to wire radius is only 2.0, which is not sufficient for the approximate results to be valid, although the error is only about 5%.]

6.1.4. A typical coaxial cable is RG-6U, which has an interior #18-gauge (radius 20.15 mils) solid wire, an interior shield radius of 90 mils, and an inner insulation of foamed polyethylene having a relative permittivity of 1.45. Determine the per-unit-length capacitance, inductance, and the velocity of propagation relative to that of free space.

6.1.5. Multilayer printed circuit boards (PCBs) consist of layers of board material that is glass-epoxy material (FR4) having a relative permittivity of 4.7 sandwiched between conducting planes. Conductors are buried midway between the conducting planes, which gives a structure resembling a stripline in Fig. 6.6a. Typical dimensions for multilayer PCBs are a plate separation of 10 mils and a conductor width of 5 mils. Determine the per-unit-length capacitance and inductance for this structure. [156.4 pF/m and 0.334 μH/m]

6.1.6. A microstrip line is constructed on an FR4 board having a relative permittivity of 4.7. The board thickness is 64 mils and the land width is 10 mils. Determine the per-unit-length capacitance and inductance as well as the effective relative permittivity.

6.1.7 A PCB shown in Fig. 6.6c has land widths of 5 mils and an edge-to-edge separation of 5 mils. The board is glass-epoxy having a relative permittivity of 4.7 and a thickness of 47 mils. Determine the per-unit-length capacitance and inductance as well as the effective relative permittivity. [0.8038 μH/m, 39.06 pF/m, and $\varepsilon_r' = 2.825$]

SECTION 6.2 TIME-DOMAIN EXCITATION OF TRANSMISSION LINES

6.2.1. Show by direct substitution that the equations in (6.13) do indeed satisfy the transmission-line equations.

6.2.2. Determine the characteristic impedance and velocity of propagation for the two-wire line in Problem 6.1.2. [122 Ω, 3 $\times$ 10^8 m/s]

6.2.3. Determine the characteristic impedance and velocity of propagation for the one-wire above a ground plane line in Problem 6.1.3.

6.2.4. Determine the characteristic impedance and velocity of propagation for the coaxial line in Problem 6.1.4. [75 Ω, 2.5 $\times$ 10^8 m/s]

6.2.5. Determine the characteristic impedance and velocity of propagation for the stripline in Problem 6.1.5.

6.2.6. Determine the characteristic impedance and velocity of propagation for the microstrip line in Problem 6.1.6. [135 Ω, 1.71 $\times$ 10^8 m/s]

6.2.7. Determine the characteristic impedance and velocity of propagation for the PCB line in Problem 6.1.7.

6.2.8. Show how the per-unit-length inductance and capacitance can be found from the characteristic impedance and velocity of propagation. [$l = Z_C/v, c = 1/vZ_C$]

6.2.9. Sketch the load voltage, $V(\mathcal{L},t)$, and the input current to the line, $I(0,t)$, for the problem depicted in Fig. P6.2.9 for $0 < t < 10$ ns. What should these plots converge to in the steady state? [$V(\mathcal{L},t),0 < t < 1$ ns, 0 V,1 ns $< t < 3$ ns, 9.375 V, 3 ns $< t < 5$ ns, 8.203 V, 5 ns $< t < 7$ ns, 8.35 V,7 ns $< t < 9$ ns, 8.331 V, 9 ns $< t < 11$ ns, 8.334 V, steady state 8.333 V, and $I(0,t),0 < t < 2$ ns, 0.125 A, 2 ns $< t < 4$ ns, 0.047 A, 4 ns $< t < 6$ ns, 0.057 A, 6 ns $< t < 8$ ns, 0.055 A, 8 ns $< t < 10$ ns, 0.056 A, steady state 0.056 A]

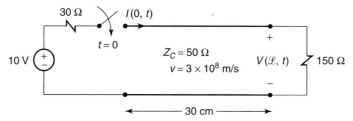

Figure P6.2.9 Problem 6.2.9.

6.2.10. Sketch the load voltage, $V(\mathcal{L},t)$, and the input voltage to the line, $V(0,t)$, for the problem depicted in Fig. P6.2.10 for $0 < t < 20$ ns. What should these plots converge to in the steady state?

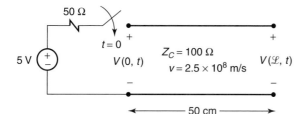

Figure P6.2.10 Problem 6.2.10.

6.2.11. Sketch the input voltage to the line, $V(0,t)$, and the load current, $I(\mathcal{L},t)$, for the problem depicted in Fig. P6.2.11 for $0 < t < 10$ μs. What should these plots converge to in the steady state? [$V(0,t),0 < t < 2$ μs, 66.67 V, 2 μs $< t < 4$ μs, 22.22 V, 4 μs $< t < 6$ μs, 7.407 V, 6 μs $< t < 8$ μs, 2.469 V, 8 μs $< t < 10$ μs, 0.823, steady state 0 V and $I(\mathcal{L},t), 0 < t < 1$ μs, 0 V, 1 μs $< t < 3$ μs, 1.33 A, 3 μs $< t < 5$ μs, 1.778 A, 5 μs $< t < 7$ μs, 1.926 A, 7 μs $< t < 9$ μs, 1.975 A, 9 μs $< t < 11$ μs, 1.992 A, steady state 2 A]

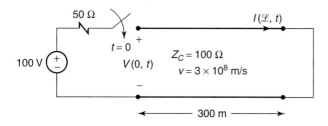

Figure P6.2.11 Problem 6.2.11.

6.2.12. Sketch the input voltage to the line, $V(0,t)$, and the load voltage, $V(\mathcal{L},t)$, for the problem depicted in Fig. P6.2.12 for $0 < t < 32$ ns. What should these plots converge to in the steady state?

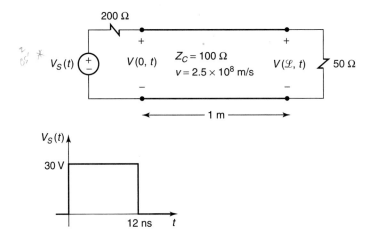

Figure P6.2.12 Problem 6.2.12.

6.2.13. A time-domain reflectometer (TDR) is an instrument used to determine properties of transmission lines. In particular, it can be used to detect the locations of imperfections such as breaks in the line. The instrument launches a pulse down the line and records the transit time for that pulse to be reflected at some discontinuity and to return to the line input. Suppose a TDR having a source impedance of 50 Ω is attached to a 50-Ω coaxial cable having some unknown length and load resistance. The dielectric of the cable is Teflon ($\varepsilon_r = 2.1$). The open-circuit voltage of the TDR is a pulse of duration 10 μs. If the recorded voltage at the input to the line is as shown in Fig. P6.2.13, determine (a) the length of the line and (b) the unknown load resistance.

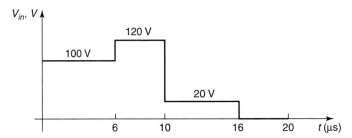

Figure P6.2.13 Problem 6.2.13.

6.2.14. A 12-V battery ($R_S = 0$) is attached to an unknown length of transmission line that is terminated in a resistance. If the input current to the line for 6 μs is as shown in Fig. P6.2.14, determine (a) the line characteristic impedance and (b) the unknown load resistance. [$Z_C = 80\ \Omega, R_L = 262.9\ \Omega$]

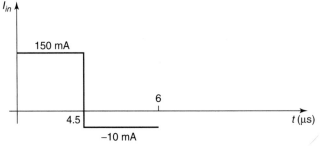

Figure P6.2.14 Problem 6.2.14.

6.2.15. Digital clock and data pulses should ideally consist of rectangular pulses. Actual clock and data pulses, however, resemble pulses having a trapezoidal shape with certain rise and fall times. Depending on the ratio of the rise/fall time to the one-way transit time of the transmission line, the received voltage may oscillate about the desired value, possibly causing a digital gate at that end to switch falsely to an undesired state and cause errors. Matching the line eliminates this problem because there are no reflections, but matching cannot always be accomplished. In order to investigate this problem, consider a line connecting two CMOS gates. The driver gate is assumed to have zero source resistance ($R_S = 0$), and the open-circuit voltage is a ramp waveform (simulating the leading edge of the clock/data pulse) given by $V_S(t) = 0$ for $t < 0$, $V_S(t) = 5(t/\tau)_r$ V for $0 \le t \le \tau_r$, and $V_S(t) = 5$ V for $t \ge \tau_r$ where τ_r is the pulse rise time. The input to a CMOS gate (the load on the line here) can be modeled as a capacitance of some 5 pF–15 pF. However, in order to simplify the problem, we will assume that the input to the load CMOS gate is an open circuit $R_L = \infty$. Sketch the load voltage of the line (the input voltage to the load CMOS gate) for line lengths having one-way transit times T such that (a) $\tau_r = T/10$, (b) $\tau_r = 2T$, (c) $\tau_r = 3T$, (d) $\tau_r = 4T$. This example shows that in order to avoid problems resulting from mismatch, one should choose line lengths short enough such that $T \ll \tau_r$, that is, the line one-way delay is much less than the rise time of the clock/data pulses being carried by the line.

6.2.16. Highly mismatched lines in digital products can cause what appears to be "ringing" on the signal output from the line. This is often referred to as "overshoot" or "undershoot" and can cause digital logic errors. To simulate this we will investigate the problem shown in Fig. P6.2.16. Two CMOS gates are connected by a transmission line as shown. A 5-V step function voltage of the first gate is applied. Sketch the output voltage of the line (the input voltage to the load CMOS gate) for $0 < t < 9T$.
$[0 < t < T, 0$ V$,T < t < 3T, 6.944$ V$, 3T < t < 5T, 3.858$ V, $5T < t < 7T, 5.23$ V, $7T < t < 9T, 4.62$ V. Steady state is 5 V$]$

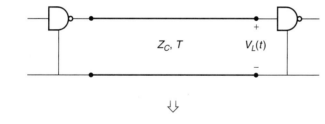

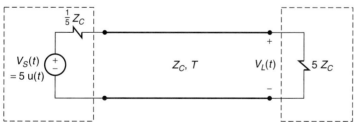

Figure P6.2.16 Problem 6.2.16.

6.2.17. A transmission line of total length 200 m and velocity of propagation of $v = 2 \times 10^8$ m/s has $Z_C = 50\ \Omega$, $R_L = 20\ \Omega$. It is driven by a source having $R_S = 100\ \Omega$ and an open-circuit voltage that is a rectangular pulse of 6-V magnitude and 3-μs duration. Sketch the input current to the line for a total time of 5 μs.

6.2.18. Confirm the results of Problem 6.2.9 using SPICE (PSPICE).

6.2.19. Confirm the results of Problem 6.2.10 using SPICE (PSPICE).

6.2.20. Confirm the results of Problem 6.2.11 using SPICE (PSPICE).

✱6.2.21. Confirm the results of Problem 6.2.12 using SPICE (PSPICE).

6.2.22. Confirm the results of Problem 6.2.13 using SPICE (PSPICE).

6.2.23. Confirm the results of Problem 6.2.14 using SPICE (PSPICE).

6.2.24. Confirm the results of Problem 6.2.15 using SPICE (PSPICE).

6.2.25. Confirm the results of Problem 6.2.16 using SPICE (PSPICE).

6.2.26. Confirm the results of Problem 6.2.17 using SPICE (PSPICE).

6.2.27. One of the important advantages in using SPICE to solve transmission-line problems is that it will readily give the solution for problems that would be difficult to solve by hand. For example, consider the case of two CMOS inverter gates connected by a 5-cm length of 100-Ω transmission line as shown in Fig. P6.2.27. The output of the driver gate is represented by a ramp waveform voltage rising from 0 V to 5 V in 1 ns and a 30-Ω internal source resistance. The receiving gate is represented at its input by 10 pF. Because of the capacitive load, this would be a difficult problem to do by hand. Use SPICE (PSPICE) to plot the output voltage of the line, $V_L(t)$, for $0 < t < 10$ ns. Observe in the solution that this output voltage varies rather drastically about the desired 5 V level going from 4.2 V to 7 V before it stabilizes to 5 V well after 10 ns.

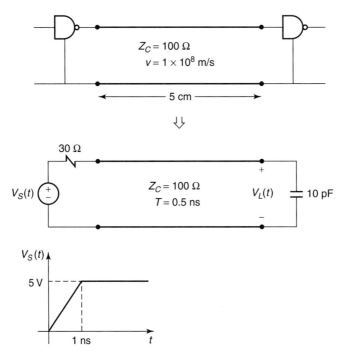

Figure P6.2.27 Problem 6.2.27.

SECTION 6.3 SINUSOIDAL (PHASOR) EXCITATION OF TRANSMISSION LINES

6.3.1. Verify by direct substitution that the solutions for the phasor voltage and current given in (6.35) satisfy the phasor transmission-line equations given in (6.33).

6.3.2. Verify by direct substitution that the time-domain solutions for the line voltage and current given in (6.37) satisfy the time-domain transmission-line equations given in (6.1).

6.3.3. For the transmission line shown in Fig. 6.19, $f = 5$ MHz, $v = 3 \times 10^8$ m/s, $\mathscr{L} = 78$ m, $Z_C = 50\ \Omega$, $\hat{V}_S = 50\angle 0°$, $\hat{Z}_S = 20 - j30\ \Omega$, $\hat{Z}_L = 200 + j500\ \Omega$. Determine (a) the line length as a fraction of a wavelength, (b) the voltage reflection coefficient at the load and at the input to the line, (c) the input impedance to the line, (d) the time-domain voltages at the input to the line and at the load, (e) the average power delivered to the load, and (f) the VSWR.

4 * **6.3.4.** For the transmission line shown in Fig. 6.19, $f = 200$ MHz, $v = 3 \times 10^8$ m/s, $\mathscr{L} = 2.1$ m,
0.5 $Z_C = 100\ \Omega$, $\hat{V}_S = 10\angle 60°$, $\hat{Z}_S = 50\Omega$, $\hat{Z}_L = 10 - j50\Omega$. Determine (a) the line length as a fraction of a wavelength, (b) the voltage reflection coefficient at the load and at the input to the line, (c) the input impedance to the line, (d) the time-domain voltages at the input to the line and at the load, (e) the average power delivered to the load, and (f) the VSWR. [(a) 1.4, (b) $0.8521\angle -126.5°$, $0.8521\angle -54.5°$, (c) $192\angle -78.83°$, (d) $9.25 \cos(4\pi \times 10^8 t + 46.33°)$, (e) $4.738 \cos(4\pi \times 10^8 t - 127°)$, (f) 43 mW, 12.52]

6.3.5. For the transmission line shown in Fig. 6.19, $f = 1$ GHz, $v = 1.7 \times 10^8$ m/s, $\mathscr{L} = 11.9$ cm, $Z_C = 100\ \Omega$, $\hat{V}_S = 5\angle 0°$, $\hat{Z}_S = 20\ \Omega$, $\hat{Z}_L = -j160\ \Omega$. Determine (a) the line length as a fraction of a wavelength, (b) the voltage reflection coefficient at the load and at the input to the line, (c) the input impedance to the line, (d) the time-domain voltages at the input to the line and at the load, (e) the average power delivered to the load, and (f) the VSWR.

6.3.6. For the transmission line shown in Fig. 6.19, $f = 600$ MHz, $v = 2 \times 10^8$ m/s, $\mathscr{L} = 53$ cm, $Z_C = 75\ \Omega$, $\hat{V}_S = 20\angle 40°$, $\hat{Z}_S = 30\ \Omega$, $\hat{Z}_L = 100 - j300\ \Omega$. Determine (a) the line length as a fraction of a wavelength, (b) the voltage reflection coefficient at the input to the line and at the load, (c) the input impedance to the line, (d) the time-domain voltages at the input to the line and at the load, (e) the average power delivered to the load, and (f) the VSWR. [(a) 1.59, (b) $0.8668\angle -25.49°$, $0.8668\angle -90.29°$, (c) $74.62\angle -81.84°$, (d) $17.71 \cos(12\pi \times 10^8 t + 19.37°)$, (e) $24.43 \cos(12\pi \times 10^8 t - 163.8°)$, (f) 0.298 W, 14.02]

6.3.7. For the transmission line shown in Fig. 6.19, $f = 1$ MHz, $v = 3 \times 10^8$ m/s, $\mathscr{L} = 108$ m, $Z_C = 300\ \Omega$, $\hat{V}_S = 100\angle 0°$, $\hat{Z}_S = 50 + j50\ \Omega$, $\hat{Z}_L = 100 - j100\ \Omega$. Determine (a) the line length as a fraction of a wavelength, (b) the voltage reflection coefficient at the load and at the input to the line, (c) the input impedance to the line, (d) the time-domain voltages at the input to the line and at the load, (e) the average power delivered to the load, and (f) the VSWR.

6.3.8. Confirm the input and load voltages for Problem 6.3.3 using SPICE.

6.3.9. Confirm the input and load voltages for Problem 6.3.4 using SPICE.

6.3.10. Confirm the input and load voltages for Problem 6.3.5 using SPICE.

6.3.11. Confirm the input and load voltages for Problem 6.3.6 using SPICE.

6.3.12. Confirm the input and load voltages for Problem 6.3.7 using SPICE.

6.3.13. A half-wavelength dipole antenna is connected to a 100-MHz source with a 3.6-m length of 300-Ω transmission line (twin lead, $v = 2.6 \times 10^8$ m/s) as shown in Fig. P6.3.13. The source

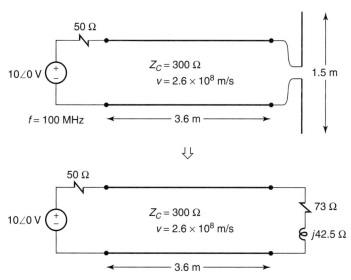

Figure P6.3.13 Problem 6.3.13.

is represented by an open-circuit voltage of 10 V and source impedance of 50 Ω, whereas the input to the dipole antenna is represented by a 73-Ω resistance in series with an inductive reactance of 42.5 Ω. The average power dissipated in the 73-Ω resistance is equal to the power radiated into space by the antenna. Determine the average power radiated by the antenna with and without the transmission line and the VSWR on the line [91.14 mW, 215.53 mW, 4.2]

6.3.14. Two identical half-wavelength dipole antennas are connected in parallel and fed from one 300-MHz source as shown in Fig. P6.3.14. Determine the average power delivered to each antenna. Hint: Determine the electrical lengths of the transmission lines.

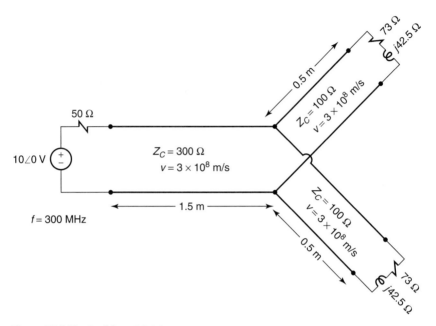

Figure P6.3.14 Problem 6.3.14.

6.3.15. Determine an expression for the input impedance to (a) a transmission line having an open-circuit load and (b) a transmission line having a short-circuit load.

$$\left[-jZ_C \frac{1}{\tan(\beta\mathcal{L})}, jZ_C \tan(\beta\mathcal{L}) \right]$$

6.3.16. Obtain an expression for the input impedance to a quarter-wavelength transmission line. If the line has an open-circuit load, what is its input impedance? If the line has a short-circuit load, what is its input impedance? [$\hat{Z}_{in} = \hat{Z}_C^2/\hat{Z}_L$, short circuit, open circuit].

SECTION 6.4 THE SMITH CHART

6.4.1. Determine, using the Smith chart, the input impedance, input reflection coefficient, load reflection coefficient, and VSWR for the line of Problem 6.3.3.
[$\hat{Z}_{in} = 2 + j12$, $\hat{\Gamma}(0) = 0.92\angle154°$, $\hat{\Gamma}_L = 0.92\angle7°$, VSWR = 30]

6.4.2. Determine, using the Smith chart, the input impedance, input reflection coefficient, load reflection coefficient, and VSWR for the line of Problem 6.3.4.

6.4.3. Determine, using the Smith chart, the input impedance, input reflection coefficient, load reflection coefficient, and VSWR for the line of Problem 6.3.5.
[$\hat{Z}_{in} = j25$, $\hat{\Gamma}(0) = 1\angle152°$, $\hat{\Gamma}_L = 1\angle-64°$, VSWR = ∞]

6.4.4. Determine, using the Smith chart, the input impedance, input reflection coefficient, load reflection coefficient, and VSWR for the line of Problem 6.3.6.

6.4.5. Determine, using the Smith chart, the input impedance, input reflection coefficient, load reflection coefficient, and VSWR for the line of Problem 6.3.7.
$[\hat{Z}_{in} = 474 - j450, \hat{\Gamma}(0) = 0.54\angle -39°, \hat{\Gamma}_L = 0.54\angle -139°, \text{VSWR} = 3.4]$

6.4.6. Determine the load impedance, VSWR, and load reflection coefficient for the following transmission lines: (a) $\hat{Z}_{in} = (30 - j100)\,\Omega$, $Z_C = 50\,\Omega$, $\mathcal{L} = 0.4\lambda$, (b) $\hat{Z}_{in} = 50\,\Omega$, $Z_C = 75\,\Omega$, $\mathcal{L} = 1.3\lambda$, (c) $\hat{Z}_{in} = (150 + j230)\,\Omega$, $Z_C = 100\,\Omega$, $\mathcal{L} = 0.6\lambda$, (d) $\hat{Z}_{in} = j250\,\Omega$, $Z_C = 100\,\Omega$, $\mathcal{L} = 0.8\lambda$.

6.4.7. Determine the shortest lengths of the following transmission lines as well as the VSWR and the load reflection coefficient: (a) $\hat{Z}_{in} = -j20\,\Omega$, $\hat{Z}_L = j50\,\Omega$, $Z_C = 100\,\Omega$, (b) $\hat{Z}_{in} = (50 - j200)\,\Omega$, $\hat{Z}_L = (12 - j50)\,\Omega$, $Z_C = 100\,\Omega$, (c) $\hat{Z}_{in} = (30 + j50)\,\Omega$, $\hat{Z}_L = (200 + j200)\,\Omega$, $Z_C = 100\,\Omega$, (d) $\hat{Z}_{in} = (135 + j0)\,\Omega$, $\hat{Z}_L = (60 - j37.5)\,\Omega$, $Z_C = 75\,\Omega$. [(a) 0.396λ, ∞, $1\angle 127°$, (b) 0.395λ, 10, $0.83\angle -126°$, (c) 0.37λ, 4.2, $0.62\angle 29.5°$, (d) 0.366λ, 1.8, $0.285\angle -96°$]

6.4.8. Using the Smith chart, show the following properties of lossless transmission lines: (a) $|\hat{\Gamma}_L| \leq 1$, (b) the input impedance replicates for distances separated by a multiple of a half wavelength, (c) the VSWR for a line having a purely reactive load is infinite, and the magnitude of the reflection coefficient is unity, (d) the input impedance at any point on a line having a purely reactive load cannot have a real part, (e) adjacent points on a line where the input impedance is purely resistive are separated by a quarter wavelength.

6.4.9. We want to determine the value of an unknown impedance, $\hat{Z}_L$, attached to a length of transmission line having a characteristic impedance of $100\,\Omega$. Removing the load yields an input impedance of $-j80\,\Omega$. With the unknown impedance attached, the input impedance is $(30 + j40)\,\Omega$. Determine the unknown impedance.

6.4.10. Shorted (or open-circuited) lengths of transmission lines can be constructed so that they appear at their input to be either capacitors or inductors. This is how microwave circuit components are constructed. Using discrete capacitors and inductors at these microwave frequencies (GHz) would be useless due to the effect of their connection leads, as we saw in Section 3.13.4 of Chapter 1. In order to demonstrate this, use the Smith chart to determine the length of a transmission line having a short-circuit load such that it appears at its input terminals as a 10-pF capacitor at 1 GHz. Assume an air-filled line ($v = 3 \times 10^8$ m/s) having a characteristic impedance of $50\,\Omega$.

SECTION 6.5 LUMPED-CIRCUIT APPROXIMATE MODELS OF TRANSMISSION LINES

6.5.1. An air-filled line ($v = 3 \times 10^8$ m/s) having a characteristic impedance of $50\,\Omega$ is driven at a frequency of 30 MHz and is 1 m in length. The line is terminated in a load of $\hat{Z}_L = (200 - j200)\,\Omega$. Determine the input impedance using (a) the transmission-line model and (b) the approximate, lumped-Pi model.

6.5.2. A coaxial cable ($v = 2 \times 10^8$ m/s) having a characteristic impedance of $100\,\Omega$ is driven at a frequency of 4 MHz and is 5 m in length. The line is terminated in a load of $\hat{Z}_L = (150 - j50)\,\Omega$ and the source is $\hat{V}_S = 10\angle 0°$ V with $\hat{Z}_S = 25\,\Omega$. Determine the input and output voltages to the line using (a) the transmission-line model and (b) the approximate, lumped-Pi model. [Exact: $\hat{V}(0) = 7.954\angle -6.578°$, $\hat{V}(\mathcal{L}) = 10.25\angle -33.6°$; Approximate: $\hat{V}(0) = 7.959\angle -5.906°$, $\hat{V}(\mathcal{L}) = 10.27\angle -35.02°$]

SECTION 6.6 LOSSY LINES

6.6.1. A low-loss coaxial cable has the following parameters: $\hat{Z}_C \cong (75 + j0)\,\Omega$, $\alpha = 0.05$, $v = 2 \times 10^8$ m/s. Determine the input impedance to a 11.175-m length of the cable at 400 MHz if the line is terminated in (a) a short circuit, (b) an open circuit, and (c) a 300-Ω resistor.

6.6.2. Lossy cables, even low-loss ones, dissipate some power as signals traverse them. Manufacturers of cables specify this loss at various frequencies in dB/100 ft or dB per some other linear dimension. The loss is the ratio of the input power to the cable over the output power extracted from it. In doing so, the cable is assumed to be matched, so there is only a forward-traveling wave (attenuated) on the line. Determine an expression for the loss in such a cable in dB as a function of the attenuation constant α. [Cable loss = $8.686\alpha\mathcal{L}$ dB]

CHAPTER 7

Antennas

Antennas are structures that launch electromagnetic waves into the surrounding air. It is useful to consider the analogy of dropping a rock into a body of water. Spherical waves (actually circular wavefronts) emanate from the point where the rock strikes the water's surface. An antenna produces spherical waves emanating from it.

Time-varying currents radiate. This is the essential mechanism by which antennas radiate electromagnetic waves. Time-varying currents on the surface of the antenna produce the radiated electromagnetic waves. When we examine the meaning of current, we get further insight into this mechanism. Current is the *rate of flow* of charge, $I = dQ/dt$. Hence the *acceleration of charge* is a time-varying current, and therefore *accelerated charges produce radiation.*

The electromagnetic fields of antennas close to the antenna are very complicated. Fortunately we are only interested in the *far fields* of the antenna: fields at a distance from the antenna that are large in terms of wavelength of the transmitted signal. In the far field of all antennas, the fields are rather simple and resemble the uniform plane waves studied in Chapter 5. Hence our knowledge of uniform plane wave characteristics can be straightforwardly carried over to the radiated fields of antennas, which will greatly simplify our understanding of antennas. We will have many occasions in this chapter to use that relation of radiated fields from antennas to uniform plane waves.

In this chapter we will examine the general properties of antennas. The ability of an antenna that is used to transmit a signal to focus or concentrate the radiated emission in a particular direction is contained in an important parameter of that antenna, its *gain*. Any energy directed away from the receiving antenna is lost and affects the efficiency of the communication system. Also, every antenna must be connected to a signal source via a transmission line (its feed line). (See Fig. 6.48a and the accompanying discussion in Section 6.7.3.) We will find that an antenna has another important parameter—its *input impedance*, whose real part is approximately its *radiation resistance*. The power dissipated in this fictitious radiation resistance represents the power radiated by the antenna. Hence the objective is to transfer power from the source along the feed line to the antenna. If the input impedance to the antenna is not matched to the characteristic impedance of the transmission line that connects it to the source, some of the power passing down the transmission line will be reflected back to the source and hence not radiated by the antenna. Therefore, another important design consideration is matching an antenna to its feed line. In a communication link we must also have a receiving antenna to convert the incident fields to voltage and current at its output. It turns out that an important principle, *reciprocity*, essentially means that the transmitting properties of antennas, such as its focusing ability, are unchanged when they are used to receive a wave.

Once we have completed this chapter, we will have enough information and insight to not only understand general antennas but also to design communication links that contain these antennas.

Chapter Learning Objectives

After completing the contents of this chapter, you should be able to

▷ understand the properties of all antennas: pattern, gain, input impedance, radiation resistance, variation with distance and phase shift of the far-field electric and magnetic fields, power density, and total radiated power

▷ compute the far-field electric and magnetic fields, power density, and total radiated power for a Hertzian dipole antenna,

▷ compute the far-field electric and magnetic fields, power density, and total radiated power for the half-wave dipole and quarter-wave monopole antennas,

▷ understand how the far-field electric fields of a two-element array combine to produce directionality of the pattern (nulls, maxima, and minima),

▷ plot the pattern of a two-element array given the separation between the two antennas and the phase of the two currents,

▷ compute the effective aperture or capture area of a receiving antenna from its gain and vice versa,

▷ compute the path loss for a two-antenna communication link using the Friis transmission equation,

▷ compute the far-field electric field using the Friis transmission equation,

▷ cite and explain some typical engineering applications of the principles of this chapter.

▷ 7.1 THE HERTZIAN DIPOLE ANTENNA

Perhaps the simplest antenna is what is called the Hertzian dipole shown in Fig. 7.1. It consists of a sinusoidal current $\hat{I}$ that is directed along the z axis of a spherical coordinate system. Use of the spherical coordinate system in characterizing antennas is standard in industry and is preferred over cylindrical or rectangular coordinate systems because the waves radiated from antennas are spherical waves. There are two important assumptions about the Hertzian dipole. First, the length of the antenna, dl, is very

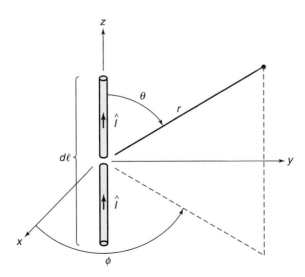

Figure 7.1 The Hertzian dipole antenna and the spherical coordinate system.

small (ideally differentially small in the calculus sense). This assumption simplifies the mathematics, allowing a closed-form solution, and is approximated in a practical sense if the physical length of the antenna is much less than a wavelength, that is, very short, electrically. The second assumption is that the current along the antenna, $\hat{I}$, is constant (magnitude and phase) along the antenna. For obvious reasons the current must go to zero at the endpoints of any practical antenna, which is not the case for this antenna. The current distribution along a wire-type antenna that is electrically short, $dl \ll \lambda_o$, is actually triangular, with the maximum at the center and linearly going to zero at the endpoints. Assumption of constant current along the Hertzian dipole also simplifies the mathematics. It turns out that despite these assumptions, the radiated fields from the Hertzian dipole are very similar in form to those of other antennas in the far field, that is, at distances that are far away from it relative to its length in wavelengths. Hence study of the far fields of a Hertzian dipole provides considerable insight into the far fields of other antennas.

From our earlier electric circuits courses, Kirchhoff's current law would indicate that the current along the Hertzian dipole must be zero since the ends are open circuited. A convenient way of seeing that this current is not zero is by visualizing capacitances between the upper and lower wires. This capacitance "completes the circuit" via its displacement current, thereby allowing currents on the wires.

For single-frequency, sinusoidal excitation, the resulting phasor electric and magnetic fields of this antenna at a distance r and angle θ from it become[1], with reference to the spherical coordinate system,

$$\hat{E}_r = \frac{2\pi f^2 \mu_o}{v_o} \hat{I} dl \cos\theta \left[\frac{1}{\left(2\pi \dfrac{r}{\lambda_o}\right)^2} - j\frac{1}{\left(2\pi \dfrac{r}{\lambda_o}\right)^3} \right] e^{-j\left(2\pi \frac{r}{\lambda_o}\right)} \tag{7.1a}$$

$$\hat{E}_\theta = \frac{\pi f^2 \mu_o}{v_o} \hat{I} dl \sin\theta \left[j\frac{1}{\left(2\pi \dfrac{r}{\lambda_o}\right)} + \frac{1}{\left(2\pi \dfrac{r}{\lambda_o}\right)^2} - j\frac{1}{\left(2\pi \dfrac{r}{\lambda_o}\right)^3} \right] e^{-j\left(2\pi \frac{r}{\lambda_o}\right)} \tag{7.1b}$$

$$\hat{E}_\phi = 0 \tag{7.1c}$$

and

$$\hat{H}_r = 0 \tag{7.1d}$$

$$\hat{H}_\theta = 0 \tag{7.1e}$$

$$\hat{H}_\phi = \frac{\pi f^2 \mu_o}{\eta_o v_o} \hat{I} dl \sin\theta \left[j\frac{1}{\left(2\pi \dfrac{r}{\lambda_o}\right)} + \frac{1}{\left(2\pi \dfrac{r}{\lambda_o}\right)^2} \right] e^{-j\left(2\pi \frac{r}{\lambda_o}\right)} \tag{7.1f}$$

where f is the frequency of excitation (and the resulting current). Equations (7.1) are for single-frequency, sinusoidal, steady-state excitation and are in phasor form. The surrounding medium is logically free space, so that $v_o = 1/\sqrt{\mu_o \varepsilon_o}$ is the speed of light in free space ($v_o \cong 3 \times 10^8$ m/s) and $\eta_o = \sqrt{\mu_o/\varepsilon_o} \cong 120\pi$ is the intrinsic impedance of free space. Because of symmetry, the fields are independent of the spherical coordinate variable, ϕ, that is, the fields are rotationally symmetric about the dipole axis.

---

[1]A derivation of these results for the interested reader can be found in C.R. Paul and S.A. Nasar, *Introduction to Electromagnetic Fields*, Second Edition, McGraw Hill, 1987.

Observe that the radial component of the electric field, $\hat{E}_r$, is a function of $\cos\theta$, and the transverse components, $\hat{E}_\theta$ and $\hat{H}_\phi$, are functions of $\sin\theta$. Hence, the radial component, $\hat{E}_r$, is a maximum off the ends, $\theta = 0°$, and zero broadside to it, $\theta = 90°$. Conversely, $\hat{E}_\theta$ and $\hat{H}_\phi$ are zero off the ends and a maximum broadside to the antenna. Observe that the radial distance from the center of the dipole, r, is written in terms of electrical distance, that is, r/λ_o, where $\lambda_o = v_o/f$ is the wavelength in free space. Observe also that the electric field has only an r and a θ component, whereas the magnetic field has only a ϕ component. The term $e^{-j(2\pi r/\lambda_o)}$ represents a phase shift of the wave as it propagates radially away from the antenna. In the time domain this represents a *time delay*, as we saw previously:

$$e^{-j\left(2\pi\frac{r}{\lambda_o}\right)} \Leftrightarrow \cos\left(\omega\left(t - \frac{r}{v_o}\right)\right)$$

It is logical to expect to find this phase shift (time delay) term in the solution since it simply says that any change in the antenna current is observed a distance r away after a time delay of r/v_o, representing the finite propagation velocity of the radiated wave. Also, the fields depend directly on the current, $\hat{I}$, and the length of the line, dl.

7.1.1 The Far Field

The complete phasor fields in (7.1) show, once again, that physical distance from the antenna is not the important dimension. The important dimension is electrical distance in wavelengths, that is, r/λ_o. Observe that the dependence on electrical distance away from the antenna occurs as the reciprocals of this electrical distance to various powers, $1/(2\pi r/\lambda_o)$, $1/(2\pi r/\lambda_o)^2$, and $1/(2\pi r/\lambda_o)^3$. At distances that are (electrically) close to the antenna, $r \ll \lambda_o$, the cubed terms dominate, that is, $1/(2\pi r/\lambda_o)^3 \gg 1/(2\pi r/\lambda_o)^2 \gg 1/(2\pi r/\lambda_o)$. This is referred to as the *near field*. The equations for the total field in (7.1) reduce in the near field to

$$\hat{E}_r = \frac{\hat{Q} dl \cos\theta}{2\pi\varepsilon_o r^3}$$

$$\hat{E}_\theta = \frac{\hat{Q} dl \sin\theta}{4\pi\varepsilon_o r^3} \qquad r \ll \lambda_o \text{ (near field)}$$

$$\hat{H}_\phi = \frac{\hat{I} dl \sin\theta}{4\pi r^2}$$

Since $I = dQ/dt$, we have substituted the phasor form of this, $\hat{I} = j\omega\hat{Q}$, into the electric field expressions. Observe that the magnetic field expression for $\hat{H}_\phi$ reduces to the Biot-Savart law for static (dc) magnetic fields for a segment of current that is given in Chapter 3, Equation (3.58). Also observe that the electric field components for $\hat{E}_r$ and $\hat{E}_\theta$ are the expressions for an electric dipole given in Chapter 3, Equation (3.14), where charges $\hat{Q}$ are separated by dl. This is logical to expect since the current of the dipole does not go to zero at the ends, indicating an accumulation of charge at the ends of the dipole. These observations once again confirm that electrical dimensions and not physical dimensions are the important parameter in electromagnetics problems. Also, if the fields are varying slowly enough, the static (dc) results of Chapter 3 may be used as a *quasistatic* approximation. For example, suppose the current of the dipole has a frequency of 60 Hz. The wavelength at 60 Hz is approximately 3000 miles or 5000 km. Hence if $r \ll \lambda_o$ or say $r < 30$ miles or 50 km, then the fields reduce to static results.

However, for use in communications, we are only interested in distances very far away from the antenna. At distances very far away from the antenna, that is, $r/\lambda_o \gg 1$, the squared and cubed terms are dominated by the $1/(2\pi r/\lambda_o)$ term. In this case, we say that the distance from the antenna is in the *far field* and the above fields reduce to

$$\hat{E}_\theta = j\frac{f\mu_o}{2}\hat{I}dl\,\sin\theta\left[\frac{e^{-j\left(2\pi\frac{r}{\lambda_o}\right)}}{r}\right] \quad \text{far field} \quad r \gg \lambda_o \tag{7.2a}$$

and

$$\hat{H}_\phi = \frac{\hat{E}_\theta}{\eta_o}$$

$$= j\frac{f\mu_o}{2\eta_o}\hat{I}dl\,\sin\theta\left[\frac{e^{-j\left(2\pi\frac{r}{\lambda_o}\right)}}{r}\right] \quad \text{far field} \quad r \gg \lambda_o \tag{7.2b}$$

The radial component, $\hat{E}_r$, is approximately zero because (7.1a) only contains squared and cubed distance terms.

Observe some important properties of the far fields:

1 the fields decay inversely as the distance, r, away from the antenna,

2 the fields contain a phase shift term, $e^{-j\left(2\pi\frac{r}{\lambda_o}\right)}$(which is equivalent to a time delay in the time domain, r/v_o), and hence the fields change phase by an angle of $\angle -2\pi r/\lambda_o$ as they propagate away from the antenna,

3 the fields are orthogonal and their cross product gives, by the right-hand-rule, the direction of propagation from the antenna: $\hat{E}_\theta \times \hat{H}_\phi \Rightarrow \mathbf{a}_r$, which is the radial direction,

4 the ratio of the electric and magnetic fields is the intrinsic impedance of free space, $\hat{E}_\theta/\hat{H}_\phi = \eta_o$, and

5 the fields depend on $\sin\theta$ and hence are zero off the ends and a maximum broadside to the antenna, $\theta = 90°$.

Properties 2, 3, and 4 are all properties of uniform plane waves. The waves are in reality spherical waves as are produced by dropping a rock in a body of water. However, to a local observer watching the approach of these waves, they appear, locally, to be uniform plane waves.

▶ EXAMPLE 7.1

A Hertzian dipole has a length of 1 cm and carries a 1-A, 100-MHz current. Determine the magnitude and phase of the electric and magnetic fields at a distance of 1000 m away and broadside to the antenna ($\theta = 90°$).

SOLUTION A wavelength at 100 MHz is 3 m. The dipole length is $\lambda_o/300$ and satisfies the requirement that it be electrically short. The distance where the fields are to be determined is $333\lambda_o$ and hence is in the far field of the dipole. Substitution into (7.2) gives

$$\hat{E}_\theta = j\frac{10^8\,\text{Hz} \times 4\pi \times 10^{-7} \times 1\text{A} \times 10^{-2}\,\text{m}}{2} \times \frac{e^{-j2\pi\frac{r\,=\,1000\,\text{m}}{\lambda_o\,=\,3\,\text{m}}}}{r\,=\,1000\,\text{m}}$$

$$= 6.28 \times 10^{-4}\angle -120{,}000° \quad \text{V/m}$$

and

$$\hat{H}_\phi = \frac{\hat{E}_\theta}{\eta_o = 120\pi}$$
$$= 1.67 \times 10^{-6}\angle -120{,}000° \qquad \text{A/m}$$

◀

▷ QUICK REVIEW EXERCISE 7.1

The magnitude of the far-field electric field of a Hertzian dipole is measured at a distance of 100 m as 1 m V/m. Determine the magnitude of the electric field at 1000 m.

ANSWER 100 μV/m.

◀

7.1.2 Power Flow and Radiation

In the far field, the time-average or simply average power density in W/m^2 is

$$\boxed{\begin{aligned} \mathbf{S}_{\text{AV}} &= \frac{1}{2}\text{Re}(\hat{\mathbf{E}} \times \hat{\mathbf{H}}^\circ) \\ &= \frac{1}{2}\frac{|\hat{E}_\theta|^2}{\eta_o}\mathbf{a}_r \end{aligned}} \qquad (7.3)$$

since $\hat{H}_\phi = \hat{E}_\theta/\eta_o$ at points in the far field and $\circ$ denotes the complex conjugate. Evaluating this for the Hertzian dipole using the far-field expressions in (7.2) gives

$$\begin{aligned} \mathbf{S}_{\text{AV}} &= \frac{1}{2}\text{Re}(\hat{\mathbf{E}} \times \hat{\mathbf{H}}^\circ) \\ &= \frac{1}{2}\text{Re}\left(j\frac{f\mu_o}{2}\hat{I}dl\sin\theta\left[\frac{e^{-j\left(2\pi\frac{r}{\lambda_o}\right)}}{r}\right]\mathbf{a}_\theta \times -j\frac{f\mu_o}{2\eta_o}\hat{I}^\circ dl\sin\theta\left[\frac{e^{j\left(2\pi\frac{r}{\lambda_o}\right)}}{r}\right]\mathbf{a}_\phi\right) \qquad (7.4) \\ &= \underbrace{\frac{\eta_o}{8}}_{15\pi}|\hat{I}|^2\left(\frac{dl}{\lambda_o}\right)^2\frac{\sin^2\theta}{r^2}\mathbf{a}_r \qquad \text{W/m}^2 \end{aligned}$$

This shows that average power is flowing away from the antenna in the radial direction. Observe that the power density decays inversely as the square of the distance, whereas the electric and magnetic fields decay inversely as distance.

▷ EXAMPLE 7.2

Determine the total average power radiated by the Hertzian dipole antenna.

SOLUTION The average power density in (7.4) has the units of W/m^2. In order to obtain the total average power, we integrate this power density over a closed surface s:

$$P_{\text{AV}} = \oint_s \mathbf{S}_{\text{AV}} \cdot d\mathbf{s} \qquad \text{W} \qquad (7.5)$$

What closed surface should we choose for the integration? Any closed surface will do but because the average power density vector is radially directed, it will simplify the integration if we choose the surface to be a sphere of radius R centered on the origin of the spherical coordinate system

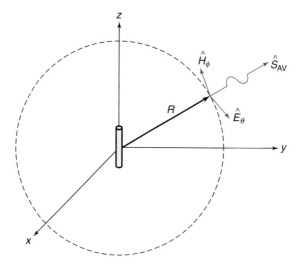

Figure 7.2 Computation of the total average power radiated by the Hertzian dipole antenna.

as shown in Fig. 7.2. The vector differential surface is, from Chapter 2, $d\mathbf{s} = R^2 \sin\theta \, d\phi \, d\theta \, \mathbf{a}_r$. Substituting (7.4) into (7.5) and performing the integration gives

$$P_{AV} = \oint_s \underbrace{\frac{\eta_o}{8}|\hat{I}|^2\left(\frac{dl}{\lambda_o}\right)^2 \frac{\sin^2\theta}{R^2}\mathbf{a}_r}_{\mathbf{S}_{AV}} \cdot \underbrace{R^2\sin\theta \, d\phi \, d\theta \, \mathbf{a}_r}_{d\mathbf{s}}$$

$$= \int_{\phi=0}^{2\pi}\int_{\theta=0}^{\pi} \frac{\eta_o}{8}|\hat{I}|^2\left(\frac{dl}{\lambda_o}\right)^2 \sin^3\theta \, d\phi \, d\theta \quad\quad (7.6)$$

$$= 80\pi^2\left(\frac{dl}{\lambda_o}\right)^2\frac{|\hat{I}|^2}{2} \quad \text{W}$$

▶ QUICK REVIEW EXERCISE 7.2

A Hertzian dipole is of length 1 cm and carries a current of 100 mA at a frequency of 10 MHz. Determine the total radiated power.

ANSWER   0.44 μW.

Antennas can be visualized as having a *radiation resistance*, R_{rad}. The radiation resistance is a fictitious resistance such that if its current is the same as the current of the dipole, then the power dissipated in the radiation resistance is equal to the total power radiated by the antenna:

$$\boxed{R_{rad} = \frac{P_{AV}}{|\hat{I}_{rms}|^2}} \quad\quad (7.7)$$

where the *rms* current is $\hat{I}_{rms} = \hat{I}/\sqrt{2}$. From (7.6) the radiation resistance is

$$\boxed{R_{rad} = 80\pi^2\left(\frac{dl}{\lambda_o}\right)^2 \quad \Omega} \quad\quad (7.8)$$

▶ **QUICK REVIEW EXERCISE 7.3**

Determine the radiation resistance of a Hertzian dipole whose length is 1 cm and frequency of operation is 100 MHz.

ANSWER 8.77 mΩ.

▶ 7.2 THE HALF-WAVE DIPOLE AND QUARTER-WAVE MONOPOLE ANTENNAS

The Hertzian dipole is not very effective in launching waves. For example, a 1-cm length antenna operated at 100 MHz has a radiation resistance of 8.77 mΩ. Hence, in order to radiate 1 W of power, the current along the dipole must be (rms) 10.7 A! Observe in (7.7) that the total power radiated depends directly on the radiation resistance. The relation for the radiation resistance in (7.7) also holds for other antennas. Hence, in order to increase the radiated power for a given current, we must increase the radiation resistance. For the Hertzian dipole the radiation resistance depends on the *electrical length* of the dipole, dl/λ_o, and not on physical length. In order to obtain a large radiation resistance, say, on the order of 100 Ω, we must increase the electrical length to a significant portion of a wavelength. But this would violate the basic assumption of the Hertzian dipole that the electrical length is very small. In this section we will investigate a more realistic and practical antenna whose length is one-half wavelength. We will find that the radiation resistance is 73 Ω and hence a significant amount of radiated power can be obtained without unreasonably large currents.

Figure 7.3a shows a general dipole antenna of total length l. It consists of two arms of equal length $l/2$. The lengths of the arms, unlike the Hertzian dipole, will not be required to be electrically short. In fact, for the most common antenna, the half-wave dipole, the total length will be one-half wavelength, $l = \lambda_o/2$. Figure 7.3b shows another common antenna, the monopole. That antenna has one arm of height h that is perpendicular to an infinite ground plane. (In practical terms, the extent of the ground plane must only be several wavelengths.) The height of a quarter-wave monopole antenna is one-quarter wavelength above the ground plane, $h = \lambda_o/4$. By the method of images, we may replace the ground plane with another arm of length h carrying a current that is the same as the arm above the ground plane. Hence, the fields of the monopole antenna, *above the position of the ground plane*, are identical to those of the dipole. So determining the fields of the dipole antenna will immediately give the fields of the monopole antenna.

We now focus on an important dipole, the half-wave dipole, whose total length is one-half wavelength, $l = \lambda_o/2$. Because the half-wave dipole is not electrically short, the current along it cannot be constant. It turns out that the current along antennas that are electrically long has a sinusoidal variation with position along the line. Hence we can assume a form of that current as

$$\hat{I}(z) = \hat{I}_m \sin\beta_o\left(\frac{l}{2} - z\right) \qquad 0 < z < \frac{l}{2} \tag{7.9a}$$

$$\hat{I}(z) = \hat{I}_m \sin\beta_o\left(\frac{l}{2} + z\right) \qquad -\frac{l}{2} < z < 0 \tag{7.9b}$$

This current distribution can be reasoned by examining the currents along a transmission line that has an open-circuit load. (See Fig. 6.23c.) The currents along the line are of the form $\hat{I}(d) \propto \sin(\beta d)$ which are sinusoidally distributed along the line, going to zero at the ends. (See Equation 6.54b.) If we spread the endpoints of the line

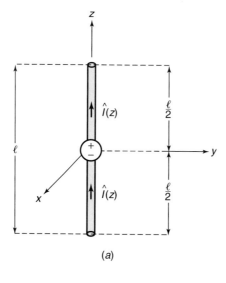

(a)

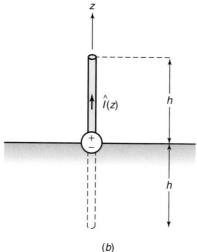

(b)

Figure 7.3 Dipole and monopole antennas. (a) The dipole antenna. (b) The monopole antenna.

apart in a "vee" fashion to the point where the line conductors are perpendicular, we obtain the dipole. Hence the form of the current distribution in (7.9). Observe that the current goes to zero at the endpoints in the expressions in (7.9) as it should.

In order to obtain the far fields of large-length dipoles, we use a simple idea. Treat this long antenna as a sequence of electrically short Hertzian dipoles and superimpose the far fields from each individual dipole as illustrated in Fig. 7.4. Hence the electric field due to a small subsection of length dz is, using (7.2a),

$$d\hat{E}_\theta = j\frac{f\mu_o}{2}\sin\theta' \frac{e^{-j\left(2\pi\frac{r'}{\lambda_o}\right)}}{r'}\hat{I}(z)dz \tag{7.10}$$

where r' and θ' are the distance and angle between the segment of current and the desired point, P, where we wish to determine the electric field. We wish to obtain the final electric field in terms of its distance and angle from the center of the dipole, r and θ. We now make an important simplifying assumption. We are only interested in

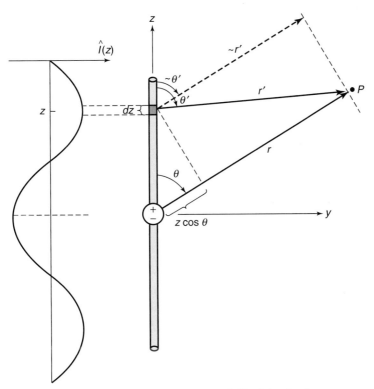

Figure 7.4 Calculating the far-field electric field of the dipole antenna as a superposition of the fields of infinitesimal Hertzian dipoles along its length.

the far fields, so we may reason that the lines from the origin to the point and from the current segment to the point are approximately parallel. Hence the angles from the origin of the spherical coordinate system to the point, θ, and from the current segment to the point, θ', are approximately the same, $\theta \cong \theta'$. The difference between r and r' is

$$r' = r - z\cos\theta \tag{7.11}$$

Substituting (7.11) into (7.10) yields

$$d\hat{E}_\theta = j\frac{f\mu_o}{2}\sin\theta\frac{e^{-j\left(2\pi\frac{(r-z\cos\theta)}{\lambda_o}\right)}}{(r-z\cos\theta)}\hat{I}(z)dz \tag{7.12}$$

The term $r - z\cos\theta$ is approximately r if (as is of interest here) the point at which we want to compute the fields is very far away from the antenna compared to its length, that is, $r \gg l$. Hence the denominator of (7.12) can be approximated as r. We cannot similarly replace the term $r - z\cos\theta$ in the numerator phase term with r because this depends not on physical distance but on electrical distance as evidenced by its division by λ_o. For example, two points that are distances 1000 m and 1000.5 m away will have a difference of phase of 180° for a frequency of 300 MHz ($\lambda_o = 1$ m). Hence (7.12) becomes

$$d\hat{E}_\theta = j\frac{f\mu_o}{2}\sin\theta\frac{e^{-j2\pi\frac{r}{\lambda_o}}}{r}e^{j\frac{2\pi z\cos\theta}{\lambda_o}}\hat{I}(z)dz \tag{7.13}$$

Integrating this along the length of the dipole and substituting the current distribution given in (7.9) gives

$$
\hat{E}_\theta = \int_{z=-\frac{l}{2}}^{\frac{l}{2}} d\hat{E}_\theta
$$
$$
= j\frac{\eta_o \hat{I}_m e^{-j2\pi\frac{r}{\lambda_o}}}{2\pi r} F(\theta)
$$

(7.14a)

where

$$
F(\theta) = \frac{\cos\left[\pi\dfrac{l}{\lambda_o}\cos\theta\right] - \cos\left(\pi\dfrac{l}{\lambda_o}\right)}{\sin\theta}
$$

(7.14b)

where $F(\theta)$ is said to be the *pattern factor* of the antenna since for a fixed distance r it gives the field as a function of angle θ, which we will plot as the *pattern* of the antenna in order to give its directional focusing ability. Since the point is in the far field of the antenna, the magnetic field is

$$
\hat{H}_\phi = \frac{\hat{E}_\theta}{\eta_o}
$$

(7.14c)

Now suppose the dipole is one-half wavelength long, $l = \lambda_o/2$. The pattern factor in (7.14b) becomes

$$
F(\theta) = \frac{\cos\left(\dfrac{\pi}{2}\cos\theta\right)}{\sin\theta} \qquad \text{half-wave dipole, } l = \frac{\lambda_o}{2}
$$

(7.15)

For the half-wave dipole, the maximum electric field intensity is broadside to the antenna, $\theta = 90°$ and $F(\theta = 90°) = 1$. Hence the magnitude of the maximum electric field intensity is obtained from (7.14) as simply

$$
|\hat{E}_\theta|_{\max} = 60\frac{|\hat{I}_m|}{r} \qquad \text{half-wave dipole, } \theta = 90°
$$

(7.16)

The *pattern* of the half-wave dipole antenna is plotted in Fig. 7.5a both in cross section and in three dimensions for a fixed distance, r, and a function of θ. Observe that the electric field decays inversely with distance. Hence at a fixed distance r the electric field depends only on angle θ through the pattern factor $F(\theta)$ given in (7.14b), which simplifies to (7.15) for the half-wave dipole antenna. The information contained in a plot of the pattern here is as follows. For a fixed value of θ, the distance from the origin to the pattern outline is the *relative magnitude of the electric field for that angle θ*. In other words, a line drawn from the origin to the pattern extremity for that angle θ is proportional to the electric field. This shows that the electric field is zero off the ends of the dipole and is a maximum broadside to it, $\theta = 90°$. For a dipole

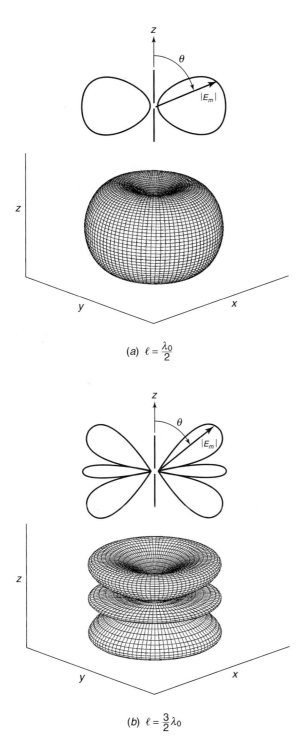

(a) $\ell = \dfrac{\lambda_0}{2}$

(b) $\ell = \dfrac{3}{2}\lambda_0$

Figure 7.5 Radiated electric field patterns of dipoles. (a) A dipole whose length is one-half wavelength, $l = \lambda_o/2$. (b) A dipole whose length is one and one-half wavelength, $l = (3/2)\,\lambda_o$.

whose length is one and one-half wavelengths, $l = (3/2) \lambda_o$, the pattern factor $F(\theta)$ in (7.14a) becomes

$$F(\theta) = \frac{\cos\left[\dfrac{3\pi}{2}\cos\theta\right]}{\sin\theta} \qquad l = \frac{3}{2}\lambda_o$$

This pattern is plotted both in cross section and in three dimensions in Fig. 7.5b. Observe that for this length of the dipole, nulls appear in the pattern off the ends of the dipole and also at other positions. There will be no transmission for these values of θ.

The average power density is

$$
\begin{aligned}
\mathbf{S}_{AV} &= \frac{1}{2}\,\mathrm{Re}(\hat{E}_\theta \hat{H}_\phi^\circ)\,\mathbf{a}_r \\
&= \frac{1}{2}\frac{|\hat{E}_\theta|^2}{\eta_o}\,\mathbf{a}_r \\
&= \underbrace{\left(\frac{\eta_o}{8\pi^2}\right)}_{4.77}\frac{|\hat{I}_m|^2}{r^2}F^2(\theta)\,\mathbf{a}_r \qquad \mathrm{W/m^2}
\end{aligned}
\tag{7.17}
$$

The total radiated average power is obtained by integrating the average power density over a closed surface (which we again choose to be a sphere of radius R):

$$
\begin{aligned}
P_{AV} &= \int\limits_{\phi=0}^{2\pi}\int\limits_{\theta=0}^{\pi} \mathbf{S}_{AV}\cdot\underbrace{R^2\sin\theta\,d\theta\,d\phi\,\mathbf{a}_r}_{d\mathbf{s}} \\
&= 4.77|\hat{I}_m|^2\int\limits_{\phi=0}^{2\pi}\int\limits_{\theta=0}^{\pi}F^2(\theta)\,\sin\theta\,d\theta\,d\phi \qquad \mathrm{W}
\end{aligned}
\tag{7.18}
$$

Unfortunately this integral cannot be evaluated in closed form. Numerical integration for a half-wave dipole gives

$$\boxed{P_{AV} = 73|\hat{I}_{rms}|^2 \qquad \mathrm{W} \quad \text{(half-wave dipole)}}\tag{7.19}$$

From this result we obtain the radiation resistance for the half-wave dipole as

$$\boxed{R_{rad} = 73 \qquad \Omega \quad \text{(half-wave dipole)}}\tag{7.20}$$

For the monopole above a ground plane, the fields above the ground plane are identical, according to the method of images, to those of the dipole and given in (7.14). However, there is one important difference between the two. The monopole radiates only half the power as the equivalent dipole (only above the ground plane). Hence the radiation resistance of the monopole is half that of the equivalent dipole. The radiation resistance of a quarter-wave monopole ($h = \lambda_o/4$) is therefore

$$\boxed{R_{rad} = 36.5 \qquad \Omega \quad \text{(quarter-wave monopole)}}\tag{7.21}$$

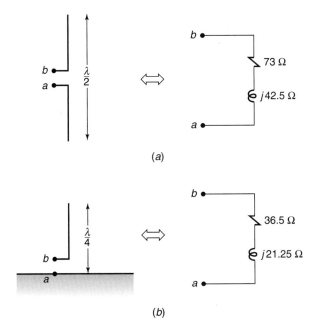

Figure 7.6 Representing the input to an antenna. (a) The input to a half-wave dipole. (b) The input to a quarter-wave monopole.

▶ **QUICK REVIEW EXERCISE 7.4**

A half-wave dipole carries a 100-MHz current whose magnitude (rms) at the center of the dipole (the excitation point) is 100 mA. Determine the total power radiated by the dipole and the power density at a distance of 1000 m away broadside to the antenna.

ANSWER   0.73 W and 95.4 nW/m². ◀

The input impedance to an antenna consists not only of a real part (the radiation resistance) but also has an imaginary part representing an inductive (or capacitive) reactance. The imaginary part of the impedance of a half-wave dipole is difficult to derive but becomes

$$jX = j42.5 \quad \Omega \qquad \text{(half-wave dipole)} \tag{7.22}$$

and for a quarter-wave monopole is

$$jX = j21.25 \quad \Omega \qquad \text{(quarter-wave monopole)} \tag{7.23}$$

Hence, the equivalent input circuits of the antennas are as shown in Fig. 7.6.

▶ **EXAMPLE 7.3**

A 1-m length dipole is driven by a 150-MHz, 50-Ω source having an open-circuit voltage of 10 V (rms) as shown in Fig. 7.7a. The dipole is constructed of #20-gauge copper wires having a radius of 4.06×10^{-4} m. Determine the total power radiated by the dipole and its efficiency.

SOLUTION   The resistance of the wires can be computed as 0.63 Ω. The total equivalent circuit of the antenna as seen at its base consists of $R_{loss} = 0.63$ Ω in series with the radiation resistance, $R_{rad} = 73$ Ω, in series with the reactance, $jX = j42.5$ Ω, as shown in Fig. 7.7b. The total power

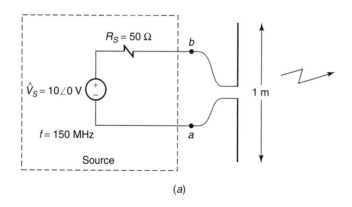

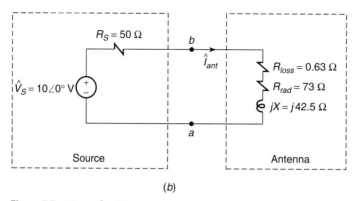

Figure 7.7 Example 7.3.

radiated is the power dissipated in the radiation resistance. To determine that radiated power, we must first determine the input current to the antenna terminals:

$$\hat{I}_{ant} = \frac{\hat{V}_S}{R_S + R_{loss} + R_{rad} + jX}$$

$$= \frac{10\angle 0°}{50 + 0.63 + 73 + j42.5}$$

$$= 76.5\angle -18.97° \qquad \text{mA}$$

The total power radiated is the power dissipated in the radiation resistance:

$$P_{rad} = |\hat{I}_{ant}|^2 R_{rad}$$

$$= 427.1 \text{ mW}$$

Since the source voltage is in rms, the current is in rms. The power lost in the wire resistance is

$$P_{loss} = |\hat{I}_{ant}|^2 R_{loss}$$

$$= 3.69 \text{ mW}$$

The efficiency of the antenna is the ratio of the total power radiated to the total power input to its terminals:

$$efficiency = \frac{P_{rad}}{P_{loss} + P_{rad}}$$

$$= 0.991$$

or approximately 99%.

▶ 7.3 ANTENNA ARRAYS

An important property of an antenna is its ability to focus the radiated power in a certain direction as illustrated in Fig. 7.8. From the standpoint of communicating with a receiving antenna, any radiated power that is not transmitted in the direction of the receiving antenna is wasted. Various antenna structures provide different degrees of focusing ability. The Hertzian dipole, half-wave dipole, and quarter-wave monopole focus their transmitted power broadside to the antenna but equally in all directions in the $\theta = 90°$ plane since the far fields depend on $\sin\theta$. Other antennas such as parabolic and horn antennas produce pencil beams such that the radiated power is maximally directed in a single direction, the main beam of the antenna.

It is possible, and frequently done, to use two or more dipoles/monopoles to provide more directional focusing than would be provided by each one alone. The principle here is that the fields of the individual antennas in the "array" combine to yield a total field that has nulls in certain directions and maxima in the desired direction. Consider two identical antennas separated a distance d as illustrated in Fig. 7.9. The antennas are placed along the x axis at $x = -d/2$ and $x = d/2$ and are directed in the z direction (perpendicular to the page). Each antenna carries a current whose magnitude I is the same for both, but the phases of the currents are different. This phase difference is routinely accomplished by placing a phase-shifter circuit at the input to the antenna. The current of the left antenna is $\hat{I}_1 = I\angle\alpha$ and the current of the right antenna is $\hat{I}_2 = I\angle 0°$. Hence the current of antenna 1 leads the current of antenna 2 by α degrees. We wish to determine the combined electric fields at a point P that is a distance r from the center of the array and an angle ϕ measured as shown in Fig. 7.9. Ordinarily we are interested in the fields at a large distance from the antenna array with respect to the separation of the antennas. Hence we may assume lines drawn from each antenna to the point are approximately parallel to each other and parallel to a line drawn from the center of the array to the point as illustrated in Fig. 7.9. Hence we have

$$r_1 \cong r - \frac{d}{2}\cos\phi \tag{7.24a}$$

and

$$r_2 \cong r + \frac{d}{2}\cos\phi \tag{7.24b}$$

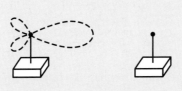

Figure 7.8 Illustration of the need to provide directionality with an array of antennas.

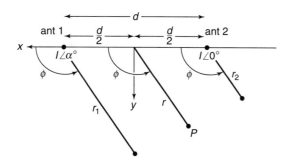

Figure 7.9 An array of two identical antennas. The antennas have an omnidirectional pattern in the plane shown.

We will assume that the two antennas are identical in structure (two half-wave dipoles, two quarter-wave monopoles, two Hertzian dipoles, etc.) so that the electric fields at point P are given by

$$\hat{E}_1 = K\hat{I}_1 \frac{e^{-j2\pi\frac{r_1}{\lambda_o}}}{r_1} \tag{7.25a}$$

and

$$\hat{E}_2 = K\hat{I}_2 \frac{e^{-j2\pi\frac{r_2}{\lambda_o}}}{r_2} \tag{7.25b}$$

The constant K contains various items that are specific to the type of antenna. For example, for the Hertzian dipole whose *far-field* electric field is given in (7.2a), $K = j(f\mu_o/2)\,dl\,\sin\theta$. The total electric field at point P is, by superposition, the sum of the electric fields due to each antenna. Adding (7.25) and substituting (7.24) yields

$$\hat{E} = \hat{E}_1 + \hat{E}_2$$

$$= KI\angle\alpha\,\frac{e^{-j2\pi\frac{r_1}{\lambda_o}}}{r_1} + KI\angle 0\,\frac{e^{-j2\pi\frac{r_2}{\lambda_o}}}{r_2}$$

$$\cong \frac{K}{r}I\left(e^{j\alpha}e^{-j2\pi\frac{\left(r-\frac{d}{2}\cos\phi\right)}{\lambda_o}} + e^{-j2\pi\frac{\left(r+\frac{d}{2}\cos\phi\right)}{\lambda_o}}\right) \tag{7.26}$$

$$= \frac{K}{r}Ie^{-j2\pi\frac{r}{\lambda_o}}e^{j\frac{\alpha}{2}}\underbrace{\left(e^{j\left(\frac{\pi d}{\lambda_o}\cos\phi + \frac{\alpha}{2}\right)} + e^{-j\left(\frac{\pi d}{\lambda_o}\cos\phi + \frac{\alpha}{2}\right)}\right)}_{\displaystyle 2\cos\left(\frac{\pi d}{\lambda_o}\cos\phi + \frac{\alpha}{2}\right)}$$

$$= 2\frac{K}{r}Ie^{-j2\pi\frac{r}{\lambda_o}}e^{j\frac{\alpha}{2}}\cos\left(\frac{\pi d}{\lambda_o}\cos\phi + \frac{\alpha}{2}\right)$$

We have substituted (7.24) into the phase term and $r_1 = r_2 \cong r$ in the denominators for reasons discussed previously. The *pattern* of this array (for a fixed r) depends on angle ϕ as

$$\hat{E} \propto \cos\left(\frac{\pi d}{\lambda_o}\cos\phi + \frac{\alpha}{2}\right) \tag{7.27}$$

Hence, the *array factor* for plotting the pattern is

$$F(\phi) = \cos\left(\frac{\pi d}{\lambda_o} \cos\phi + \frac{\alpha}{2}\right) \qquad (7.28)$$

► **EXAMPLE 7.4**

Suppose the separation of the two antennas in the array is one-half wavelength, $d = \lambda_o/2$, and the currents are in phase, $\alpha = 0°$. Sketch the pattern of the array.

SOLUTION The array factor in (7.28) becomes

$$F(\phi) = \cos\left(\frac{\pi}{2} \cos\phi\right)$$

which is sketched in Fig. 7.10a. This pattern can be determined with a physical argument with the following observations. The wave propagated from each antenna suffers a phase shift of $\angle -2\pi r/\lambda_o$ as it propagates away from the antenna. First examine the direction $\phi = 0°$. A wave propagating to the left from antenna 1 has a relative amplitude of 1 arriving at the point where we desire to determine the field. A wave propagating to the left from antenna 2 suffers a phase shift of $\angle -2\pi(d/\lambda_o) = \angle -\pi$ or 180° in propagating from antenna 2 to antenna 1. Hence it arrives at the field point 180° out of phase with the wave propagating from antenna 1. Thus the two contributions cancel for the $\phi = 0°$ direction. Similarly for the $\phi = 180°$, we also have a null

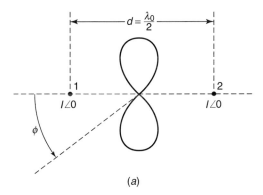

(a)

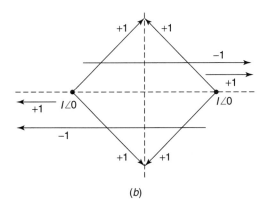

(b)

Figure 7.10 Example 7.4.

in the pattern because the two waves are out of phase when they arrive at the field point. Now consider points broadside to the array, $\phi = 90°$ and $\phi = 270°$. The waves from the two antennas in this case travel the same distances and hence arrive in phase, thus reinforcing each other, giving a maximum in the pattern.

▶ EXAMPLE 7.5

Suppose the separation of the two antennas in the array is one-half wavelength, $d = \lambda_o/2$, and the currents are $180°$ out of phase, $\alpha = \pi$. Sketch the pattern of the array.

SOLUTION The array factor in (7.28) becomes

$$F(\phi) = \cos\left(\frac{\pi}{2}\cos\phi + \frac{\pi}{2}\right)$$

The pattern is sketched in Fig. 7.11a. Again, this pattern is relatively simple to sketch using a physical argument. The wave propagated from each antenna suffers a phase shift of $\angle -2\pi(r/\lambda_o)$ as it propagates away from the antenna. In addition, the current of the antenna may have, as is the case for antenna 1, a phase angle that further shifts the phase of the wave arriving at the field point. First, examine the direction $\phi = 0°$. A wave propagating to the left from antenna 1 has a relative amplitude of -1 arriving at the point where we desire to determine the field. This is due to the fact that the wave starts out with $180°$ phase angle due to the phase angle of its current. A wave propagating to the left from antenna 2 suffers a phase shift of $\angle -2\pi(d/\lambda_o) = \angle -\pi$ or $-180°$ in propagating from antenna 2 to antenna 1. Since the phase angle of the current is zero

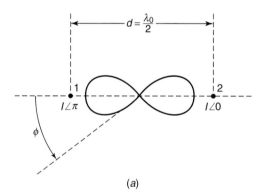

(a)

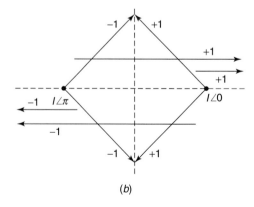

(b)

Figure 7.11 Example 7.5.

degrees, there is no additional phase shift incurred. Hence it arrives at the field point in phase with the wave propagating from antenna 1. Thus the two contributions add for the $\phi = 0°$ direction. Similarly for the $\phi = 180°$, we also have a maxima in the pattern because the two waves are in phase when they arrive at the field point. The wave arriving from antenna 1 suffers $-180°$ due to its having to propagate a distance $d = \lambda_o/2$ and another $180°$ by virtue of the phase angle of the current being $180°$, giving it a net phase shift of $0°$. Hence it arrives in phase with the wave transmitted from antenna 2. Now consider points broadside to the array, $\phi = 90°$ and $\phi = 270°$. The waves from the two antennas in this case travel the same distances but are $180°$ out of phase by virtue of the current of antenna 1 starting out $180°$ out of phase. Hence they arrive $180°$ out of phase, thus giving a null in the pattern. ◀

▶ **EXAMPLE 7.6**

Suppose the separation of the two antennas in the array is one-quarter wavelength, $d = \lambda_o/4$, and the currents are $90°$ out of phase, $\alpha = \pi/2$. Sketch the pattern of the array.

SOLUTION   The array factor in (7.28) becomes

$$F(\phi) = \cos\left(\frac{\pi}{4}\cos\phi + \frac{\pi}{4}\right)$$

The pattern is sketched in Fig. 7.12a. Again, this pattern is relatively simple to sketch using a physical argument. The wave propagated from each antenna suffers a phase shift of $\angle -2\pi(r/\lambda_o)$ as it propagates away from the antenna. In addition, the current of the antenna may have, as is the case for antenna 1, a phase angle that further shifts the phase of the wave arriving at the field point. First examine the direction $\phi = 0°$. A wave propagating to the left from antenna 1 has a

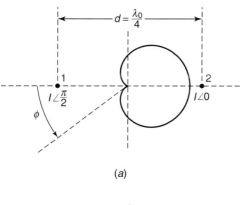

(a)

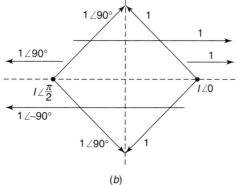

(b)

Figure 7.12   Example 7.6.

phase angle of 90° when it arrives at the point where we desire to determine the field due to the phase angle of the current. A wave propagating to the left from antenna 2 suffers a phase shift of $\angle -2\pi(d/\lambda_o) = \angle -(\pi/2)$ or −90° in propagating from antenna 2 to antenna 1. Since the phase angle of the current is zero degrees, there is no additional phase shift incurred. Hence it arrives at the field point with a phase angle of −90°, which is out of phase with the wave propagating from antenna 1, which arrives with a phase of 90°. Hence the two waves cancel, giving a null for $\phi = 0°$. Now consider $\phi = 180°$. A wave propagating to the right from antenna 1 starts with a phase angle of 90°. As it propagates from antenna 1 to antenna 2, it suffers a phase shift of $\angle -2\pi(d/\lambda_o) = \angle -(\pi/2)$ or −90°. This gives a net phase of 90° − 90° = 0° when it arrives at the field point. The wave from antenna 2 arrives at the field point with a relative phase of 0° since it starts out with a phase angle of 0° due to the phase of the current. Hence the two fields add for $\phi = 180°$. Now consider points broadside to the array, $\phi = 90°$ and $\phi = 270°$. The waves from the two antennas in this case travel the same distances but are 90° out of phase by virtue of the currents starting out 90° out of phase. Hence they arrive 90° out of phase, thus giving a total of $1 + j1 = \sqrt{2}\angle 45°$. ◀

The location of nulls in the pattern can be methodically determined by setting the array factor equal to zero:

$$F(\phi) = \cos\left(\frac{\pi d}{\lambda_o}\cos\phi + \frac{\alpha}{2}\right)$$
$$= 0$$

Thus the location of nulls in the pattern can be found by solving for

$$\frac{\pi d}{\lambda_o}\cos\phi + \frac{\alpha}{2} = \pm\frac{\pi}{2} \tag{7.29}$$

For example, in Example 7.4, where $d = \lambda_o/2$ and $\alpha = 0°$, we have

$$\frac{\pi}{2}\cos\phi = \pm\frac{\pi}{2}$$

giving nulls at $\cos\phi = \pm1$ or $\phi = 0°, 180°$. For Example 7.5, where $d = \lambda_o/2$ and $\alpha = 180°$, we have

$$\frac{\pi}{2}\cos\phi + \frac{\pi}{2} = \pm\frac{\pi}{2}$$

giving nulls at $\cos\phi = 0, -2$ or $\phi = \pm90°$ and $\cos\phi = -2$ does not represent a physical angle. In Example 7.6, where $d = \lambda_o/4$ and $\alpha = 90°$, we have

$$\frac{\pi}{4}\cos\phi + \frac{\pi}{4} = \pm\frac{\pi}{2}$$

giving nulls at $\cos\phi = 1, -3$ or $\phi = 0°$ and $\cos\phi = -3$ does not represent a physical angle.

▶ 7.4 PROPERTIES OF ANTENNAS

We have studied the properties of important wire-type antennas, the Hertzian dipole and the half-wave dipole and quarter-wave monopole. The important properties of focusing of the radiated power in certain directions and the input impedance to those antennas were studied. Other types of antennas such as parabolic and horn antennas share these and several other important properties. In this section we will study these important parameters that are shared by all other antennas. For more complicated antennas these properties are often measured rather than being calculated directly.

7.4.1 Directivity and Gain

The ability to focus the radiated power in certain directions is an important property of an antenna and is contained in the *gain* of that antenna. The *directivity* in a particular θ,ϕ direction, $D(\theta,\phi)$, is the ratio of the power density radiated in that direction to the power density at this point *if* the total power radiated by the antenna were radiated equally in all directions:

$$D(\theta,\phi) = \frac{S_{AV}(\theta,\phi)}{S_{AV,\text{ total power radiated equally in all directions}}} \tag{7.30}$$

The power densities in this expression are at the same distance from the antenna. Hence, the directivity gives a measure of the focusing ability of the antenna in a particular direction. The *gain*, $G(\theta,\phi)$, is related to the directivity of the antenna by the efficiency of the antenna, e:

$$G(\theta,\phi) = eD(\theta,\phi) \tag{7.31}$$

For example, in Example 7.3, some of the power supplied to the terminals of the antenna is dissipated in the ohmic resistance of the wires, giving it an efficiency of 99.1%. Hence, 99.1% of the total power supplied to the terminals is radiated and the efficiency is $e = 0.991$. For most useful antennas, the efficiency is close to 100%, so that the directivity and gain are approximately equal. For inefficient antennas such as the electrically small Hertzian dipole, the efficiency can be quite small, so that the directivity and gain are not the same. Antennas tend to focus their radiated power in certain directions. Hence, it is common to simply specify the gain of the antenna for the θ and ϕ direction that results in the maximum radiation:

$$G = eD$$
$$= e\frac{S_{AV}(\theta_{max},\phi_{max})}{S_{AV,\text{ total power radiated equally in all directions}}} \tag{7.32}$$

If no θ,ϕ direction is specified, it is to be assumed that the stated gain is in the direction of a maximum, that is, the main beam. For example, for the Hertzian dipole as well as the half-wave dipole, the direction of maximum radiation is broadside to the dipole, so that $\theta_{max} = 90°$. We will assume that the efficiency of the antenna is 100% in the following, so that directivity and gain are the same, and will use the symbol G to denote that gain.

An ideal antenna that radiates equally in all directions, that is, has no focusing ability, is the *isotropic point source* shown in Fig. 7.13a. An isotropic point source is nonphysical, since all radiating structures have some focusing ability. The efficiency of an isotropic point source is 100%, that is, all the power applied to its input is radiated. Hence its directivity, D, and gain, G, are the same. If it is transmitting a total power of P_T W, then the power at a point r away is spread equally over the surface of a sphere of surface area $4\pi r^2$ m^2 as illustrated in Fig. 7.13a. Hence the power density at the point is

$$S_{AV} = \frac{P_T}{4\pi r^2} \quad \text{W/m}^2, \text{ isotropic point source} \tag{7.33}$$

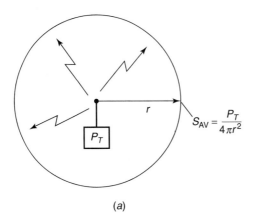

(a)

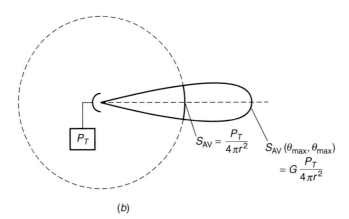

(b)

Figure 7.13 Illustration of the meaning of antenna directivity. (a) The isotropic point source. (b) Directivity of a general antenna.

Therefore, the gain of an antenna that is transmitting a total power of P_T W is

$$G = \frac{S_{AV}(\theta_{max}, \phi_{max})}{S_{AV,\text{ total power radiated equally in all directions}}}$$

$$= 4\pi r^2 \frac{S_{AV}(\theta_{max}, \phi_{max})}{P_T} \tag{7.34}$$

Consequently, the gain of an antenna gives its ability to focus the radiated power in a particular direction relative to an isotropic point source as illustrated in Fig. 7.13b, so that

$$\boxed{S_{AV}(\theta_{max}, \phi_{max}) = G\frac{P_T}{4\pi r^2} \quad \text{W/m}^2} \tag{7.35}$$

▷ **EXAMPLE 7.7**

Determine the gain of the Hertzian dipole.

SOLUTION The power density of the Hertzian dipole is given in (7.4) as

$$S_{AV} = 15\pi \, |\hat{I}|^2 \left(\frac{dl}{\lambda_o}\right)^2 \frac{\sin^2\theta}{r^2} \quad \text{W/m}^2$$

which is a maximum in the direction $\theta_{\max} = 90°$. The total power radiated is given by (7.6) as

$$P_T = 80\pi^2 \left(\frac{dl}{\lambda_o}\right)^2 \frac{|\hat{I}|^2}{2} \qquad \text{W}$$

Hence the gain is given by (7.34) as

$$G = 1.5$$

◀

EXAMPLE 7.8

Determine the gain of a half-wave dipole.

SOLUTION   The power density of a half-wave dipole is given by (7.17) as

$$S_{AV} = 4.77 \frac{|\hat{I}_m|^2}{r^2} F^2(\theta) \qquad \text{W/m}^2$$

which is a maximum in the direction $\theta_{\max} = 90°$, so that $F(\theta_{\max}) = 1$. The total power radiated is given by

$$P_T = 73 \frac{|\hat{I}_m|^2}{2} \qquad \text{W}$$

Hence the gain is given by (7.34) as

$$G = 1.64$$

According to this, the half-wave dipole is only slightly better than the Hertzian dipole in its ability to focus the radiated power. However, there is a large difference between the two antennas. The radiation resistance of the half-wave dipole is considerably larger than that of the Hertzian dipole and hence power can be transmitted using a much smaller input current. The quarter-wave monopole radiates half the power of a half-wave dipole (in the space above the location of the ground plane). Hence the gain of the quarter-wave monopole is twice that of the corresponding half-wave dipole or 3.28.

◀

Gain of an antenna is often specified in decibels (dB). The gain of an antenna is specified in dB as

$$G_{\text{dB}} = 10 \log_{10}(G) \tag{7.36}$$

The gain of the Hertzian dipole is $G = 10 \log_{10}(1.5) = 1.76$ dB. The gain of the half-wave dipole is $G = 10 \log_{10}(1.64) = 2.15$ dB, and the gain of a quarter-wave monopole is $G = 10 \log_{10}(3.28) = 5.17$ dB. Antennas that have considerable focusing ability such as the parabolic or horn antennas have very large gains. A gain of 40 dB is equivalent to a gain of 10,000, that is, the power density in the direction of maximum radiation is 10,000 times the power density if the total transmitted power were radiated equally in all directions, as with an isotropic point source. Satellite antennas such as are used to receive television signals rely on these very large gains (as much as 60 dB) to compensate for the very small power density received on the earth from an orbiting satellite, which can be on the order of 10^{-12} W/m^2.

Gain of an antenna is specified with respect to the gain of some reference antenna. The reference antenna used here is the isotropic point source, which has a gain of unity, $G = 1$. In some instances in industry, the gain of an antenna may be specified with respect to a half-wave dipole.

7.4.2 Effective Aperture

In the previous section we discussed the focusing ability of an antenna when it is used to transmit power. Of equal importance is the ability of a receiving antenna to extract power from a passing wave that has been transmitted from another antenna. This ability to extract power from a passing wave is given by the *effective aperture* of the antenna, which is sometimes referred to its *effective area* or *capture area*.

Consider an antenna, depicted as a horn antenna, shown in Fig. 7.14a which has an incident wave impinging on it. This receiving antenna extracts some of the power in the passing wave and converts it to received power, P_R W, at its base. In the far field of antennas, the waves radiated by them resemble uniform plane waves. Hence we can characterize the incident wave as having a power density of

$$S_{AV,inc} = \frac{1}{2} \frac{|\hat{E}|^2}{\eta_o} \quad \text{W/m}^2 \tag{7.37}$$

The *effective aperture* of an antenna is the ratio of the power extracted and received at its base to the power density of the incident (uniform plane) wave:

$$\boxed{A_{eff} = \frac{P_R}{S_{AV,inc}} \quad \text{m}^2} \tag{7.38}$$

Observe that the effective aperture is the ratio of a power in W to a power density in W/m^2, giving the units of effective aperture as square meters, a surface area. Hence we can think of the effective aperture of an antenna as the effective capture area of it. Surface-type antennas such as parabolic and horn antennas have an effective aperture that is very close to their physical aperture or opening. Wire-type antennas such as the half-wave dipole have no physical capture area but do extract power from a passing wave and therefore have an effective aperture also.

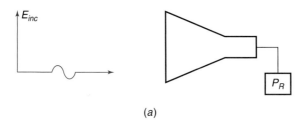

(a)

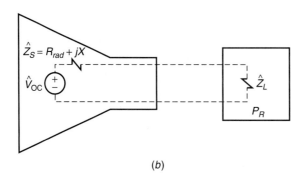

(b)

Figure 7.14 Receiving properties of antennas. (a) Illustration of effective aperture (area). (b) The equivalent circuit of a receiving antenna.

There are some ideal assumptions inherent in the above definition and concept of effective aperture. The ability of the antenna to extract power from the passing wave depends on its orientation with respect to the wave front. For example, a surface-type antenna such as a horn antenna must be oriented such that the area of its mouth is parallel to the wave front of the incoming uniform plane wave in order to maximally extract the power from it. Similarly, a wire-type antenna such as a half-wave dipole must have its wire axis oriented parallel to the electric field vector of the incident wave in order to maximally extract power from it. If a wire antenna is oriented with its axis perpendicular to the electric field vector of the incoming wave, the wave will evoke no response in the antenna. Hence the definition of effective aperture assumes that the receiving antenna is oriented for maximal extraction of the incident wave power. Also, this receiving antenna can be thought of or modeled as a typical source with an open-circuit voltage $\hat{V}_{OC}$ and a source impedance $\hat{Z}_S$ as illustrated in Figure 7.14b. In order for the extracted power to be maximally delivered to the receiver, represented by $\hat{Z}_L$, the source and load impedances must be matched, that is, $\hat{Z}_L = \hat{Z}_S^*$. The principle of reciprocity gives the source impedance of an antenna when it is used for reception as being the same as its input impedance when it is used for transmission, $\hat{Z}_S = R_{rad} + jX$. Hence a matched load would have its impedance as $\hat{Z}_L = \hat{Z}_S^* = R_{rad} - jX$. The definition of effective aperture also assumes this matched condition between the antenna and the receiver at its base.

It turns out that the reception ability of an antenna when used as a receiving antenna is directly related to its gain or focusing ability when it is used for transmission. This important relationship is

$$A_{eff}(\theta,\phi) = \frac{\lambda_o^2}{4\pi} G(\theta,\phi) \tag{7.39}$$

Hence, the effective aperture of an antenna can be determined from its gain. Strictly speaking, the gain in (7.39) should be the directivity, $D(\theta,\phi)$, but we will henceforth assume that the antenna efficiency is 100%, so that the two are the same.

► **EXAMPLE 7.9**

Demonstrate the relation between effective aperture and gain in (7.39) for a Hertzian dipole.

SOLUTION   An incident uniform plane wave has power density given by (7.37). If the antenna axis is oriented parallel to the electric field vector of this incident wave, the open-circuit voltage induced at its base is

$$\hat{V}_{OC} = |\hat{E}|\, dl$$

where dl is the length of the dipole. This follows from the observation that the passing wave will induce a voltage between the two ends of the dipole, but since the dipole is assumed to be electrically short, this induced voltage is approximately the product of the electric field tangent to the dipole and its length. With reference to Fig. 7.14b and assuming the load is matched, $\hat{Z}_L = \hat{Z}_S^* = R_{rad} - jX$, the power received is

$$P_R = \frac{|\hat{V}_{OC}|^2}{8R_{rad}}$$
$$= \frac{|\hat{E}|^2\, dl^2}{8R_{rad}}$$

and we assume that the antenna is lossless. Substituting the value of radiation resistance for the Hertzian dipole given in (7.8) yields

$$P_R = \frac{|\hat{E}|^2 \lambda_o^2}{640\pi^2}$$

The average power density in the passing wave is given by (7.37). Thus the effective aperture is

$$A_{eff} = \frac{P_R}{S_{AV,inc}}$$
$$= 1.5\frac{\lambda_o^2}{4\pi}$$

The gain of the Hertzian dipole was shown to be 1.5 previously.   ◀

7.4.3   The Friis Transmission Equation[2]

An important use of antennas is in a communication link illustrated in Fig. 7.15. A transmitter is transmitting a total power P_T W via a transmitting antenna. A receiver is at a distance d from it and is attached to a receiving antenna. From our previous results, the power density at the receiving antenna is

$$S_{AV} = \frac{P_T}{4\pi d^2}G_T(\theta_T,\phi_T) \tag{7.40}$$

where θ_T,ϕ_T is the angular orientation between the main beam of the transmitting antenna and the direction toward the receiving antenna. The received power is

$$P_R = S_{AV}A_R(\theta_R,\phi_R) \tag{7.41}$$

where θ_R,ϕ_R is the angular orientation between the main beam of the receiving antenna and the direction toward the transmitting antenna and A_R is the effective aperture of the receiving antenna in that direction. Substituting yields

$$\frac{P_R}{P_T} = \frac{G_T(\theta_T,\phi_T)A_R(\theta_R,\phi_R)}{4\pi d^2} \tag{7.42}$$

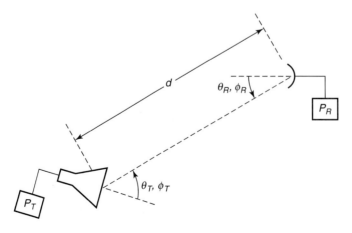

Figure 7.15   A general communication link.

[2]H.T. Friis, "A Note on a Simple Transmission Formula," *Proc. IEEE*, Vol. 34, pp. 254–256, May 1946.

Ordinarily, the antennas are characterized in terms of their gain rather than effective aperture. Substituting the relation between gain and effective aperture given in (7.39) for the receiving antenna:

$$A_R(\theta_R,\phi_R) = \frac{\lambda_o^2}{4\pi} G_R(\theta_R,\phi_R) \tag{7.43}$$

yields the Friis transmission equation:

$$\frac{P_R}{P_T} = G_T(\theta_T,\phi_T) G_R(\theta_R,\phi_R) \left(\frac{\lambda_o}{4\pi d}\right)^2 \tag{7.44}$$

This important result allows the calculation of the received power in terms of the separation distance between the two antennas and their gains (in the direction of transmission). Alternatively, (7.44) is said to be the *path loss* of the transmission path.

A second useful result is the calculation of the electric field intensity at the receiving antenna. The power density of the transmitted wave is that of a uniform plane wave:

$$S_{AV} = \frac{1}{2} \frac{|\hat{E}|^2}{\eta_o} \tag{7.45}$$

From (7.40)

$$S_{AV} = \frac{P_T}{4\pi d^2} G_T(\theta_T,\phi_T) \tag{7.46}$$

Equating these two gives the electric field at the receiving antenna:

$$|\hat{E}| = \frac{\sqrt{60 P_T G_T(\theta_T,\phi_T)}}{d} \tag{7.47}$$

The Friis transmission equation or path loss can also be written in dB as

$$10 \log_{10}\left(\frac{P_R}{P_T}\right) = G_{T,dB} + G_{R,dB} - 20 \log_{10}(f) - 20 \log_{10}(d) + 148 \tag{7.48}$$

▶ **EXAMPLE 7.10**

A microwave communication link is designed to transmit 3-GHz signals over a distance of 3000 m. The antennas are parabolic antennas having (main beam) gains of 20 dB. The received power must be at least 1 μW for proper operation. Determine the required transmitter power and the electric field intensity at the receiving antenna.

SOLUTION   The gains of the two antennas are 100, and the wavelength at 3 GHz is 0.1 m. Using the Friis transmission equation gives

$$P_R = 10^{-6}$$

$$= P_T \times 100 \times 100 \times \left(\frac{0.1}{12{,}000\pi}\right)^2$$

Solving for the required transmitter power gives $P_T = 14.2$ W. The electric field at the receiving antenna is obtained from (7.47):

$$|\hat{E}| = \frac{\sqrt{60 \times 14.2 \times 100}}{3000}$$

$$= 97.3 \qquad \text{mV/m}$$

◀

▶ **QUICK REVIEW EXERCISE 7.5**

An electronic countermeasures (ECM) system is designed to jam an enemy radar that is operating at 6 GHz. The enemy radar is 5 miles away from the ECM system. To provide successful jamming requires that the electric field at the enemy radar site must be 1 V/m. If the ECM antenna has a gain of 30 dB, determine the required transmitting power of the ECM transmitter.

ANSWER 1.08 kW.   ◀

▶ 7.5 ENGINEERING APPLICATIONS

In this final section we will discuss several engineering applications of the principles of this chapter. The first application examines the radiated electric field from a transmission line. An approximate but simple model is derived that allows the prediction of the radiated field. The second application examines the electronic steering of an antenna array by electronically changing the phase of the currents to the antennas. The third application examines the design of a satellite communication link for a certain signal-to-noise ratio. The background noise inherent in any communication system is incorporated into that design by use of the Friis transmission equation. The fourth application examines the converse to the first application, namely that an incident electromagnetic wave can induce noise or interference signals at the ends of a transmission line. Again, a simple, approximate model is obtained that allows an estimate of the induced terminal voltages. Moreover, this simple model shows the underlying factors that can be controlled in order to reduce the levels of those induced voltages.

7.5.1 Radiated Emissions from Transmission Lines

The Federal Communications Commission (FCC) in the United States requires that all digital devices (digital computers, laser printers, electronic typewriters, etc.) be tested for their radiated electric fields over the frequency band of 30 MHz to over 1 GHz.[3]

Similar restrictions are imposed by governmental bodies in other countries. Limits on the allowable radiated electric field are provided by the FCC, and any digital device whose radiated electric field exceeds those limits cannot, by law, be sold in the United States! The intent of the regulations is to prevent digital devices from causing interference with radios, televisions, and other communications. Hence manufacturers of digital devices must be able to design digital products so that, in addition to the desired functional performance, their radiated emissions do not exceed the legal limits set by the FCC. In this section we will obtain a simple model that can be used to estimate the radiated electric field produced by transmission lines (wire-type cables or printed circuit board lands) in the product. The model is based on the superposition of two Hertzian dipoles.

Consider a source and load connected by a two-conductor transmission line as shown in Fig. 7.16a. The length of the line is $\mathscr{L}$ and the separation between the two conductors is denoted as s. The two conductors can represent a variety of transmission lines such as two wires in a ribbon cable or two lands on a printed circuit board. Current I passes down one conductor and returns on the other conductor. Currents radiate.

---

[3]C.R. Paul, *Introduction to Electromagnetic Compatibility*, John Wiley Interscience, 1992.

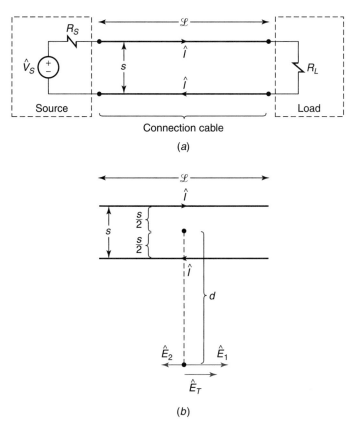

Figure 7.16 Illustration of a model to predict the electric field radiated by a transmission line. (a) The problem specification. (b) Treating each wire as a Hertzian dipole and superimposing the fields.

Hence we expect that their combination will radiate an electric field at some point a distance d away from their center as illustrated in Fig. 7.16b. We will determine that electric field in the plane of the wires as a worst case. In order to build a model to predict this, the simple idea is to treat each of the two conductors as Hertzian dipoles and superimpose the two electric fields from them. In essence we have a two-element array with $\alpha = 180°$ (the currents in the two wires are equal and oppositely directed) and want to determine the pattern in the broadside plane. In order to determine the total radiated electric field, we will make some simplifying assumptions. We will assume that (a) the point at which we wish to compute the field is in the *far field* of the two conductors and (b) the line length is electrically short at the frequency of the source, $\mathcal{L} \ll \lambda_o$. The first restriction is imposed so that we can use only the far-field electric field given in (7.2a) to model the electric field produced by the currents of each conductor rather than the complete but complicated fields in (7.1). The second restriction is imposed so that we can assume that the currents along the conductors are approximately constant along their length, that is, do not vary in magnitude or phase along their lengths. This is imposed so that the basic assumption of the Hertzian dipole is satisfied. For an electrically short transmission line, the current does not change substantially down the line. For lines that are electrically long, the current does change substantially along the line. (See Fig. 6.23.)

The current of the top conductor will produce an electric field at the point that is maximum broadside to the conductor and directed opposite to the current direction [see (7.2a)]:

$$\hat{E}_2 = j\frac{f\mu_o}{2}\hat{I}\,\mathcal{L}\frac{e^{-j2\pi\frac{(d+s/2)}{\lambda_o}}}{d+\dfrac{s}{2}}$$

$$\cong j\frac{f\mu_o}{2}\hat{I}\,\mathcal{L}\frac{e^{-j2\pi\frac{d}{\lambda_o}}}{d}e^{-j\pi\frac{s}{\lambda_o}}$$

(7.49a)

Similarly, the current of the bottom conductor produces an electric field that is directed opposite to the field of the other current:

$$\hat{E}_1 = j\frac{f\mu_o}{2}\hat{I}\,\mathcal{L}\frac{e^{-j2\pi\frac{(d-s/2)}{\lambda_o}}}{d-\dfrac{s}{2}}$$

$$\cong j\frac{f\mu_o}{2}\hat{I}\,\mathcal{L}\frac{e^{-j2\pi\frac{d}{\lambda_o}}}{d}e^{j\pi\frac{s}{\lambda_o}}$$

(7.49b)

Clearly $\hat{E}_1 > \hat{E}_2$ since the lower conductor is closer to the measurement point. Observe in these expressions we have approximated $d + (s/2) \cong d - (s/2) \cong d$ in the denominators, since we are assuming that the conductors are very close together and the measurement point is considerably farther away than the separation, that is, $d \gg s$. We cannot similarly make this approximation in the phase terms since electrical length and not physical length is involved in those terms. The total electric field is the sum of these individual contributions:

$$\hat{E}_T = \hat{E}_1 - \hat{E}_2$$

$$= j\frac{f\mu_o}{2}\hat{I}\,\mathcal{L}\frac{e^{-j2\pi\frac{d}{\lambda_o}}}{d}e^{j\pi\frac{s}{\lambda_o}} - j\frac{f\mu_o}{2}\hat{I}\,\mathcal{L}\frac{e^{-j2\pi\frac{d}{\lambda_o}}}{d}e^{-j\pi\frac{s}{\lambda_o}}$$

$$= j\frac{f\mu_o}{2}\hat{I}\,\mathcal{L}\frac{e^{-j2\pi\frac{d}{\lambda_o}}}{d}\underbrace{\left(e^{j\pi\frac{s}{\lambda_o}} - e^{-j\pi\frac{s}{\lambda_o}}\right)}_{2j\sin\left(\pi\frac{s}{\lambda_o}\right)}$$

(7.50)

$$= -\frac{f\mu_o\hat{I}\,\mathcal{L}}{d}\sin\left(\pi\frac{s}{\lambda_o}\right)e^{-j2\pi\frac{d}{\lambda_o}}$$

The magnitude of the electric field is therefore

$$|\hat{E}_T| = \frac{f\mu_o|\hat{I}|\mathcal{L}}{d}\left|\sin\left(\pi\frac{s}{\lambda_o}\right)\right|$$

(7.51)

We will now make one final observation to simplify this result. Since the line length must be electrically short, the separation between the two conductors, s, must also be very small, electrically, since $s \ll \mathcal{L}$. Therefore, using the small-angle approximation for the sin function gives

$$|\hat{E}_T| \cong \frac{f\mu_o |\hat{I}| \mathcal{L}}{d}\left(\pi \frac{s}{\lambda_o}\right)$$

$$= 1.316 \times 10^{-14} \frac{|\hat{I}| f^2 \, Area}{d}$$

(7.52)

Observe that the radiated electric field depends on

1 the magnitude of the current at that frequency, $|\hat{I}|$,

2 the square of the frequency of the current, f^2, and

3 the area of the loop formed by the two conductors, $Area = \mathcal{L}s$.

Hence, in order to reduce the radiated electric field, we either reduce the level of the current and/or reduce the loop area of the line, that is, make it shorter or bring the two conductors closer together. Large loop areas create larger radiated electric fields and hence designers should, for example, route the two traces of a digital clock signal on a printed circuit board as close together as possible and place the clock as close as possible to the integrated circuit (IC) that it serves.

As an example of the application, the FCC limit on a radiated electric field for digital devices at 300 MHz is 200 μV/m. This is to be measured at a distance of 3 m away from the device. Let us determine the maximum allowable current on a ribbon cable that will just pass this requirement. Suppose the cable length is 10 ins (0.254 m) and the typical wire separation is 50 mils (1.27×10^{-3} m). Observe that the wavelength at 300 MHz is 1 m, so that the measurement distance of 3 m is in the far field and the line length is electrically short. Hence the two basic restrictions on the validity of our model are satisfied. Substituting into (7.52) yields

$$200 \times 10^{-6} = 1.316 \times 10^{-14} \frac{|\hat{I}| \, (f = 3 \times 10^8)^2 (Area = 1.27 \times 10^{-3} \text{ m} \times 0.254 \text{ m})}{d = 3}$$

Solving for the current gives a maximum of 1.57 mA. Hence if the 300-MHz component of current (perhaps a harmonic of a digital clock or data signal being carried by the conductors) exceeds a level of 1.57 mA, the radiated emission from this cable will exceed the FCC limit and the product cannot be sold in the United States!

7.5.2 Phased Array Radars

Radars determine the distance to an object such as an airplane by transmitting a signal (a pulse) and determining the time (round trip) delay of the signal reflected off the object. Many radar applications require rapid steering of the antenna to point the beam at the object. Mechanically steering the antenna is relatively slow. Phased array antennas electronically steer the beam without having to mechanically point the antenna. The principle of phased array antennas is to electronically change the phases of the currents delivered to the antennas in the array so that the far fields of each antenna of the array combine constructively or destructively, depending on their relative phase, to give maxima (main beam) and minima (nulls) in the overall pattern in the desired directions. The phases of the currents are shifted by placing electronic phase shifters at the input to each antenna. This, then, is a simple extension of the principles of arrays studied in Section 7.3.

To illustrate this principle, consider a two-element array shown in Fig. 7.9, where the amplitudes of the two currents are equal but the phase of the left antenna (1) leads that of the right antenna (2) by α. The pattern of the array is determined in terms of the angle, ϕ, separation between the two antennas, d, and relative phase of the currents, α, by the array factor obtained in Section 7.3:

$$F(\phi) = \cos\left(\pi \frac{d}{\lambda_o} \cos\phi + \frac{\alpha}{2}\right) \tag{7.53}$$

Suppose we wish to have the main beam pointing in direction ϕ_m. Define the relative phase of the currents in terms of this desired direction of the main beam as

$$\alpha = -2\pi \frac{d}{\lambda_o} \cos\phi_m \tag{7.54}$$

Substituting (7.54) into (7.53) gives the array factor as

$$F(\phi) = \cos\left(\pi \frac{d}{\lambda_o} (\cos\phi - \cos\phi_m)\right) \tag{7.55}$$

The maximum ($F(\phi) = 1$) occurs when $\cos\phi - \cos\phi_m = 0$ or $\phi = \phi_m$. Hence we can determine for a desired location of the main beam, ϕ_m, the desired phase of the current from (7.54).

For example, suppose we choose the spacing between the two antennas as one-half wavelength, $d = \lambda_o/2$. Hence the desired phase to produce a main beam at $\phi = \phi_m$ is

$$\alpha = -\pi \cos\phi_m \tag{7.56}$$

The required phase to produce several main beam directions is given in Table 7.1: The patterns for each of these phasings of the currents are shown in Fig. 7.17. The pattern of the array must be symmetric about a line drawn between the two antennas. This requires that there be two maxima symmetrically disposed about this line. For example, consider the case for which $\phi_m = 45°$ as shown in Fig. 7.17c. The maximum of 1 occurs at $\phi = 45°$ and also at $\phi = 315°$. Hence, there are two main beams located symmetrically about the array axis. However, for two elements in the array, the beams are rather broad and the "steering of the beam" is not particularly evident. Arrays of several elements give more narrow, well-defined beams. Figure 7.18 shows a military phased array search radar. The panels contain a very large number of antennas, and the currents applied to them are shifted in phase to electronically steer the main beam.

TABLE 7.1

ϕ_m	α
0°	$-\pi = -180°$
30°	$-0.866\pi = -155.9°$
45°	$-0.707\pi = -127.3°$
75°	$-0.259\pi = -46.59°$
90°	$0°$

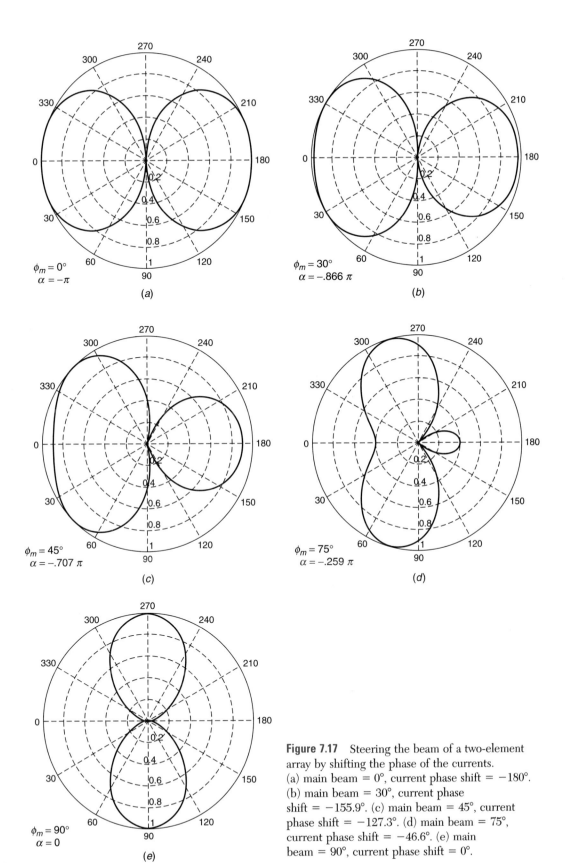

Figure 7.17 Steering the beam of a two-element array by shifting the phase of the currents. (a) main beam = 0°, current phase shift = −180°. (b) main beam = 30°, current phase shift = −155.9°. (c) main beam = 45°, current phase shift = −127.3°. (d) main beam = 75°, current phase shift = −46.6°. (e) main beam = 90°, current phase shift = 0°.

Figure 7.18 A military phased array radar antenna at the Clear Air Force Station in central Alaska. Its purpose is to detect and provide early warning of a ballistic missle attack on the United States and Canada and to provide space surveillance for satellites and space objects. The radar antennas are housed in a 105-ft-high building having three sides. The two array faces are 102 ft wide and tilted back 20 degrees from the vertical. Each array face contains 1792 active antenna elements. The main beams of each can be electronically steering between 3 and 85 degrees above the horizontal. (Photo courtesy of the U.S. Air Force. For more details, see the website http://www.pavepaws.org/.)

7.5.3 Design of a Satellite Communication Link

Satellite "downlinks" are an increasingly standard method of communicating with all parts of the world. Orbiting satellites receive the signal to be rebroadcast from an earth station. This signal is then broadcast to multiple receivers on the ground. One of the problems in this and other types of communication links is noise. Receivers generate noise internally in their electronics as a part of the amplification process. In addition, if we point the receiving antenna to a portion of the sky that contains no intentional transmitters, we will nevertheless receive background or cosmic noise. The amount of this combined noise is characterized by *noise temperature*, T_{noise}, that is measured in kelvins (K). The noise power in a certain bandwidth, B, measured in hertz (Hz), is given by

$$P_{noise} = kT_{noise}B \qquad (7.57)$$

where k is Boltzman's constant, $k = 1.38 \times 10^{-23}$ J/K. The desired signal, P_{signal}, is received in the presence of the noise. In order to discern the signal from that noise, we must have a certain signal-to-noise ratio

$$SN = \frac{P_{signal}}{P_{noise}} \qquad (7.58)$$

Consider the design of a satellite downlink to achieve a certain signal-to-noise ratio. The transmitter on the satellite is transmitting a certain power, P_T W. The antenna of the satellite has a gain of G_T. Using the Friis transmission equation, the received power is

$$P_R = P_T G_T G_R \left(\frac{\lambda_o}{4\pi d}\right)^2 \tag{7.59}$$

where G_R is the gain of the receiving antenna on the earth and d is the distance from the satellite to the earth receiver. Hence the signal-to-noise ratio in the receiver output is

$$\boxed{SN = \frac{P_T G_T G_R}{kT_{noise}B}\left(\frac{\lambda_o}{4\pi d}\right)^2} \tag{7.60}$$

Consider the design of a high-quality direct broadcast satellite link operating at 4 GHz transmitting a television signal having a bandwidth of 5 MHz. A typical noise temperature is on the order of $T_{noise} = 550$ K. Suppose the transmitter is transmitting 120 W of power and the gain of the transmitting antenna is 40 dB. The satellite is in orbit 22,370 miles above the earth. Determine the required gain of the receiving antenna to achieve a signal-to-noise ratio of 30 dB (10^3). The wavelength at 4 GHz is 7.5 cm, the distance above the earth is 3.6×10^7 m (36,000 km), and the gain of the transmitting antenna is 10,000. Substituting into (7.60) yields

$$SN = 10^3 = \frac{120 \times 10^4 \times G_R}{1.38 \times 10^{-23} \times 550 \times 5 \times 10^6}\left(\frac{.075}{4\pi \times 3.6 \times 10^7}\right)^2$$

Solving this gives a required gain of the receiving antenna of

$$G_R = 1.15 \times 10^3 \qquad (30.6\,\text{dB})$$

If a parabolic antenna is used to construct the receiving antenna, the area of its opening is approximately its effective aperture, A_R. But the effective aperture of an antenna and its gain are related by

$$A = \frac{\lambda_o^2}{4\pi} G$$

Hence the area of the required parabolic antenna is 0.515 m^2 and the required diameter of the parabolic antenna is 0.81 m or 2.7 ft.

7.5.4 Radiated Susceptibility of Transmission Lines

In Section 7.5.1 we investigated the ability of a transmission line to radiate emissions that could potentially cause interference with other electronic devices. The converse is also possible: an electromagnetic wave can induce noise signals into a transmission line, thereby causing interference with the devices the line connects. This is illustrated in Fig. 7.19a. A pair of parallel conductors (wires, PCB lands, etc.) connect two devices which are represented by the termination resistors, R_S and R_L. The line is of length $\mathcal{L}$ and the conductors are separated by s. An electromagnetic wave from some distant transmitter is incident on this line. We assume that the line is in the far field of the transmitter and hence the wave is a uniform plane wave. The component of the magnetic field that is normal or perpendicular to the area of the loop formed by the two conductors is denoted as H_n, whereas the component of the electric field that is transverse to (in the plane of and perpendicular to the conductors) the line is denoted as E_t. These

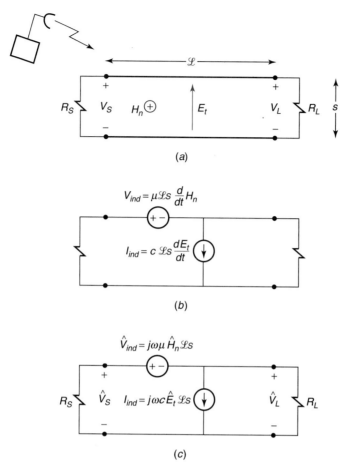

Figure 7.19 Development of a model for predicting the susceptibility of a transmission line to an incident plane wave. (a) The problem specification. (b) A simple equivalent circuit representing the effects of the incident wave as induced voltage and current sources. (c) The phasor equivalent circuit for single-frequency incident waves.

two components of the incident wave will induce sources in the line which will cause noise voltages, V_S and V_L, to be produced at the endpoints of the line (the inputs to the devices represented by R_S and R_L).

We will derive a simple but approximate model for predicting these induced voltages. The model will assume that the line is electrically short, that is, $\mathcal{L} \ll \lambda$. First consider the contribution due to the normal magnetic field, H_n. The magnetic flux penetrating the loop between the two conductors is

$$\psi = \int \mu \mathbf{H} \cdot d\mathbf{s}$$

$$= \mu H_n \underbrace{\mathcal{L}s}_{\text{Area}}$$

We have employed the assumption that the line is electrically short so that the field is approximately constant along it, and hence the H field can be removed from the integral. According to Faraday's law, a voltage source will be induced into the loop formed by the line and its terminations which is equal to the time rate-of-change of the magnetic flux penetrating the loop as shown in Fig. 7.19b:

$$V_{ind} = \frac{d\psi}{dt}$$

$$= \mu \frac{dH_n}{dt} \underbrace{\mathcal{L}s}_{\text{Area}}$$

Similarly, the transverse electric field, E_t, will induce a voltage between the two conductors, given by

$$V = - \int_c E \cdot dl$$

$$= -E_t s$$

which is positive at the lower conductor. This voltage will be in series with the total line capacitance of value $c\mathcal{L}$ F and will result in a current directed downward of

$$I_{ind} = c\mathcal{L} \frac{dV}{dt}$$

$$= c \underbrace{\mathcal{L}s}_{\text{Area}} \frac{dE_t}{dt}$$

The phasor equivalent circuit is shown in Fig. 7.19c.

Consider an example shown in Fig. 7.20. A 10-m length of "twin lead" that is commonly used to carry TV signals from a rooftop antenna to a TV set is immersed in an AM broadcast wave whose electric field is 1 V/m and has a frequency of 1 MHz. The twin lead has a characteristic impedance of 300 Ω and velocity of propagation of $v = 2.82 \times 10^8$ m/s. Hence the per-unit-length capacitance is $c = 1/vZ_C = 11.8$ pF/m.

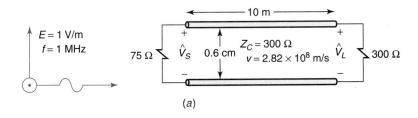

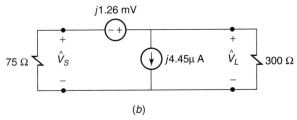

Figure 7.20 An example illustrating the prediction of induced terminal voltages. (a) The problem specification. (b) The model showing the effect of the incident wave as induced sources.

We assume that the incident wave is traveling parallel to the line, with the electric field vector completely transverse to the line. The magnetic field intensity vector must be directed out of the page in order for power flow to be in the direction of propagation of the wave, left to right. Its value is $(1 \text{ V/m})/377 \ \Omega = 2.65 \text{ mA/m}$. Hence the phasor sources in the model are

$$\hat{V}_{ind} = j\omega\mu \ \underbrace{\mathcal{L}s}_{\text{Area}} \ \hat{H}_n$$

$$= j1.26 \text{ mV}$$

The current source is

$$\hat{I}_{ind} = j\omega c \ \underbrace{\mathcal{L}s}_{\text{Area}} \ \hat{E}_t$$

$$= j4.45 \ \mu\text{A}$$

Observe that according to Lenz's law, the polarity of the induced Faraday law source has the positive on the right because the magnetic field, H_n, is directed out of the page. This circuit may be solved by superposition to give

$$\hat{V}_S = -\frac{75}{75 + 300}\hat{V}_{ind} - \frac{75 \times 300}{75 + 300}\hat{I}_{ind}$$

$$= -j0.519 \text{ mV}$$

and

$$\hat{V}_L = \frac{300}{75 + 300}\hat{V}_{ind} - \frac{75 \times 300}{75 + 300}\hat{I}_{ind}$$

$$= -j0.741 \text{ mV}$$

In addition to giving estimates of the induced interference voltages, this simple model gives further insight into the coupling of an electromagnetic wave to a transmission line. Observe that the two induced sources are due to (1) the component of the incident magnetic field that is normal to the loop formed by the transmission line and (2) the component of the incident electric field that is transverse to the transmission line. If either one is zero, then the associated induced source is not present. For example, suppose that the wave is propagating in the plane of the line perpendicular to the conductors with the electric field polarized parallel to the conductors as shown in Fig. 7.21a. In this case the H field is normal to the loop and induces source V_{ind}, but the electric field is parallel to the conductors and hence the current source is zero, $I_{ind} = 0$. Next, suppose that the wave is propagating perpendicular to the plane of the line as shown in Fig. 7.21b. Here the H field is parallel to the loop and hence the voltage source is zero, $V_{ind} = 0$. The electric field is completely transverse, so that the current source, I_{ind}, is nonzero. Finally, consider the situation shown in Fig. 7.21c wherein the wave is propagating along the line but the electric field vector is perpendicular to the loop and the magnetic field vector is transverse to the loop. In this case both sources are zero, $V_{ind} = 0$ and $I_{ind} = 0$, and no interference voltage is induced across the terminals of the line. This model shows that reorienting the line (where this is feasible) such that no component of the incident magnetic field is perpendicular to the loop and no component of the incident electric field is transverse to the line will (ideally) eliminate any induced interference voltages across the terminations.

An example of where this has been useful in the past has to do with susceptibility of electronic typewriters and laser printers to electrostatic discharge (ESD). The industry

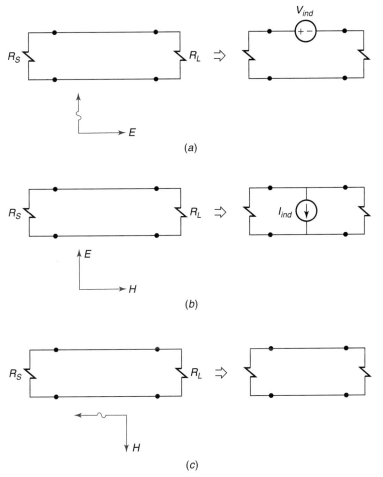

Figure 7.21 Illustration of the effect of the direction of incidence of the wave and the polarization of its fields with respect to the transmission line. (a) Propagation in the plane of the line with the electric field parallel to the line. Only the voltage source is induced. (b) Propagation perpendicular to the plane of the line with the electric field transverse to the line. Only the current source is induced. (c) Propagation along the line with the electric field perpendicular to the plane of the line. No sources are induced.

standard test for susceptibility to ESD consists of placing the product on a metal table and discharging an ESD gun to the table as shown in Fig. 7.22. This creates a wave that propagates across the table and induces noise in the product that may cause improper operation of it. Manufacturers of digital electronic devices such as electronic typewriters, laser printers, and computers routinely test their products for susceptibility to ESD in this fashion before shipment to customers. A particular product was having trouble passing the ESD test. In the presence of the ESD discharge, the product would "lock up" and would only operate when rebooted. That product had one printed circuit board containing the electronics which was mounted vertically at the rear. Because the ESD table had a metal top, the boundary conditions require that at the table surface, the electric field generated by the spark must be perpendicular to the table and the associated magnetic field must be parallel to it. The pairs of lands on the PCB were, because of the placement of the PCB in the vertical plane at the rear of the

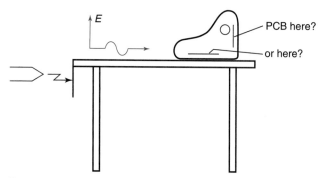

Figure 7.22 Illustration of using these principles to design an electronic product to withstand the field of an electrostatic discharge (ESD) pulse.

product, such that the electric field was totally transverse to the transmission lines on the PCB formed by the lands. Hence a current source was induced in those circuits, causing interference. This problem was solved in future designs by placing the PCB flat on the bottom of the product. This had the effect of the incident electric field being perpendicular to the circuits on the PCB and not transverse to them, and the magnetic field, being parallel to the ESD table, was not normal to the plane of the circuits on the PCB. Once this design attitude was adopted, most of the ESD problems disappeared.

See C.R. Paul, *Introduction to Electromagnetic Compatibility*, John Wiley Interscience, 1992, for a comparison of the predictions of this model to experimental results. Those results show that so long as the line is electrically very short, predictions match experimental results very well.

▶ SUMMARY OF IMPORTANT CONCEPTS AND FORMULAE

1. **Far fields of the Hertzian dipole:**

$$\hat{E}_\theta = j\frac{f\mu_o}{2}\hat{I}dl\,\sin\theta\left[\frac{e^{-j\left(2\pi\frac{r}{\lambda_o}\right)}}{r}\right] \qquad \text{far field} \quad r \gg \lambda_o$$

$$\hat{H}_\phi = \frac{\hat{E}_\theta}{\eta_o}$$

$$= j\frac{f\mu_o}{2\eta_o}\hat{I}dl\,\sin\theta\left[\frac{e^{-j\left(2\pi\frac{r}{\lambda_o}\right)}}{r}\right] \qquad \text{far field} \quad r \gg \lambda_o$$

2. **Properties of the far fields of the Hertzian dipole (and all antennas):**

1. the fields decay inversely as the distance, r, away from the antenna,

2. the fields contain a phase shift term, $e^{-j2\pi r/\lambda_o}$ (which is equivalent to a time delay in the time domain, r/v_o), and hence the fields change phase by an angle of $\angle -2\pi r/\lambda_o$ as they propagate away from the antenna,

3. the fields are orthogonal and their cross product gives, by the right-hand-rule, the direction of propagation from the antenna: $\hat{E}_\theta \times \hat{H}_\phi \Rightarrow \mathbf{a}_r$, which is the radial direction, and

4. the ratio of the electric and magnetic fields is the intrinsic impedance of free space, $\hat{E}_\theta/\hat{H}_\phi = \eta_o$.

3. **Power density of Hertzian dipole:**

$$\mathbf{S}_{AV} = \frac{1}{2}\text{Re}(\hat{\mathbf{E}} \times \hat{\mathbf{H}}^{\circ})$$

$$= \frac{1}{2}\frac{|\hat{E}|^2}{\eta_o}\mathbf{a}_r$$

$$= \underbrace{\frac{\eta_o}{8}}_{15\pi}|\hat{I}|^2\left(\frac{dl}{\lambda_o}\right)^2\frac{\sin^2\theta}{r^2}\mathbf{a}_r \qquad \text{W/m}^2$$

4. **Radiation resistance of Hertzian dipole:** The radiation resistance is a fictitious resistance such that if its current is the same as the current at the input to the dipole, then the power dissipated in the radiation resistance is equal to the total power radiated by the antenna:

$$R_{rad} = \frac{P_{AV}}{|\hat{I}_{rms}|^2}$$

where the *rms* input current is $\hat{I}_{rms} = \hat{I}/\sqrt{2}$. The radiation resistance of a Hertzian dipole is

$$R_{rad} = 80\pi^2\left(\frac{dl}{\lambda_o}\right)^2 \qquad \Omega$$

5. **The far fields of a half-wave dipole:**

$$\hat{E}_\theta = j\frac{\eta_o\hat{I}_m e^{-j2\pi\frac{r}{\lambda_o}}}{2\pi r}F(\theta)$$

where the pattern factor is

$$F(\theta) = \frac{\cos\left[\pi\frac{l}{\lambda_o}\cos\theta\right] - \cos\left(\pi\frac{l}{\lambda_o}\right)}{\sin\theta}$$

$$F(\theta) = \frac{\cos\left(\frac{\pi}{2}\cos\theta\right)}{\sin\theta} \qquad \text{half-wave dipole, } l = \frac{\lambda_o}{2}$$

and

$$|\hat{E}_\theta|_{\max} = 60\frac{|\hat{I}_m|}{r} \qquad \text{half-wave dipole, } \theta = 90°$$

6. **Power density of a half-wave dipole:**

$$\mathbf{S}_{AV} = \frac{1}{2}\text{Re}(\hat{E}_\theta\hat{H}_\phi^{\circ})\mathbf{a}_r$$

$$= \frac{1}{2}\frac{|\hat{E}_\theta|^2}{\eta_o}\mathbf{a}_r$$

$$= \underbrace{\left(\frac{\eta_o}{8\pi^2}\right)}_{4.77}\frac{|\hat{I}_m|^2}{r^2}F^2(\theta)\mathbf{a}_r \qquad \text{W/m}^2$$

7. **Input impedance of dipoles:**

half-wave dipole:

$$\hat{Z}_{in} = \underbrace{73}_{R_{rad}} + \underbrace{j42.5}_{jX} \qquad \Omega$$

quarter-wave monopole:

$$\hat{Z}_{in} = \underbrace{36.5}_{R_{rad}} + \underbrace{j21.25}_{jX} \quad \Omega$$

radiated power:

$$P_{AV,radiated} = R_{rad}|\hat{I}_{rms}|^2.$$

8. **Array factor of two-element array:**

$$F(\phi) = \cos\left(\frac{\pi d}{\lambda_o}\cos\phi + \frac{\alpha}{2}\right)$$

9. **Directivity and gain of all antennas:**

$$D(\theta,\phi) = \frac{S_{AV}(\theta,\phi)}{S_{AV,\text{ total power radiated equally in all directions}}},$$
$$G(\theta,\phi) = eD(\theta,\phi),$$

Hertzian dipole: $G = 1.5$,
half-wave dipole: $G = 1.64$,
quarter-wave monopole: $G = 3.28$

10. **Power density:**

isotropic point source:

$$S_{AV} = \frac{P_T}{4\pi r^2} \quad \text{W/m}^2$$

general antenna:

$$S_{AV}(\theta_{max},\phi_{max}) = G\frac{P_T}{4\pi r^2} \quad \text{W/m}^2.$$

11. **Effective aperture:**

$$A_{eff} = \frac{P_R}{S_{AV,inc}} \quad \text{m}^2$$

relation to gain:

$$A_{eff}(\theta,\phi) = \frac{\lambda_o^2}{4\pi}G(\theta,\phi)$$

12. **Friis transmission equation:**
path loss:

$$\frac{P_R}{P_T} = G_T(\theta_T,\phi_T)G_R(\theta_R,\phi_R)\left(\frac{\lambda_o}{4\pi d}\right)^2$$

electric field:

$$|\hat{E}| = \frac{\sqrt{60P_TG_T(\theta_T,\phi_T)}}{d}.$$

► PROBLEMS

SECTION 7.1 THE HERTZIAN DIPOLE ANTENNA

7.1.1. A Hertzian dipole antenna of length 1 cm is carrying a 100-MHz phasor current of $\hat{I} = 10\angle30°$. Determine the electric and magnetic fields at a distance of 10 cm away from the dipole and $\theta = 45°$. Compute the ratios $|\hat{E}_\theta|/|\hat{E}_r|$ and $|\hat{E}_\theta|/|\hat{H}_\phi|$ at this distance. Repeat for distances of 1 m and 10 m and $\theta = 45°$. Determine these distances in wavelengths. Is the result for the 10-m distance expected? [(a) 10 cm, $\hat{H}_\phi = 0.575\angle 29.8°$ A/m, $\hat{E}_r = 2069.67\angle -60.17°$ V/m, $\hat{E}_\theta = 991.4\angle -59.64°$ V/m, $|\hat{E}_\theta|/|\hat{E}_r| = 0.479$, $|\hat{E}_\theta|/|\hat{H}_\phi| = 1724.65$ Ω, (b) 1 m, $\hat{H}_\phi = 1.306 \times 10^{-2}\angle -25.5°$ A/m, $\hat{E}_r = 4.701\angle -115.5°$ V/m, $\hat{E}_\theta = 4.033\angle -31.74°$ V/m, $|\hat{E}_\theta|/|\hat{E}_r| = 0.858$, $|\hat{E}_\theta|/|\hat{H}_\phi| = 308.8$ Ω, (c) 10 m, $\hat{H}_\phi = 1.18 \times 10^{-3}\angle -2.7°$ A/m, $\hat{E}_r = 4.247 \times 10^{-2}\angle -92.73°$ V/m, $\hat{E}_\theta = 0.444\angle -2.74°$ V/m, $|\hat{E}_\theta|/|\hat{E}_r| = 10.45$, $|\hat{E}_\theta|/|\hat{H}_\phi| = 376.08$ Ω $\cong \eta_o$]

7.1.2. Compute the radiation resistance of and the total average power radiated by the Hertzian dipole of Problem 7.1.1.

7.1.3. A Hertzian dipole is carrying a 500-MHz current. An observer at a point 100 m broadside to the dipole, $\theta = 90°$, measures the magnitude of the electric field as 100 mV/m. Determine the magnitude of the electric field at a distance of 1000 m. Determine the magnitudes of the magnetic field at both distances. Determine the phase difference between the fields at 100 m and 1000 m. (Give this answer as an angle whose magnitude is less than or equal to 360°, and state whether the field vectors at 1000 m lead or lag those at 100.) Determine the average power densities at these two points.

SECTION 7.2 THE HALF-WAVE DIPOLE AND QUARTER-WAVE MONOPOLE ANTENNAS

7.2.1. Determine the magnitudes of the electric and magnetic fields of a half-wave dipole operated at a frequency of 300 MHz at a distance of 100 m in the broadside plane, that is, $\theta = 90°$. The input current to the terminals is $100\angle 0°$ mA. Determine the total power radiated and the power density in the wave. [60 mV/m, 159.15 μA/m, 365 mW, 4.775 μW/m²]

7.2.2. A lossless quarter-wave monopole antenna is situated above a perfectly conducting ground plane and is driven by a 100-V 300-MHz source that has a source impedance of 50 Ω. Determine the total average power radiated. Also determine the magnitude of the electric field broadside to the antenna ($\theta = 90°$) and the power density at a distance of 100 m. What is the direction of this electric field vector with respect to the ground plane?

7.2.3. The quarter-wave monopole antenna of Problem 7.2.2 is replaced by a $(1/5)\,\lambda_o$ lossless monopole that has an input impedance of $(20 - j50)$ Ω. Determine the total average power radiated. [13.51 W]

7.2.4. The quarter-wave monopole antenna of Problem 7.2.2 is replaced by a $(1/10)\,\lambda_o$ lossless monopole that has an input impedance of $(4 - j180)$ Ω. Determine the total average power radiated.

7.2.5. A lossless dipole antenna is attached to a source with a length of lossless 50-Ω coaxial cable. The source has an open-circuit voltage of 100 V (rms) and a source impedance of 50 Ω. If the frequency of the source is such that the dipole length is one-half wavelength and the transmission-line length is 1.3λ, determine the total average power radiated by the antenna and the VSWR on the cable. [43.1 W, 2.18]

SECTION 7.3 ANTENNA ARRAYS

7.3.1. For an array of two antennas as in Fig. 7.9, the electric field is proportional to $E \propto \cos[(\pi d/\lambda_o)\cos\phi + (\alpha/2)]$. Nulls in the pattern are where the argument of the cosine is an odd multiple of 90°, that is, $[(\pi d/\lambda_o)\cos\phi + (\alpha/2)] = \pm 90°, \pm 270°$, etc. Points in the pattern which exhibit peaks and valleys (minima not necessarily zero) are called maxima and minima. Show that the locations of maxima and minima in the pattern can be determined by setting the argument of the cosine to a multiple of 180°, that is, $[(\pi d/\lambda_o)\cos\phi + (\alpha/2)] = 0°, \pm 180°$.

7.3.2. Two identical monopole antennas are perpendicular to the earth. The antennas are separated by d and fed with currents of equal amplitude as shown in Fig. 7.9. Sketch the pattern of the array in a plane parallel to the earth for the following conditions: (a) $d = \lambda_o/2$, $\alpha = 90°$, (b) $d = 5\lambda_o/8$, $\alpha = 45°$, (c) $d = \lambda_o$, $\alpha = 180°$, (d) $d = \lambda_o/4$, $\alpha = 180°$.

7.3.3. A standard AM broadcast band transmitting station consists of two vertical monopoles above the earth. The two antennas are separated by 164 ft, and the transmitting frequency is 1500 kHz. The antennas are fed with signals of equal amplitude and a phase difference of 135°. Sketch the electric field pattern at the surface of the earth. Show the location of all maxima and minima and their relative values. [Nulls at $\phi = \pm 60°$ and maxima/minima at $\phi = 0°, 180°$]

7.3.4. Two dipoles are separated by one wavelength. The terminal currents are of equal magnitude but are out of phase by 90°. Sketch the electric field pattern in a plane perpendicular to the dipoles. Show the location of all maxima and minima and their relative values.

SECTION 7.4 PROPERTIES OF ANTENNAS

7.4.1. An aircraft transmitter is designed to communicate with a ground station. The ground receiver must receive at least 1 μW for proper reception. Assume that both antennas are omnidirectional. After takeoff, the airplane flies over the station at an altitude of 5000 ft. When the airplane is directly over the station, a signal of 500 mW is received by the station. Determine the maximum communication range of the airplane. [670 miles]

7.4.2. A telemetry transmitter placed on the moon is to transmit data to the earth. The transmitter power is 100 mW and the gain of the transmitting antenna in the direction of transmission is 12 dB. Determine the minimum gain of the receiving antenna in order to receive 1 nW. The distance from the moon to the earth is 238,857 miles and the transmitter frequency is 100 MHz.

7.4.3. A microwave relay link is to be designed. The transmitting and receiving antennas are separated by 30 miles, and the gain in the direction of transmission for both antennas is 45 dB. If both antennas are lossless and matched and the frequency is 3 GHz, determine the minimum transmitter power if the received power is to be 1 mW. [36.81 W]

7.4.4. An antenna on an aircraft is being used as a electronic countermeasures (ECM) to jam an enemy radar. If the antenna has a gain of 12 dB in the direction of transmission and the transmitted power is 5 kW, determine the electric field intensity in the vicinity of the enemy radar, which is 2 miles away. The frequency of transmission is 7 GHz.

7.4.5. A lossless half-wave dipole is being driven by a 10-V, 50-Ω source. Determine the electric field intensity at a distance of 10 km in a plane perpendicular to the antenna. Compute your result by using the Friis transmission equation, and check you result by using Equation (7.14). [0.461 mV/m]

Oblique Incidence of Uniform Plane Waves on Plane Boundaries

In Section 5.5 of Chapter 5 we examined the reflection and transmission of uniform plane waves (UPWs) that are incident normal (perpendicular) to a boundary between two media. In Section 5.6 we briefly examined the incidence of UPWs that are incident on that boundary at some arbitrary angle, and determined Snell's laws relating the angle of reflection to the angle of incidence and the angle of transmission to the angle of incidence. In this appendix we will examine the structure of the electric and magnetic fields of UPWs that are incident at arbitrary angles to this boundary.

▷ A.1 PERPENDICULAR POLARIZATION

Consider Fig. A.1a. A UPW is incident on a boundary between two lossless media at some angle of incidence θ_i that is measured with respect to a perpendicular to that boundary. A reflected wave will be generated at an angle of reflection θ_r, and a wave transmitted across the boundary will be generated at an angle of transmission θ_t, both of which are again measured with respect to a perpendicular to the boundary. In the case of *perpendicular polarization* shown in Fig. A.1a, the incident electric field is perpendicular to the page, as are the electric fields of the reflected and transmitted waves. The associated magnetic fields are in the plane of the page with their directions being such that $\mathbf{E} \times \mathbf{H}$ yields the direction of propagation of the wave.

First, we must construct the general form of these waves and then apply the boundary conditions to determine the relationships between the incident, reflected, and transmitted electric fields. In order to construct the general forms of these waves, we note the following important points.

1 The phase of the waves will have components in the x and z directions. These are simply determined by the projections of the propagation vector on the x and z axes. For example, the projections of the incident wave propagation vector on these axes are

$$\beta_1 \sin\theta_i \quad \text{(in the } x \text{ direction)} \tag{A.1a}$$

$$\beta_1 \cos\theta_i \quad \text{(in the } z \text{ direction)} \tag{A.1b}$$

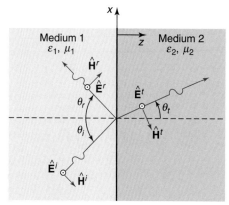

(a) Perpendicular polarization

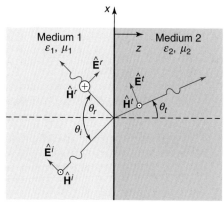

(b) Parallel polarization

Figure A.1 Oblique incidence of uniform plane waves on plane boundaries. (a) Perpendicular polarization. (b) Parallel polarization.

Similarly, the reflected wave projections are

$$\beta_1 \sin\theta_r \qquad \text{(in the } x \text{ direction)} \qquad \text{(A.2a)}$$

$$-\beta_1 \cos\theta_r \qquad \text{(in the } z \text{ direction)} \qquad \text{(A.2b)}$$

and the transmitted wave projections are

$$\beta_2 \sin\theta_t \qquad \text{(in the } x \text{ direction)} \qquad \text{(A.3a)}$$

$$\beta_2 \cos\theta_t \qquad \text{(in the } z \text{ direction)} \qquad \text{(A.3b)}$$

where the phase constants in the appropriate medium are given by $\beta_1 = \omega\sqrt{\mu_1\varepsilon_1}$ and $\beta_2 = \omega\sqrt{\mu_2\varepsilon_2}$.

2   The electric field vectors are in the y direction, so that we may write their form as

$$\hat{\mathbf{E}}^i = E_m^i e^{-j\beta_1(\sin\theta_i x + \cos\theta_i z)}\mathbf{a}_y \qquad \text{(A.4a)}$$

$$\hat{\mathbf{E}}^r = E_m^r e^{j\beta_1(-\sin\theta_r x + \cos\theta_r z)}\mathbf{a}_y \qquad \text{(A.4b)}$$

$$\hat{\mathbf{E}}^t = E_m^t e^{-j\beta_2(\sin\theta_t x + \cos\theta_t z)}\mathbf{a}_y \qquad \text{(A.4c)}$$

3 The phases of the magnetic field vectors will have the same projections as the electric fields but the vectors themselves will have projections on the x and z axes. Hence, from Fig. A.1a we may write these as

$$\hat{\mathbf{H}}^i = \frac{E_m^i}{\eta_1}(-\cos\theta_i\mathbf{a}_x + \sin\theta_i\mathbf{a}_z)e^{-j\beta_1(\sin\theta_i x + \cos\theta_i z)} \tag{A.5a}$$

$$\hat{\mathbf{H}}^r = \frac{E_m^r}{\eta_1}(\cos\theta_r\mathbf{a}_x + \sin\theta_r\mathbf{a}_z)e^{j\beta_1(-\sin\theta_r x + \cos\theta_r z)} \tag{A.5b}$$

$$\hat{\mathbf{H}}^t = \frac{E_m^t}{\eta_2}(-\cos\theta_t\mathbf{a}_x + \sin\theta_t\mathbf{a}_z)e^{-j\beta_2(\sin\theta_t x + \cos\theta_t z)} \tag{A.5c}$$

where $\eta_1 = \sqrt{\mu_1/\varepsilon_1}$ and $\eta_2 = \sqrt{\mu_2/\varepsilon_2}$ are the intrinsic impedances of the two media.

The magnitude of the incident electric field, E_m^i, is presumed known, leaving two undetermined items, the magnitudes of the reflected and transmitted electric fields E_m^r and E_m^t. We will determine these with the boundary conditions at $z = 0$. The tangential components of the electric fields (the y components) obtained from (A.4) at $z = 0$ must be continuous, giving

$$E_m^i e^{-j\beta_1(\sin\theta_i x)} + E_m^r e^{j\beta_1(-\sin\theta_r x)} = E_m^t e^{-j\beta_2(\sin\theta_t x)} \tag{A.6a}$$

and the tangential components (the x components) of the magnetic fields obtained from (A.5) at $z = 0$ must be continuous, giving

$$\frac{E_m^i}{\eta_1}(-\cos\theta_i)e^{-j\beta_1(\sin\theta_i x)} + \frac{E_m^r}{\eta_1}(\cos\theta_r)e^{j\beta_1(-\sin\theta_r x)} = \frac{E_m^t}{\eta_2}(-\cos\theta_t)e^{-j\beta_2(\sin\theta_t x)} \tag{A.6b}$$

From (A.6a) we observe that because these must be satisfied for all x, the phase terms require that

$$\beta_1 \sin\theta_i = \beta_1 \sin\theta_r$$
$$= \beta_2 \sin\theta_t \tag{A.7}$$

This requirement is referred to as "phase matching." The first requirement in (A.7) yields

$$\boxed{\theta_i = \theta_r} \tag{A.8a}$$

which is Snell's law of reflection obtained in an alternative fashion in Section 5.6 of Chapter 5, and the second requirement in (A.7) yields

$$\boxed{\begin{aligned} \sin\theta_t &= \frac{\beta_1}{\beta_2}\sin\theta_i \\[1ex] &= \sqrt{\frac{\mu_1\varepsilon_1}{\mu_2\varepsilon_2}}\sin\theta_i \end{aligned}} \tag{A.8b}$$

which is Snell's law of refraction, also obtained in an alternative fashion in Section 5.6 of Chapter 5. Substituting these into (A.6) leaves

$$E_m^i + E_m^r = E_m^t \tag{A.9a}$$

$$\frac{E_m^i}{\eta_1}(-\cos\theta_i) + \frac{E_m^r}{\eta_1}(\cos\theta_r) = \frac{E_m^t}{\eta_2}(-\cos\theta_t) \tag{A.9b}$$

Solving these gives the reflection and transmission coefficients as

$$\Gamma_\perp = \frac{E_m^r}{E_m^i}$$
$$= \frac{\eta_2 \cos\theta_i - \eta_1 \cos\theta_t}{\eta_2 \cos\theta_i + \eta_1 \cos\theta_t} \tag{A.10a}$$

and

$$T_\perp = \frac{E_m^t}{E_m^i}$$
$$= \frac{2\eta_2 \cos\theta_i}{\eta_2 \cos\theta_i + \eta_1 \cos\theta_t} \tag{A.10b}$$

These are referred to as the Fresnel coefficients and allow us to obtain the reflected and transmitted fields from the incident field. We have added a subscript $\perp$ to denote that these are for perpendicular polarization, since we investigate parallel polarization in the next section. Observe that for normal incidence, $\theta_i = \theta_r = \theta_t = 0°$, these expressions reduce to those obtained in Section 5.5 of Chapter 5, as they should. Also observe that if the second medium is a perfect conductor, $\eta_2 = 0$, then $\Gamma_\perp = -1$ and $T_\perp = 0$, again as they should.

▶ A.2 PARALLEL POLARIZATION

For the case of parallel polarization shown in Fig. A.1b, the electric field vectors are parallel to the page, and the magnetic field vectors are perpendicular to it with their direction such that $\mathbf{E} \times \mathbf{H}$ gives the direction of propagation of the wave. The field vectors may be written in a fashion identical to the previous derivation (only the projections of the field components are different; the projections of the propagation vectors remain the same) as

$$\hat{\mathbf{E}}^i = E_m^i(\cos\theta_i\mathbf{a}_x - \sin\theta_i\mathbf{a}_z)e^{-j\beta_1(\sin\theta_i x + \cos\theta_i z)} \tag{A.11a}$$

$$\hat{\mathbf{E}}^r = E_m^r(\cos\theta_r\mathbf{a}_x + \sin\theta_r\mathbf{a}_z)e^{j\beta_1(-\sin\theta_r x + \cos\theta_r z)} \tag{A.11b}$$

$$\hat{\mathbf{E}}^t = E_m^t(\cos\theta_t\mathbf{a}_x - \sin\theta_t\mathbf{a}_z)e^{-j\beta_2(\sin\theta_t x + \cos\theta_t z)} \tag{A.11c}$$

and

$$\hat{\mathbf{H}}^i = \frac{E_m^i}{\eta_1}e^{-j\beta_1(\sin\theta_i x + \cos\theta_i z)}\mathbf{a}_y \tag{A.12a}$$

$$\hat{\mathbf{H}}^r = -\frac{E_m^r}{\eta_1}e^{j\beta_1(-\sin\theta_r x + \cos\theta_r z)}\mathbf{a}_y \tag{A.12b}$$

$$\hat{\mathbf{H}}^t = \frac{E_m^t}{\eta_2}e^{-j\beta_2(\sin\theta_t x + \cos\theta_t z)}\mathbf{a}_y \tag{A.12c}$$

The boundary conditions again require at $z = 0$ that the components of the electric and magnetic fields tangent to the boundary be continuous. Evaluating (A.11) and (A.12) at $z = 0$ yields

$$E_m^i(\cos\theta_i)e^{-j\beta_1(\sin\theta_i x)} + E_m^r(\cos\theta_r)e^{j\beta_1(-\sin\theta_r x)} = E_m^t(\cos\theta_t)e^{-j\beta_2(\sin\theta_t x)} \tag{A.13a}$$

and

$$\frac{E_m^i}{\eta_1}e^{-j\beta_1(\sin\theta_i x)} - \frac{E_m^r}{\eta_1}e^{j\beta_1(-\sin\theta_r x)} = \frac{E_m^t}{\eta_2}e^{-j\beta_2(\sin\theta_t x)} \tag{A.13b}$$

Again matching the phase terms, we obtain Snell's laws, leaving

$$E_m^i(\cos\theta_i) + E_m^r(\cos\theta_i) = E_m^t(\cos\theta_t) \tag{A.14a}$$

and

$$\frac{E_m^i}{\eta_1} - \frac{E_m^r}{\eta_1} = \frac{E_m^t}{\eta_2} \tag{A.14b}$$

Solving these gives the reflection and transmission coefficients for parallel polarization as

$$\boxed{\begin{aligned} \Gamma_{\parallel} &= \frac{E_m^r}{E_m^i} \\ &= -\frac{\eta_1\cos\theta_i - \eta_2\cos\theta_t}{\eta_1\cos\theta_i + \eta_2\cos\theta_t} \end{aligned}} \tag{A.15a}$$

and

$$\boxed{\begin{aligned} T_{\parallel} &= \frac{E_m^t}{E_m^i} \\ &= \frac{2\eta_2\cos\theta_i}{\eta_1\cos\theta_i + \eta_2\cos\theta_t} \end{aligned}} \tag{A.15b}$$

Observe that for normal incidence, $\theta_i = \theta_r = \theta_t = 0°$, these expressions reduce to those obtained in Section 5.5 of Chapter 5, as they should. Also observe that if the second medium is a perfect conductor, $\eta_2 = 0$, then $\Gamma_{\parallel} = -1$ and $T_{\parallel} = 0$, again as they should.

▶ A.3 BREWSTER ANGLE OF TOTAL TRANSMISSION

Observe that the numerators of the reflection coefficients in (A.10a) and (A.15a) involve the difference between two terms. Hence there is a possibility that for certain angles of incidence, these reflection coefficients may become zero, indicating no reflection of the incident wave and total transmission of it through the boundary. These angles of incidence are called the Brewster angles.

First, investigate the case of perpendicular polarization in (A.10a). For $\Gamma_{\perp} = 0$, this requires that the numerator be zero, giving

$$\eta_2\cos\theta_i = \eta_1\cos\theta_t \tag{A.16}$$

Squaring both sides gives

$$\begin{aligned} \cos^2\theta_i &= 1 - \sin^2\theta_i \\ &= \left(\frac{\eta_1}{\eta_2}\right)^2(1 - \sin^2\theta_t) \end{aligned} \tag{A.17}$$

Substituting Snell's law of refraction given in (A.8b) gives

$$\boxed{\sin\theta_i = \sqrt{\frac{1 - (\mu_1\varepsilon_2/\mu_2\varepsilon_1)}{1 - (\mu_1/\mu_2)^2}} \qquad \Gamma_{\perp} = 0} \tag{A.18}$$

Observe that for materials such as dielectrics that are not magnetic, $\mu_{r1} = \mu_{r2} = 1$, there is no physical angle of incidence satisfying this.

For the case of perpendicular polarization in (A.15a), $\Gamma_{\parallel} = 0$ requires that the numerator be zero, giving

$$\eta_1 \cos\theta_i = \eta_2 \cos\theta_t \tag{A.19}$$

Squaring both sides and substituting Snell's law of refraction given in (A.8b) gives

$$\sin\theta_i = \sqrt{\frac{1 - (\mu_2\varepsilon_1/\mu_1\varepsilon_2)}{1 - (\varepsilon_1/\varepsilon_2)^2}} \qquad \Gamma_{\parallel} = 0 \tag{A.20}$$

Unlike the case of perpendicular polarization, for typical dielectrics, $\mu_{r1} = \mu_{r2} = 1$, such an angle may exist. For example, suppose the two materials are dielectrics with $\varepsilon_{r1} = 1$ (free space) and $\varepsilon_{r2} = 4$ (glass). Evaluating (A.20) gives the Brewster angle of total transmission as 63.4°. Hence if a parallel polarized wave is incident from air onto the surface of glass at an angle of incidence of 63.4°, it will be totally transmitted across the barrier, and none of the incident wave will be reflected.

Index